UNIVERSITY LIBRARY
UW-STEVENS POINT

W9-BIY-581

ADVANCES IN CHEMICAL PHYSICS

VOLUME 104

EDITORIAL BOARD

BRUCE J. BERNE, Department of Chemistry, Columbia University, New York, New York, U.S.A.

KURT BINDER, Institut für Physik, Johannes Gutenberg-Universität Mainz, Mainz, Germany

A. WELFORD CASTLEMAN, JR., Department of Chemistry, The Pennsylvania State University, University Park, Pennsylvania, U.S.A.

DAVID CHANDLER, Department of Chemistry, University of California, Berkeley, California, U.S.A.

M. S. CHILD, Department of Theoretical Chemistry, University of Oxford, Oxford, U.K.

WILLIAM T. COFFEY, Department of Microelectronics and Electrical Engineering, Trinity College, University of Dublin, Dublin, Ireland

F. FLEMING CRIM, Department of Chemistry, University of Wisconsin, Madison, Wisconsin, U.S.A.

ERNEST R. DAVIDSON, Department of Chemistry, Indiana University, Bloomington, Indiana, U.S.A.

GRAHAM R. FLEMING, Department of Chemistry, The University of Chicago, Chicago, Illinois, U.S.A.

KARL F. FREED, The James Franck Institute, The University of Chicago, Chicago, Illinois, U.S.A.

PIERRE GASPARD, Center for Nonlinear Phenomena and Complex Systems, Brussels, Belgium

ERIC J. HELLER, Institute for Theoretical Atomic and Molecular Physics, Harvard-Smithsonian Center for Astrophysics, Cambridge, Massachusetts, U.S.A.

ROBIN M. HOCHSTRASSER, Department of Chemistry, The University of Pennsylvania, Philadelphia, Pennsylvania, U.S.A.

R. KOSLOFF, The Fritz Haber Research Center for Molecular Dynamics and Department of Physical Chemistry, The Hebrew University of Jerusalem, Jerusalem, Israel

RUDOLPH A. MARCUS, Department of Chemistry, California Institute of Technology, Pasadena, California, U.S.A.

G. NICOLIS, Center for Nonlinear Phenomena and Complex Systems, Université Libre de Bruxelles, Brussels, Belgium

THOMAS P. RUSSELL, Department of Polymer Science, University of Massachusetts, Amherst, Massachusetts, U.S.A.

DONALD D. TRUHLAR, Department of Chemistry, Univesity of Minnesota, Minneapolis, Minnesota, U.S.A.

JOHN D. WEEKS, Institute for Physical Science and Technology and Department of Chemistry, University of Maryland, College Park, Maryland, U.S.A.

PETER G. WOLYNES, Department of Chemistry, School of Chemical Sciences, University of Illinois, Urbana, Illinois, U.S.A.

Advances in
CHEMICAL PHYSICS

Edited by

I. PRIGOGINE

Center for Studies in Statistical Mechanics and Complex Systems
The University of Texas
Austin, Texas
and
International Solvay Institutes
Université Libre de Bruxelles
Brussels, Belgium

and

STUART A. RICE

Department of Chemistry
and
The James Franck Institute
The Univesity of Chicago
Chicago, Illinois

VOLUME 104

AN INTERSCIENCE® PUBLICATION
JOHN WILEY & SONS, INC.
NEW YORK · CHICHESTER · WEINHEIM · BRISBANE · SINGAPORE · TORONTO

This book is printed on acid-free paper. ☉

Copyright © 1998 by John Wiley & Sons, Inc. All rights reserved.

Published simultaneously in Canada.

No part of this publication may be reproduced, stored in a retrieval system or transmitted in any form or by any means, electronic, mechanical, photocopying, recording, scanning or otherwise, expect as permitted under Sections 107 or 108 of the 1976 United States Copyright Act, without either the prior written permission of the Publisher, or authorization through payment of the appropriate per-copy fee to the Copyright Clearance Center, 222 Rosewood Drive, Danvers, MA 01923, (978) 750-8400, fax (978) 750-4744. Requests to the Publisher for permission should be addressed to the Permissions Department, John Wiley & Sons, Inc., 605 Third Avenue, New York, NY 10158-0012, (212) 850-6011, fax (212) 850-6008, E-mail: PERMREQ@WILEY.COM.

Library of Congress Catalog Number: 58-9935

ISBN 0-471-29338-5

Printed in the United States of America.

10 9 8 7 6 5 4 3 2 1

QD
453
.A27
v. 104

CONTRIBUTORS TO VOLUME 104

DAVID M. BISHOP, Department of Chemistry, University of Ottawa, Ottawa, Canada

P. C. FANNIN, Department of Electronic and Electrical Engineering, Trinity College, Dublin, Ireland

LAURI HALONEN, Laboratory of Physical Chemistry, University of Helsinki, Finland

INTRODUCTION

Few of us can any longer keep up with the flood of scientific literature, even in specialized subfields. Any attempt to do more and be broadly educated with respect to a large domain of science has the appearance of tilting at windmills. Yet the synthesis of ideas drawn from different subjects into new, powerful, general concepts is as valuable as ever, and the desire to remain educated persists in all scientists. This series, *Advances in Chemical Physics*, is devoted to helping the reader obtain general information about a wide variety of topics in chemical physics, a field that we interpret very broadly. Our intent is to have experts present comprehensive analyses of subjects of interest and to encourage the expression of individual points of view. We hope that this approach to the presentation of an overview of a subject will both stimulate new research and serve as a personalized learning text for beginners in a field.

I. PRIGOGINE
STUART A. RICE

CONTENTS

MOLECULAR VIBRATION AND NONLINEAR OPTICS

DAVID M. BISHOP

Department of Chemistry, University of Ottawa, Ottawa, Canada K1N 6N5

CONTENTS

I. INTRODUCTION

In principle the discipline of nonlinear optics was born in 1875 when John Kerr discovered that a material became birefringent when it was placed in an electric field: the refractive index was different in the directions parallel and perpendicular to the field. Subsequently this eponymous effect and others like it were understood in terms of the induced polarization which atoms and molecules sustain in the presence of static or dynamic

Advances in Chemical Physics, Volume 104, Edited by I. Prigogine and Stuart A. Rice.
ISBN 0-471-29338-5 © 1998 John Wiley & Sons, Inc.

(oscillating) electric fields. In a static field the dipole moment can be written as

$$\mu = \mu^0 + \alpha F + \tfrac{1}{2}\beta F^2 + \tfrac{1}{6}\gamma F^3 + \cdots \tag{1.1}$$

where μ^0 is the permanent dipole moment (if any), α is the dipole polarizability and governs the linear effect, and β, γ, ... are atomic or molecular parameters which govern the nonlinear effect and were dubbed dipole hyperpolarizabilities by Coulson et al. [1]. They are tensor quantities. The optics in nonlinear optics arises when any of the fields or properties are optical in nature. For optical fields the induced polarization is very small unless the radiation is very intense, a situation which was only put into practice with the invention of the laser in the 1950s. This led to commercial possibilities such as all-optical switching: a light beam turning another one on and off. The basic theory was quickly developed, by and large, by physicists, and it is only in the last 20 years that chemists, and in particular quantum chemists, have taken an active interest in the calculation of the hyperpolarizabilities (β, γ, ...) which modulate nonlinear optical processes.

The initial approach was to compute the effects which electric fields have on the electronic motions of atoms and molecules, that is, electronic hyperpolarizabilities, and to leave aside the parallel effects on nuclear motions. This has changed in the last decade and this review is an attempt to describe the role of nuclear vibration in nonlinear optics. That there will be such effects is easily envisioned when one realizes that under an electric field the nuclear geometry and potential surface are changed. The species' equilibrium structure and zero-point vibrational energy will be altered, and this should all be reflected in corrections to the electronic hyperpolarizabilities. These corrections are called the vibrational dipole hyperpolarizabilities: $\beta^{\text{vibrational}}$, $\gamma^{\text{vibrational}}$, ... As we will see later, these quantities can be interpreted either as the effect of changes to the geometry and potential surface of a molecule when an external field is applied or as the sum of the zero-point vibrational averaging (ZPVA) correction to the electronic property and a "pure" vibrational correction. It has been argued that the ZPVA correction should be included in the electronic property since it exists in the absence of an external field. In this chapter we do not subscribe to this view since it would negate correspondences which can be made between different theoretical formulations. It might be noted that the linear polarizability counterpart, $\alpha^{\text{vibrational}}$, was investigated as long ago as 1924 [2]; at that time the pure vibrational component went under the misnomer atomic polarizability. This name remained around for many years afterward.

The first question which must be addressed is are these vibrational effects important? In my opinion, and for the reasons which follow, the answer is unequivocally yes. In certain circumstances they dominate their electronic cousins, for example, in the hydrogen molecular ion (H_2^+), by an order of magnitude [3]. When calculated values of hyperpolarizabilities are to be used to calibrate gas-phase nonlinear optical measurements, and H_2 has been nominated for this role [4] (though helium is the more common reference), it is essential that, in the interests of accuracy, vibration be taken into account. For the electronic hyperpolarizabilities there exist universal dispersion formulas [5] which link different nonlinear optical experiments, but to make use of these formulas to intercompare experimental data, it is essential to subtract the calculated vibrational contributions [6]. Under-lying this last statement is the fact that vibrational contributions, for which no universal dispersion formula exists, differ significantly between one non-linear optical (NLO) process and another [7], just as the dielectric constant differs from the refractive index. In the same vein, it is often necessary to extrapolate an observed NLO property to the static optical frequency in order to compare with calculation; this cannot be done without knowledge of $\beta^{\text{vibrational}}$ or $\gamma^{\text{vibrational}}$ if the experimental values are at frequencies where, say, vibrational contributions are different from the static limit. So far most vibrational calculations have been carried out for relatively small molecules which, to be honest do not have much practical potential in the nonlinear optical field. But progress is being made with respect to larger systems which do have such a potential. In reality, the requirement for great preci-sion is not necessary; we need only know the order of magnitude of the vibrational properties for such systems and less effete calculations than those described later for small molecules are sufficient.

The objectives of this review are to describe in a transparent way for the nonspecialist the relevant formulas and equations that are needed to deter-mine vibrational hyperpolarizabilities; to make the connections between the different approaches which have been used; and to establish a depository, by way of references, of where values of $\beta^{\text{vibrational}}$ and $\gamma^{\text{vibrational}}$ are to be found. Since Stark vibrational shifts are related to this subject, this area will also be briefly touched upon.

Given only a finite amount of space, certain boundary conditions have been imposed. Very little will be said about $\alpha^{\text{vibrational}}$; most references will postdate 1990 (when an earlier review on this subject was written [8]) and predate October 1997. Only the effects of uniform static or dynamic electric fields will be discussed since little attention has been paid to nonuniform ones. Hence only dipole hyperpolarizabilities and not, for example, quadru-pole hyperpolarizabilities will be considered. For the same reason, mol-ecules in the condensed phase will be excluded. As in electronic calculations, the problems of basis-set size and electronic correlation are all-important; however, for these matters the reader is directed to the original references. The other type of nuclear contribution is rotational. In practically all cases this has been taken on board by considering classical orientational averaging, and this leads to particular combinations of the tensor components of β and γ which make up the appropriate quantities to compare with the observables. This is the approach we take here. The non-classical quantum-mechanical study of rotation has been addressed in Refs. [8–13].

A number of reviews and special journal issues have been devoted to nonlinear optics; largely they are restricted to the electronic side of things. References [14] and [15] are concerned with both theory and experiment and complement each other; Refs. [16–18] are of a theoretical nature but have little to say on vibrational matters. The subject of this review is to be found in Ref. [8] and, to a degree, in Refs. [14] and [19–21].

In line with the objectives above, this review, after a brief section on conventions, definitions, and units, is divided into two sections: theory and applications. Before a final section on conclusions, wrapping up where we are now, a few words are said about the controversial topic of the relationship between electronic and vibrational hyperpolarizabilities.

II. CONVENTIONS, DEFINITIONS, AND UNITS

The matter of conventions in nonlinear optics is rife with confusion. This includes the very way in which one expands the induced polarization as in Eq. (1.1), for example, to include or not to include the factors of $\frac{1}{2}$ and $\frac{1}{6}$. This leads to problems in comparing theory and experiment when different conventions may have been used, and the reader is advised to be extremely vigilant. The problem similarly exists in comparing electronic and vibrational hyperpolarizabilities. Fortunately an article by Willetts et al. [22] has done much to clarify the situation. And, in line with that work, we use the Taylor series definition (T-convention) for the polarization expansion such as that given in Eq. (1.1).

When the tensor components of the first hyperpolarizability (β) are combined, under the aegis of classical orientational averaging, to produce the measurable quantity, there is again more than one convention and, as well, more than one symbol for the pertinent quantity ($\bar{\beta}$, $\beta_{\parallel}$, β_{vec}). These problems exist for the vibrational β as much as for its electronic counterpart. The symbol $\bar{\beta}$, which we use, might imply an isotropic value; but this is not so since β, like μ, has no isotropic value. It should be appreciated that in the experiment it is the product $\mu \cdot \beta$ which occurs. Sometimes this contribution is written as $\mu \cdot \beta/3kT$ and sometimes as $\mu \cdot \beta/5kT$, and this effectively changes the value of β by a numerical factor. To start with, we take the $\mu \cdot \beta/3kT$ formulation. Further, since we are dealing with frequency-dependent fields, the frequencies (ω_1, ω_2) of the fields associated with the β of a particular NLO process must be specified as arguments of β, and these are preceded by (with a semicolon in between) the negative of the frequency of the observed induced polarization $\omega_\sigma = \omega_1 + \omega_2$. The tensor component must also be specified by the subscripts x, y, and z, which define, in order, the molecular axes along which the polarization occurs and the directions of the pertinent applied fields; we have thus

$$\beta_{\alpha\beta\gamma}(-\omega_\sigma; \omega_1, \omega_2)$$

where α, β, and γ stand for any choice of x, y, and z. Different NLO processes are recognized by certain optical frequencies (ω_1, ω_2, ...) being zero.

For the moment we shall, for the sake of compactness, drop the frequency arguments, but it is understood that they are there. We now define

$$\bar{\beta} = \sum_{i=x, y, z} \frac{\mu_i \beta_i}{\|\mu\|} \tag{2.1}$$

where μ is the dipole moment and

$$\beta_i = \sum_{j=x, y, z} \frac{\beta_{ijj} + \beta_{jij} + \beta_{jji}}{5} \tag{2.2}$$

If the z Cartesian coordinate (molecular axis) is chosen to coincide with the dipole axis, then

$$\bar{\beta} = \sum_{j=x, y, z} \frac{\beta_{zjj} + \beta_{jzj} + \beta_{jjz}}{5} \tag{2.3}$$

If all the applied fields are static ($\omega_\sigma = \omega_1 = \omega_2 = 0$) and we are considering $\beta(0; 0, 0)$, then the β components are invariant to the permutation of x, y, and z and Eq. (2.2) becomes

$$\beta_i = \frac{3}{5} \sum_{j=x, y, z} \beta_{ijj} \tag{2.4}$$

and Eq. (2.3) becomes

$$\bar{\beta} = \frac{3}{5} \sum_{j=x, y, z} \beta_{zjj} \tag{2.5}$$

Sometimes the invariance of β (as well as the other hyperpolarizabilities) to permutation of x, y, and z is invoked even when the frequencies are not zero; this is referred to as using Kleinman symmetry [23,24], and it is often a good approximation. Using Kleinman symmetry, fewer tensor components need to be calculated. It is the same as using Eqs. (2.4) and (2.5) but with frequency-dependent β's or arbitrarily choosing in Eq. (2.3), say, one of β_{zjj}, β_{jzj}, β_{jjz} to stand for the whole [25].

If the measured quantity is instead written as $\mu \cdot \beta/5kT$, then the factors of 5 in Eqs. (2.2)–(2.5) must be changed to 3. In this context, a less frequently used definition (an example is Ref. [26]) is

$$\beta_{vec} = (\beta_x^2 + \beta_y^2 + \beta_z^2)^{1/2} \tag{2.6}$$

with

$$\beta_i = \sum_{j=x,y,z} \frac{\beta_{ijj} + \beta_{jij} + \beta_{jji}}{3} \tag{2.7}$$

however, β_{vec} does not define any measurable quantity. It is also important to note that whatever sign convention is used for μ should be maintained for β, so that the observed product $\mu \cdot \beta$ is of the correct sign; some common computer packages do not meet this requirement.

In the Kerr experiment it is an anisotropy which is measured and the parameter of interest is

$$\beta^K = \tfrac{3}{2}(\beta_\| - \beta_\perp) \tag{2.8}$$

where the symbols $\|$ and $\perp$ indicate the light beam is parallel or perpendicular to the electric field, respectively. The $\beta_\|$ equals the $\bar{\beta}$ of the previous equations. We then have, in the $\mu \cdot \beta/3$ context,

$$\beta^K(-\omega; \omega, 0) = \frac{3}{10} \sum_{j=x,y,z} [3\beta_{jzj}(-\omega; \omega, 0) - \beta_{jjz}(-\omega; \omega, 0)] \tag{2.9}$$

where z is the dipole axis. It should always be remembered that the number of independent frequency-dependent components of any hyperpolarizability tensor is limited by invariance to the simultaneous permutation of the spatial indices (x, y, z) and their corresponding frequency arguments. As well, they are invariant to the reversal of all the signs of the frequency arguments.

For the second hyperpolarizability γ things are much simpler and $\bar{\gamma}$ (or $\gamma_{\parallel}$) is a true isotropic quantity; it is defined by

$$\bar{\gamma} = \sum_{i=x, y, z} \sum_{j=x, y, z} \frac{\gamma_{iijj} + \gamma_{ijij} + \gamma_{ijji}}{15} \tag{2.10}$$

and for static fields (or using Kleinman symmetry)

$$\bar{\gamma} = \sum_{i=x, y, z} \sum_{j=x, y, z} \frac{\gamma_{iijj}}{5} \tag{2.11}$$

Again, the number of nonzero independent components of γ is constrained by the permutation rule, sign reversal, and molecular symmetry. For the Kerr anisotropy the following definition is used:

$$\gamma^K(-\omega; \omega, 0, 0) = \sum_{i=x, y, z} \sum_{j=x, y, z} \frac{3\gamma_{ijij}(-\omega; \omega, 0, 0) - \gamma_{iijj}(-\omega; \omega, 0, 0)}{10} \tag{2.12}$$

In theoretical studies the usual units used are atomic units and in experimental work in the recent literature Système International (SI) units are used; the older literature often gives values in the electrostatic unit (esu) system. The conversion factors are as follows:

$$\beta: \quad au = e^3 a_0^3 E_h^{-2} = 3.20636 \times 10^{-53} \text{ C}^3 \text{ m}^3 \text{ J}^{-2}$$
$$= 8.63922 \times 10^{-33} \text{ esu} \tag{2.13}$$

$$\gamma: \quad au = e^4 a_0^4 E_h^{-3} = 6.23538 \times 10^{-65} \text{ C}^4 \text{ m}^4 \text{ J}^{-3}$$
$$= 5.03670 \times 10^{-40} \text{ esu} \tag{2.14}$$

III. THEORY

With one exception, the hydrogen molecular ion [3], all calculations of nonlinear optical vibrational phenomena have been carried out within the framework of the Born–Oppenheimer approximation. This is not so much a question of expediency as that it allows us to divide the hyperpolarizability into two familiar distinct parts: electronic and vibrational. Since there have already been many electronic hyperpolarizability calculations carried out, we can now, within the Born–Oppenheimer approximation, find the vibrational counterparts which should be added to them. Since the review in

1990 by the present author [8] there has been a great deal of methodologi-cal work. Much of it is interrelated and some of it has, in a way, been a rediscovery of earlier work. The intention of this part of the review is to make the different approaches transparent.

Since we are looking at the perturbation of vibrational motion by exter-nal electric fields, all calculations are, in that sense, based on perturbation theory. Nonetheless, two distinct approaches, as mentioned in the Intro-duction, are to be discerned: (a) the dissection of the vibrational hyperpo-larizabilities into a part emanating from zero-point vibrational averaging and a part, which is the remainder, called the pure vibrational hyperpolar-izability, and we can write

$$\beta^{\text{vibration}} = \Delta\beta^{\text{ZPVA}} + \beta^{\text{v}} \qquad (3.1)$$

$$\gamma^{\text{vibration}} = \Delta\gamma^{\text{ZPVA}} + \gamma^{\text{v}} \qquad (3.2)$$

or (b) the vibrational hyperpolarizability can be thought of as coming about from (i) the change of the equilibrium geometry in the presence of the elec-tric fields (nuclear relaxation) and (ii) the change in the shape (curvature) of the potential surface in the presence of the fields; the two quantities in Eqs. (3.1) and (3.2) are then written as

$$\beta^{\text{vibration}} = \beta^{\text{nr}} + \beta^{\text{curv}} \qquad (3.3)$$

$$\gamma^{\text{vibration}} = \gamma^{\text{nr}} + \gamma^{\text{curv}} \qquad (3.4)$$

Within both schemes (a) and (b), further development is possible. In Section A the fundamental formulas based on sum-over-states expressions, which are derived from standard perturbation theory, are presented. In Section B these formulas are further simplified by approximations to the vibrational wave functions and electric properties which are involved. The evaluation of the ZPVA correction in scheme (a) is discussed in Section C. The second general approach, (b), which is, overall, equivalent to that of (a), simply another way of cutting the vibrational hyperpolarizability cake, is outlined in Section D, along with the relationship between the two schemes. The connection between the Stark vibrational shift and the curvature contribu-tion of (b) is also discussed. Further approximations to the formulas of Section B, including the use of experimental data, are given in Sections E and F. In the last section, G, the finite-field application of approach (b) is described.

One extremely important difference between the two methodologies is that (a) applies to both static and dynamic external fields, whereas (b) is strictly only applicable to static fields. This is certainly a drawback when dealing with nonlinear optical properties, since both qualitatively and quantitatively vibrational nonlinear optical terms are highly dependent on whether the external fields are oscillating or not. In third-harmonic generation, for example, where all external fields are dynamic, γ^v is essentially null, which is often far from the case when the fields are static.

A. Conventional Perturbation Theoretic Approach

The most general sum-over-states formula (derived from perturbation theory) for a hyperpolarizability tensor component $X_{\alpha\beta\dots}^n$ ($-\omega_\sigma$; ω_1, ω_2, ... ω_n), where $\omega_\sigma = \sum_i \omega_i$, is

$$X_{\alpha\beta\dots}^n(-\omega_\sigma; \omega_1, \dots, \omega_n) = \hbar^{-n} \sum P_{\alpha,\beta,\dots} \sum_{a_1} \sum_{a_2} \cdots \sum_{a_n} \langle g|\hat{\mu}_\alpha|a_1\rangle$$

$$\times \langle a_1|\hat{\mu}_\beta|a_2\rangle \cdots \langle a_n|\hat{\mu}_i|g\rangle$$

$$\times [(\omega_{a_1} - \omega_\sigma)(\omega_{a_2} - \omega_\sigma + \omega_1) \cdots (\omega_{a_n} - \omega_n)]^{-1}$$

$$(3.5)$$

In this very general formula we are using the notation of the author [27]. More conventionally $X_{\alpha\beta}^1 = \alpha_{\alpha\beta}$ (dipole polarizability), $X_{\alpha\beta\gamma}^2 = \beta_{\alpha\beta\gamma}$ (first dipole hyperpolarizability), and $X_{\alpha\beta\gamma\delta}^3 = \gamma_{\alpha\beta\gamma\delta}$ (second dipole hyperpolarizability). The symbol $\sum P_{\alpha,\beta,\dots}$ represents summation over all the terms obtained by permuting pairs of dipole moment operators ($\hat{\mu}$) and optical frequencies (ω_i). The pairs are ($\hat{\mu}_\alpha$, $-\omega_\sigma$), ($\hat{\mu}_\beta$, ω_1), ... This summation may also be denoted by $\sum P_{-\sigma,1,\dots}$. The other summations in Eq. (3.5) are over vibronic (or electronic) states, including the ground state, of the molecule, and each state is identified by a wave function $|a_j\rangle$ and a corresponding energy E_j; the circular transition frequency between the ground state (wave function $|g\rangle$) and the jth excited state is denoted by $\omega_j = (E_j - E_g)/\hbar$. In practice Eq. (3.5), as written, is difficult to use because of apparent divergences arising from cases when a term in the denominator becomes zero. The method of handling this problem has been discussed in detail in Ref. [27] and leads to nondivergent formulas for α, β, ... which can be exploited without problem.

If the wave functions and energies in Eq. (3.5) are read as electronic ones, the result will be a purely electronic hyperpolarizability. If they are read as true nonadiabatic vibronic ones, the result will be a total nonadiabatic hyperpolarizability. Likewise, if the wave functions are written as products of an electronic and nuclear (vibrational) wave function, the adiabatic approximation [28], and furthermore, if the nuclear wave functions are defined by a Hamiltonian which does not include the adiabatic correction

[29], then the Born–Oppenheimer electronic plus vibrational hyperpolarizabilities will appear. We take this last approach.

With the sum-over-states formulas of Ref. [27] and writing $|a_i\rangle = \psi_K \phi_k^K = |k, K\rangle$, where K and k refer, respectively, to electronic and vibra-
tional states, and $\hbar\omega_{a_i} = \hbar\omega_{kK}$, which is the energy of the state (K, k) rela-
tive to the ground state $(0, 0)$, we obtain the following expression for the
first hyperpolarizability:

$$\beta_{\alpha\beta\gamma}(-\omega_\sigma; \omega_1, \omega_2) = \hbar^{-2} \sum P_{-\sigma, 1, 2} \sum_{k, K}' \sum_{l, L}' (\omega_{kK} - \omega_\sigma)^{-1}(\omega_{lL} - \omega_2)^{-1}$$

$$\times \langle 0, 0| \hat{\mu}_\alpha | K, k\rangle\langle k, K| \overline{\hat{\mu}_\beta} | L, l\rangle\langle l, L| \hat{\mu}_\gamma | 0, 0\rangle$$

(3.6)

In this equation $\overline{\hat{\mu}}$ is the fluctuation dipole moment operator
$\hat{\mu} - \langle 0, 0| \hat{\mu}|0, 0\rangle$ and the alternative symbol for the permutation operator
has been employed. The primes indicate omission of the vibronic ground
state. A number of approximations are now introduced which reduce this
expression to the conventional (or canonical) one for the total (electronic
plus vibrational) first hyperpolarizability. The validity of these approximations has been discussed in detail by Bishop et al. [30]. The most critical is that the vibrational energy differences are assumed to be small
compared to the electronic energy differences. The underlying physical distinction between Eq. (3.6) and the conventional formulation which follows
is the difference between (i) the simultaneous application of the external
fields to both electronic and vibrational motions and (ii) the sequential
application: first to the electronic motions and then to the nuclear motions.
This question has also been discussed with respect to the linear polarizability [31]. It is clear that the sequential approach falls within the spirit of the
Born–Oppenheimer approximation.

For the first hyperpolarizability we then obtain a zero-point vibrational-
averaged electronic component and a pure vibrational component; the
latter is expressed as

$$\beta^v(-\omega_\sigma; \omega_1, \omega_2) = [\mu\alpha] + [\mu^3]$$

(3.7)

with

$$[\mu\alpha] = \tfrac{1}{2}\hbar^{-1} \sum P_{-\sigma, 1, 2} \sum_k' (\mu_\alpha)_{0k}(\alpha_{\beta\gamma})_{k0}(\omega_k \pm \omega_\sigma)^{-1}$$

(3.8)

$$[\mu^3] = \hbar^{-2} \sum P_{-\sigma, 1, 2} \sum_k' \sum_l' (\mu_\alpha)_{0k}(\overline{\mu}_\beta)_{kl}(\mu_\gamma)_{l0}(\omega_k - \omega_\sigma)^{-1}(\omega_l - \omega_2)^{-1}$$

(3.9)

where

$$(\omega_k \pm \omega_\sigma)^{-1} = (\omega_k + \omega_\sigma)^{-1} + (\omega_k - \omega_\sigma)^{-1} \tag{3.10}$$

and $(\mu_\alpha)_{0k} = \langle 0 | \mu_\alpha | k \rangle$, $(\alpha_{\alpha\beta})_{k0} = \langle k | \alpha_{\alpha\beta} | 0 \rangle$, and so on, ω_k are the vibrational frequencies. The primes indicate omission of the vibrational ground state. The $[\mu\alpha]$ term corresponds to two terms in Eq. (3.6): those where $K = 0$, $L \neq 0$ or $K \neq 0$, $L = 0$; the $[\mu^3]$ term corresponds to the term in Eq. (3.6) where $K = 0$ and $L = 0$ [32]. That is, we are extracting and naming as the pure vibrational contribution those terms in Eq. (3.6) which result from summations which involve an intermediate vibronic state which contains the electronic ground state. The electric properties μ_α and $\alpha_{\alpha\beta}$ which appear in Eqs. (3.8) and (3.9) are electronic and pertain to the electronic ground state. Within this scheme the frequency dependence of the polarizability $\alpha_{\beta\gamma}$ in Eq. (3.8) is undefined. In the derivation it has been implicitly ignored, and consequently a consistent interpretation would be that it and all electric properties that occur in subsequent vibrational hyperpolarizabilities are static quantities. There has been little discussion of this question, but it does not have significant quantitative or qualitative importance [13].

For the second hyperpolarizability a similar treatment [32] gives γ^v of Eq. (3.2) as

$$\gamma^v_{\alpha\beta\gamma\delta}(-\omega_\sigma; \omega_1, \omega_2, \omega_3) = [\alpha^2] + [\mu\beta] + [\mu^2\alpha] + [\mu^4] \tag{3.11}$$

where

$$[\alpha^2] = \tfrac{1}{4}\hbar^{-1} \sum P_{-\sigma, 1, 2, 3} \sum_k{}' (\alpha_{\alpha\beta})_{0k}(\alpha_{\gamma\delta})_{k0}(\omega_k - \omega_2 - \omega_3)^{-1} \tag{3.12}$$

$$[\mu\beta] = \tfrac{1}{6}\hbar^{-1} \sum P_{-\sigma, 1, 2, 3} \sum_k{}' (\mu_\alpha)_{0k}(\beta_{\beta\gamma\delta})_{k0}(\omega_k \pm \omega_\sigma)^{-1} \tag{3.13}$$

$$[\mu^2\alpha] = \tfrac{1}{2}\hbar^{-2} \sum P_{-\sigma, 1, 2, 3} \sum_k{}' \sum_l{}' \{(\mu_\alpha)_{0k}(\bar\mu_\beta)_{kl}(\alpha_{\gamma\delta})_{l0}$$

$$\times [(\omega_k + \omega_\sigma)^{-1}(\omega_l + \omega_2 + \omega_3)^{-1} + (\omega_k - \omega_\sigma)^{-1}(\omega_l - \omega_2 - \omega_3)^{-1}]$$

$$+ (\mu_\alpha)_{0k}(\bar\alpha_{\beta\gamma})_{kl}(\mu_\delta)_{l0}(\omega_k - \omega_\sigma)^{-1}(\omega_l - \omega_3)^{-1}\} \tag{3.14}$$

$$[\mu^4] = \hbar^{-3} \sum P_{-\sigma, 1, 2, 3}\left[\sum_k{}' \sum_l{}' \sum_m{}' (\mu_\alpha)_{0k}(\bar\mu_\beta)_{kl}(\bar\mu_\gamma)_{lm}(\mu_\delta)_{m0} \right.$$

$$\times (\omega_k - \omega_\sigma)^{-1}(\omega_l - \omega_2 - \omega_3)^{-1}(\omega_m - \omega_3)^{-1}$$

$$\left. - \sum_k{}' \sum_l{}' (\mu_\alpha)_{0k}(\mu_\beta)_{k0}(\mu_\gamma)_{0l}(\mu_\delta)_{l0}(\omega_k - \omega_\sigma)^{-1}(\omega_l - \omega_3)^{-1}(\omega_l + \omega_2)^{-1} \right] \tag{3.15}$$

An alternative derivation of Eq. (3.12) is to be found in Ref. [33], and there are similarities to the much-simplified formulation of Ref. [34]. In the next section the method of evaluating the square-bracketed terms is reviewed.

B. Working Formulas for β^v and γ^v

For diatomic molecules the evaluation of the terms denoted by square brackets in Eqs. (3.7) and (3.11) is straightforward if the electronic properties (μ, α, β) are known for a range of internuclear separations. The numerical vibrational wave functions and vibrational energies can be found from the standard Numerov–Cooley procedure and the required transition integrals determined by numerical integration. Examples of this are Ref. [32], where the vibrational hyperpolarizabilities of HF are calculated, and Refs. [35] and [36], where Cl_2, Br_2, and N_2 are investigated.

For polyatomic molecules perturbation theory has to be applied in order to find the square-bracketed contributions. This is the approach taken by Bishop and Kirtman [32,37,38]; note that some errors in Refs. [32] and [37] are corrected in Ref. [38]. First, the electrical properties are expanded in a power series in the normal coordinates (Q_a, Q_b, ...), for example,

$$\mu_\alpha = \mu_\alpha^0 + \sum_a \frac{\partial \mu_\alpha}{\partial Q_a} Q_a + \frac{1}{2} \sum_a \sum_b \frac{\partial^2 \mu_\alpha}{\partial Q_a \partial Q_b} Q_a Q_b + \cdots \qquad (3.16)$$

Quadratic (and higher) order terms in this expansion account for electrical anharmonicity. Second the vibrational wave functions are expressed as perturbed harmonic oscillator functions found by solving the vibrational Schrödinger equation with the anharmonic potential:

$$V = V^0 + \frac{1}{2} \sum_a \omega_a^2 Q_a^2 + \frac{1}{6} \sum_a \sum_b \sum_c F_{abc} Q_a Q_b Q_c + \cdots \qquad (3.17)$$

The perturbation terms linear in the cubic force constants F_{abc} account for mechanical anharmonicity in first order.

Using these two equations, analytical expressions can be obtained for the vibrational matrix elements and thus for the right-hand sides of Eqs. (3.8), (3.9), and (3.12)–(3.15). In their work Bishop and Kirtman retained contributions through second order in electrical and first order in mechanical anharmonicity (except for second-order terms involving third derivatives of the electrical property). The terms were labeled as, for example, $[\mu^2]^{n,m}$ where n is the order of electrical anharmonicity and m is the order of mechanical anharmonicity. This leads to the following results for α^v, β^v, and γ^v:

$$\alpha^v = [\mu^2]^{0,\,0} + [\mu^2]^{2,\,0} + [\mu^2]^{1,\,1} \tag{3.18}$$

$$\beta^v = [\mu\alpha]^{0,\,0} + [\mu\alpha]^{2,\,0} + [\mu\alpha]^{1,\,1} + [\mu^3]^{1,\,0} + [\mu^3]^{0,\,1} \tag{3.19}$$

$$\gamma^v = [\alpha^2]^{0,\,0} + [\alpha^2]^{2,\,0} + [\alpha^2]^{1,\,1} + [\mu\beta]^{0,\,0} + [\mu\beta]^{2,\,0}$$
$$+ [\mu\beta]^{1,\,1} + [\mu^2\alpha]^{1,\,0} + [\mu^2\alpha]^{0,\,1} + [\mu^4]^{2,\,0} + [\mu^4]^{1,\,1} \tag{3.20}$$

Explicit expressions for the $[A]^{n,\,m}$ in these equations, which involve the electrical properties, the harmonic vibrational frequencies, and the anharmonic (cubic) force constants, are to be found in Ref. [38]. They were derived after a good deal of tedious and arduous algebraic manipulation. However, when written in a slightly different format, certain correspondences become apparent and some higher order terms can be obtained from the lower order ones simply by use of analogy. This is what we will do now. Let us define

$$\lambda_a^{\pm\sigma} = \lambda_a^{+\sigma}\lambda_a^{-\sigma}$$

$$\lambda_a^{\pm(\sigma-1)} = \lambda_a^{+(\sigma-1)}\lambda_a^{-(\sigma-1)}$$

$$\lambda_{ab}^{\pm\sigma} = \lambda_{ab}^{+\sigma}\lambda_{ab}^{-\sigma}$$

where

$$\lambda_a^{+\sigma} = (\omega_a + \omega_\sigma)^{-1}$$

$$\lambda_a^{+(\sigma-1)} = (\omega_a + \omega_\sigma - \omega_1)^{-1}$$

$$\lambda_{ab}^{+\sigma} = (\omega_a + \omega_b + \omega_\sigma)^{-1}$$

and $\omega_\sigma = \omega_1$, or $\omega_\sigma = \omega_1 + \omega_2$, or $\omega_\sigma = \omega_1 + \omega_2 + \omega_3$, depending, in an obvious way, on the term type; we will also use the notation $\partial\mu_\alpha/\partial Q_a = (\alpha/a)$, $\partial^2\mu_\alpha/\partial Q_a\,\partial Q_b = (\alpha/ab)$, and so on. Then the expressions in Ref. [38] become

$$[\mu^2]^{0,\,0} = \frac{1}{2}\sum P_{-\sigma,\,1}\sum_a (\alpha/a)(\beta/a)\lambda_a^{\pm\sigma} \tag{3.21}$$

$$[\mu^3]^{1,\,0} = \frac{1}{2}\sum P_{-\sigma,\,1,\,2}\sum_{a,\,b} (\alpha/a)(\beta/ab)(\gamma/b)\lambda_a^{\pm\sigma}\lambda_b^{\pm(\sigma-1)} \tag{3.22}$$

$$[\mu^4]^{2,\,0} = \frac{1}{2}\sum P_{-\sigma,\,1,\,2,\,3}\sum_{a,\,b,\,c} (\alpha/a)(\beta/ab)(\gamma/bc)(\delta/c)\lambda_a^{\pm\sigma}\lambda_b^{\pm(\sigma-1)}\lambda_c^{\pm(\sigma-1-2)}$$

$$\tag{3.23}$$

and we notice systematic changes as we add a μ and increase the order of electrical anharmonicity: an extra differentiation with respect to Q_b or Q_c, an extra (γ/b) or (δ/c), and an additional λ corresponding to the extra denominator. In Eqs. (3.21)–(3.23), $\omega_\sigma = \omega_1$, $\omega_\sigma = \omega_1 + \omega_2$, and $\omega_\sigma = \omega_1 + \omega_2 + \omega_3$, respectively. We also have

$$[\mu^3]^{0,\,1} = -\frac{1}{6} \sum P_{-\sigma,\,1,\,2} \sum_{a,\,b,\,c} F_{abc}(\alpha/a)(\beta/b)(\delta/c)$$
$$\times \lambda_a^{\pm\sigma} \lambda_b^{\pm[\sigma-(\sigma-1)]} \lambda_c^{\pm(\sigma-1)} \tag{3.24}$$

$$[\mu^4]^{1,\,1} = -\frac{1}{6} \sum P_{-\sigma,\,1,\,2,\,3} \sum_{a,\,b,\,c,\,d} F_{abc}(\alpha/a)(\beta/b)(\delta/cd)(\delta/d)$$
$$\times \lambda_a^{\pm\sigma} \lambda_b^{\pm[\sigma-(\sigma-1)]} \lambda_c^{\pm(\sigma-1)} \lambda_d^{\pm(\sigma-1-2)} \tag{3.25}$$

where it is seen that $[\mu^4]^{1,\,1}$ is simply an extension of $[\mu^3]^{0,\,1}$ with the same systematics which showed up in $[\mu^{n+2}]^{n,\,0}$. As well, $[\mu^2]^{2,\,0}$ and $[\mu^2]^{1,\,1}$ are given by

$$[\mu^2]^{2,\,0} = \frac{\hbar}{8} \sum P_{-\sigma,\,1} \sum_{a,\,b} (\alpha/ab)(\beta/ab)\lambda_b^{\pm\sigma}(\omega_a^{-1} + \omega_b^{-1}) \tag{3.26}$$

$$[\mu^2]^{1,\,1} = -\frac{\hbar}{4} \sum P_{-\sigma,\,1} \sum_{a,\,b,\,c} [F_{abc}(\alpha/ab)(\beta/c)(\omega_a^{-1} + \omega_b^{-1})\lambda_{ab}^{\pm\sigma}\lambda_c^{\pm\sigma}$$
$$+ F_{bcc}(\alpha/ab)(\beta/a)\omega_b^{-2}\omega_c^{-1}\lambda_a^{\pm\sigma}] \tag{3.27}$$

and

$$[\mu^2\alpha]^{1,\,0} = \frac{1}{4} \sum P_{-\sigma,\,1,\,2,\,3} \sum_{a,\,b} [(\alpha/a)(\beta\gamma/ab)(\delta/b)\lambda_a^{\pm\sigma}\lambda_b^{\pm(\sigma-1-2)}$$
$$+ 2(\alpha/a)(\beta/ab)(\gamma\delta/b)\lambda_a^{\pm\sigma}\lambda_b^{\pm(\sigma-1)}] \tag{3.28}$$

$$[\mu^2\alpha]^{0,\,1} = -\frac{1}{12} \sum P_{-\sigma,\,1,\,2,\,3} \sum_{a,\,b,\,c} F_{abc}[(\alpha/a)(\beta\gamma/b)(\delta/c)$$
$$\times \lambda_a^{\pm\sigma} \lambda_b^{\pm[\sigma-(\sigma-1-2)]}\lambda_c^{\pm(\sigma-1-2)} + 2(\alpha/a)(\beta/b)(\gamma\delta/c)$$
$$\times \lambda_a^{\pm\sigma} \lambda_b^{\pm[\sigma-(\sigma-1)]}\lambda_c^{\pm(\sigma-1)}]$$
$$= -\frac{1}{4} \sum P_{-\sigma,\,1,\,2,\,3} \sum_{a,\,b,\,c} F_{abc}(\alpha/a)(\beta/b)(\gamma\delta/c)$$
$$\times \lambda_a^{\pm\sigma} \lambda_b^{\pm[\sigma-(\sigma-1)]}\lambda_c^{\pm(\sigma-1)} \tag{3.29}$$

where $[\mu^2\alpha]^{1,\,0}$ can be found from $[\mu^3]^{1,\,0}$, though there are now two terms depending on whether α comes first or last (same contribution) or in the middle of the general expression, Eq. (3.14), and an appropriate denominator. Similarly, $[\mu^2\alpha]^{0,\,1}$ can be deduced from $[\mu^3]^{0,\,1}$. All other required $[A]^{n,\,m}$ can be easily found from the ones above and the relations in Table V of Ref. [32]. Recently, the $[\mu^2]^{0,\,2}$ and $[\mu^4]^{0,\,2}$ terms have also been developed. These formulas have been applied, as we shall see, to several small molecules for the determination of the frequency-dependent β^v and γ^v. Finally, it is interesting to note that the hyperpolarizability derivatives that we have introduced can be related to higher order hyperpolarizability densities; for example, $\partial\beta/\partial Q$ is related to the γ hyperpolarizability density [39].

C. Working Formulas for Zero-Point Vibrational Averaging

The difference between the value of the electronic hyperpolarizability (β^e or γ^e) determined for the equilibrium geometry and the value of the property averaged over the zero-point vibration is the ZPVA correction $\Delta\beta^{ZPVA}$ or $\Delta\gamma^{ZPVA}$; see Eqs. (3.1) and (3.2). This correction is interesting in that its frequency dependence (dispersion) follows more or less the dispersion in β^e or γ^e, and this is quite unlike the dispersion character of β^v or γ^v. The universal dispersion formulas for β^e and γ^e apply equally well to the ZPVA quantities. This implies that though for some nonlinear optical processes such as third-harmonic generation β^v or γ^v are negligible at optical frequencies, the ZPVA correction will have the same significance at all frequencies (including the static).

Like β^v and γ^v, $\Delta\beta^{ZPVA}$ and $\Delta\gamma^{ZPVA}$ for diatomic molecules can be found accurately by use of the Numerov–Cooley vibrational wave functions and numerical integration:

$$\Delta\beta^{ZPVA} = \langle 0 | \beta^e(R) | 0 \rangle - \beta^e(R_e) \tag{3.30}$$

$$\Delta\gamma^{ZPVA} = \langle 0 | \gamma^e(R) | 0 \rangle - \gamma^e(R_e) \tag{3.31}$$

where $|0\rangle$ is the ground-state vibrational wave function, R is the internuclear separation, and R_e is the equilibrium value of R.

An alternative approach is to once again use perturbation theory, and with few exceptions [40,41], this is the method of choice for polyatomic molecules. The explicit theory has been developed by Kern and co-workers [42–44], Raynes and co-workers [45,46], Fowler [47,48], and Russell and Spackman [49–51]. First-order expressions for the linear polarizability have been used by Werner and Meyer [52] and for linear molecules were first introduced by Schlier [53].

To get the flavor of the expressions, we will give those pertaining to first order. These are derived in the same manner as the vibrational matrix elements in the previous section, that is, expansion of β^e and γ^e and the use of perturbed harmonic oscillator wave functions. We then have

$$\Delta P^{\text{ZPVA}} = [P]^{0,\,1} + [P]^{1,\,0} \tag{3.32}$$

where the first term is first order in mechanical anharmonicity and the second is first order in electrical anharmonicity and P represents β^e or γ^e. It is straightforward to show that

$$[P]^{0,\,1} = -\frac{\hbar}{4} \sum_a \frac{\left(\sum_b F_{abb}/\omega_b\right)(\partial P/\partial Q_a)}{\omega_a^2} \tag{3.33}$$

and

$$[P]^{1,\,0} = \frac{\hbar}{4} \sum_a \frac{\partial^2 P/\partial Q_a^2}{\omega_a} \tag{3.34}$$

It seems to have only recently been recognized [54] that the factor $\sum_b F_{abb}/\omega_b$ can be written as $(4/\hbar)(\partial E^{\text{ZP}}/\partial Q_a)$, where E^{ZP} is the zero-point vibrational energy; this leads to computational simplification.

The calculation of the ZPVA corrections is more computationally demanding than that of the pure vibrational corrections, since we require derivatives of higher order electrical properties. It is noteworthy that the Cambridge group [55], in their study on HF and H_2O, did not determine $\Delta\gamma^{\text{ZPVA}}$. But since these corrections can easily be of the order of 10–20% for polyatomics and since they do not become muted at the optical frequencies used in nonlinear optical processes as with β^v and γ^v; it is essential that they be obtained.

An approximate alternative for diatomic molecules is to calculate β^e and γ^e at $R_0 = \langle 0 | R | 0 \rangle$ rather than at R_e; see, for examples, Ref. [56]. It can be shown [57] that, to first order,

$$\langle 0 | P | 0 \rangle = P(R_0) + \frac{1}{2}\left[\frac{B_e}{\omega_e} - \left(1 - \frac{R_0}{R_e}\right)^2\right]\frac{\partial^2 P}{\partial \xi^2} \tag{3.35}$$

where $\xi = (R - R_e)/R_e$ and B_e is the rotational constant. Thus the validity of this approximation rests on the degree of cancellation between the two terms in square brackets.

D. Alternative Formulation of the Conventional Perturbation Theoretic Approach

The alternative interpretation of vibrational effects in nonlinear optical processes is to consider, first, the effect that an electric field has by altering the equilibrium nuclear configuration of the molecule (nuclear relaxation) and, second, the effect it has by changing the zero-point vibrational energy which follows from changes to the potential surface shape (curvature). Both of these effects can be expressed separately through equations similar to Eq. (1.1), in terms of the vibrational polarizability and hyperpolarizabilities. The subdivisions in Eqs. (3.3) and (3.4) show a transparent notation for the two contributions. It must be emphasized that this approach is incapable, in principle, of handling dynamic fields and consequently can lead only to static vibrational properties. In a rigorous sense it is therefore irrelevant to nonlinear optics, since the static vibrational hyperpolarizabilities cannot be obtained (by extrapolation or any other means) from the experimental data. Nonetheless, it is interesting, in an academic sense, to see how the treatment relates to that given in Section A. The concept can be applied in either an analytic manner [58–60] or in a numeric (finite-field) manner. The former will be discussed in this section and the latter left to Section G. The relationship of this treatment to the vibrational Stark effect [61] will be described in both sections.

The first analytic development of the concept was given by Martí and Bishop [58] in 1993 and it is their theory which is outlined here. For simplicity, we will initially consider diatomic molecules. The first step (nuclear relaxation) begins by expressing the potential curve of the diatomic molecule in the presence of a uniform static field (F) as a power series expansion in the field and the normal coordinate (Q):

$$V(Q, F) = \sum_n \sum_m a_{nm} Q^n F^m \tag{3.36}$$

There are obvious identities between the coefficient a_{nm} in Eq. (3.36) and the electric property derivatives in Sections B and C, for example, $a_{11} = -(\partial\mu/\partial Q)$, $a_{12} = -\frac{1}{2}(\partial\alpha/\partial Q)$, and $a_{22} = -\frac{1}{4}(\partial^2\alpha/\partial Q^2)$. First, we consider the vibrational polarizability. If we differentiate $V(Q, F)$ with respect to Q and set the result equal to zero, solution of the resulting equation leads to a field-dependent equilibrium geometry:

$$Q_e(F) = -(2a_{20})^{-1}\left[a_{11}F + \left(a_{12} + \frac{3a_{30}a_{11}^2}{4a_{20}^2} - \frac{a_{11}a_{21}}{a_{20}}\right)F^2 + \cdots\right]$$

$$\tag{3.37}$$

Inserting this formula into $V(Q, F)$, we arrive at

$$V(Q_e, F) = a_{00} + a_{01}F + \left(a_{02} - \frac{a_{11}^2}{4a_{20}}\right)F^2 + \cdots \tag{3.38}$$

and we can identify α^{nr} by

$$\alpha^{nr} = \frac{a_{11}^2}{2a_{20}} \tag{3.39}$$

or, using the notation of Section B,

$$\alpha^{nr} = [\mu^2]_{\omega=0}^{0,\,0} \tag{3.40}$$

The same treatment, but with terms in F^3 consistently retained, gives

$$\beta^{nr} = [\mu\alpha]_{\omega=0}^{0,\,0} + [\mu^3]_{\omega=0}^{1,\,0} + [\mu^3]_{\omega=0}^{0,\,1} \tag{3.41}$$

and with terms in F^4 retained [62]:

$$\gamma^{nr} = [\alpha^2]_{\omega=0}^{0,\,0} + [\mu\beta]_{\omega=0}^{0,\,0} + [\mu^2\alpha]_{\omega=0}^{0,\,1} + [\mu^2\alpha]_{\omega=0}^{1,\,0} + [\mu^4]_{\omega=0}^{2,\,0} + [\mu^4]_{\omega=0}^{0,\,2}$$
$$+ [\mu^4]_{\omega=0}^{1,\,1} \tag{3.42}$$

It is interesting to note that the $[A]^{n,\,m}$ terms that appear in the nuclear relaxation contribution to the (hyper) polarizabilities are the lowest order $(n + m)$ ones for each allowable and nonzero type of $[A]$. This is another way of defining nuclear relaxation and makes a clear-cut link with the approach taken in Section A.

The change in curvature by the applied field affects the zero-point vibrational energy which, if we ignore third derivatives of electric properties, we can take as $\frac{1}{2}\hbar\omega_e(F)$, where $\omega_e(F)$ is the field-perturbed harmonic frequency and is related to the curvature by the equations

$$\omega_e(F) = [k(F)]^{1/2} \tag{3.43}$$

$$k(F) = \left[\frac{\partial^2 V(Q, F)}{\partial Q^2}\right]_{Q_e(F)} \tag{3.44}$$

and k is the harmonic force constant. Combining Eqs. (3.36), (3.44), and (3.37), we obtain

$$k(F) = 2a_{20} + \left(2a_{21} - \frac{3a_{30}a_{11}}{a_{20}}\right)F$$

$$+ \left[2a_{22} - \frac{3a_{30}(a_{12} + 3a_{30}a_{11}^2/4a_{20}^2 - a_{11}a_{21}/a_{20})}{a_{20}}\right]F^2 + \cdots \quad (3.45)$$

And, using Eq. (3.43) and abstracting the F^2 term in $\frac{1}{2}\hbar\omega_e(F)$, we deduce that

$$\alpha^{\text{curv}} = -\frac{\hbar\omega(2a_{22}a_{20} - 3a_{30}a_{12})}{4a_{20}^2}$$

$$+ \frac{\hbar\omega(4a_{21}^2a_{20}^2 - 36a_{30}a_{11}a_{21}a_{20} + 27a_{30}^2a_{11}^2)}{32a_{20}^4} \quad (3.46)$$

or, in the language of Section B,

$$\alpha^{\text{curv}} = \Delta\alpha^{\text{ZPVA}} + [\mu^2]_{\omega=0}^{2,0} + [\mu^2]_{\omega=0}^{1,1} + [\mu^2]_{\omega=0}^{0,2} \quad (3.47)$$

The same procedure, but with terms in F^3 consistently retained, produces

$$\beta^{\text{curv}} = \Delta\beta^{\text{ZPVA}} + [\mu\alpha]_{\omega=0}^{2,0} + [\mu\alpha]_{\omega=0}^{1,1} + [\mu\alpha]_{\omega=0}^{0,2}$$

$$+ [\mu^3]_{\omega=0}^{3,0} + [\mu^3]_{\omega=0}^{2,1} + [\mu^3]_{\omega=0}^{1,2} + [\mu^3]_{\omega=0}^{0,3} \quad (3.48)$$

It is apparent that though we have more terms than were actually evaluated in Section B, the two methods will, to the same order of perturbation theory, produce identical results; compare Eqs. (3.19) and (3.48). It is also clear that extension of the above derivation to polyatomic molecules just requires reinterpretation of the square-bracketed terms.

The Girona group [59,60] have exploited this method further and taken higher order perturbation theory for the curvature terms α^{curv} and β^{curv} (which they call α_{vib} and β_{vib}). To solve the relevant equations, they used an iterative method. Since the potential surface is still described in terms of an harmonic force constant, it is not clear how consistent this extension is. They have also gone to higher orders of perturbation theory than Martí and Bishop and made an extension to polyatomic molecules [60]. In neither Ref. [59] nor [60] have they given an expression for γ^{curv}. It should be noted that the formulas in Ref. [59] are based on an unstated approximation to the expression, given in Eq. (5) of that reference, for the field-dependent equilibrium coordinate. They are not, therefore, the same as those in Ref. [58]. In Ref. [60], as well as discarding anharmonic terms in the zero-point vibrational energy, the authors also, unwittingly, ignored the off-diagonal terms in the field-dependent harmonic force constants; the latter were expressed as a Taylor series in the nuclear coordinates. The formulas in Ref. [60] cannot, therefore, be compared with Bishop and Kirtman [38], and the numerical results for H_2O are not strictly comparable to either the finite-field values in Refs [55], [97] or those in Ref. [104].

Since it was shown in 1993 [58] that the two approaches to determining the vibrational hyperpolarizabilities are completely equivalent (two ways to cut the same pie), it is perhaps strange that the work in Ref. [60] is called "a third alternate way." As is apparent from the previous paragraphs, simply replacing the coefficients a_{nm} by the corresponding derivatives of the electric properties with respect to the normal coordinates is all that is required to make the link. The treatment in Ref. [60] is also claimed to be a "feasible method." This is not borne out by the authors' own investigation. The attempt was made to find all the required derivatives of the energy (a_{nm}) analytically. This became quite impracticable since only second- and third-order derivatives of the energy were available and, in general, third-order (and higher) derivatives had to be found by numerical differencing. The extension of the formulas, therefore, became irrelevant and the imputed analytic nature was destroyed. If we consider γ^{curv}, one of the contributions will contain $\partial^2\gamma^e/\partial Q^2$, which, analytically, implies four differentiations of the energy with respect to the field and two with respect to the normal coordinate, overall a sixth-order differentiation, which is not a very realistic proposition at the present time.

The curvature term in this formulation is directly related to the vibrational Stark effect [61]. This effect covers changes to the vibrational frequencies and intensities which are caused by an external electric field. For a change to the fundamental vibrational frequency of a diatomic molecule, in the harmonic oscillator approximation, the change will be simply related to the change the field makes to the zero-point vibrational energy:

$$\hbar \Delta\omega = -2(\mu^{curv}F + \tfrac{1}{2}\alpha^{curv}F^2 + \tfrac{1}{6}\beta^{curv}F^3 + \cdots) \tag{3.49}$$

where $\mu^{curv} = \Delta\mu^{ZPVA} = \langle 0|\mu^e|0\rangle$. Therefore, theoretical Stark vibrational studies can be used to validate formulas for the "curvature" polarizability and hyperpolarizabilities. Hush and co-workers [63–65] have thus verified [64] the α^{curv} expression of Ref. [58] as well as going to higher orders of perturbation theory.

E. Further Approximations to the Conventional Perturbation Theoretic Approach

There are many ways in which the formulas for the frequency-dependent β^v and γ^v given in Sections A and B can be approximated. This is over and above the frequent neglect of the ZPVA corrections, which, as has already been mentioned, are corrections to be ignored at one's peril. The approximations for β^v and γ^v fall into three categories: (i) ignoring certain terms in Eqs. (3.19) and (3.20), (ii) treating the frequency dependence in a simplistic fashion, and (iii) doing both. Many of these approximations for γ^v have been assessed by Bishop and Dalskov [66] for CH_4, NH_3, H_2O, HF, and CO_2, and they concluded that in many cases the relaxation/infinite-frequency approximation, to be described shortly, is satisfactory.

Let us first consider the evaluation of the static values of β^v and γ^v, when obviously frequency dependence is not an issue. One might then approximate β^v and γ^v as simply the nuclear relaxation terms β^{nr} and γ^{nr} as given in Eqs. (3.41) and (3.42). This is usually done in the context of the finite-field method and therefore it will be discussed in Section G. The one exception to this is Ref. [66] where the relation

$$\gamma^v \simeq [\alpha^2]^{0,\,0} + [\mu\beta]^{0,\,0} + [\mu^2\alpha]^{0,\,1} + [\mu^2\alpha]^{1,\,0} + [\mu^4]^{2,\,0} + [\mu^4]^{1,\,1}$$

$$(3.50)$$

was used. This differs from the nuclear relaxation formula in Eq. (3.42) by the omission of $[\mu^4]^{0,\,2}$; this is also true of Eq. (3.20), for which no analytic formula has yet been published [38].

The next level of approximation is to neglect all mechanical and electrical anharmonicities: the so-called double-harmonic-oscillator approximation. This has been widely used, particularly for the static ($\omega = 0$) quantities. The pertinent equations are then

$$\beta^v \simeq [\mu\alpha]^{0,\,0}$$

$$(3.51)$$

$$\gamma^v \simeq [\alpha^2]^{0,\,0} + [\mu\beta]^{0,\,0}$$

$$(3.52)$$

For diatomic molecules the $[\mu\beta]$ term is zero, and in fact, without any use of perturbation theory for evaluating the $[A]$ for these molecules, $\beta^v = 0$ and $\gamma^v = [\alpha^2]$ [9,35,36]. The Milan group, which uses the double-harmonic-oscillator approximation [67–74], has, however, often neglected $[\mu\beta]^{0,\,0}$ in Eq. (3.52) even for polyatomic molecules. This practice has been criticized by the present author [75] (a rebuttal appears in Ref. [76]). The formulas of the Milan group were a "rediscovery" of those already in the literature, as is also the case for Ref. [77]. Champagne and co-workers have also made frequent use of the double-harmonic-oscillator approximation [78–83], as have Kirtman and Hasan [84]. Comparison of Eq. (3.41) with Eq. (3.51) and Eq. (3.42) with Eq. (3.52) reveals that in the double-harmonic-oscillator approximation, $\beta^v = \beta^{nr}$ and $\gamma^v = \gamma^{nr}$.

For calculations of the dynamic vibrational hyperpolarizabilities Elliott and Ward [85] suggested that for optical frequencies (laser wavelengths) it was sufficient to treat only those terms which they called "enhanced." In essence this means those terms which survive as all-optical frequencies are allowed to go to infinity. Thus we call this the infinite-frequency approximation. It is normally used in conjunction with the double-harmonic-oscillator approximation. The sole exception appears, again, to be Ref. [66], where its use in conjunction with just the nuclear relaxation terms in Eq. (3.50) was found to be a very reasonable approximation in many cases. Rowlands [86] and Champagne and co-workers (see, e.g., Ref. [82]) have used the double-harmonic-oscillator infinite-frequency approximation.

It is instructive to relate the formulas for $\bar{\gamma}^v$, the mean (isotropic) second vibrational hyperpolarizability [Eq. (2.10)] obtained in the infinite-frequency approximation, to those for static frequencies ($\omega_1 = \omega_2 = \omega_3 = 0$). For third-harmonic generation (THG), electric-field-induced second-harmonic generation (ESHG), the Kerr effect, parallel component (KERR), and intensity-dependent refractive index (IDRI) we find [66]

THG: $\qquad [\bar{\gamma}^v(-3\omega; \omega, \omega, \omega)]_{\omega \to \infty} = 0$ $\hfill (3.53)$

ESHG: $\qquad [\bar{\gamma}^v(-2\omega; \omega, \omega, 0)]_{\omega \to \infty} = \frac{1}{4}[\overline{\mu\beta}]_{\omega=0}$ $\hfill (3.54)$

KERR: $\qquad [\bar{\gamma}^v(-\omega; \omega, 0, 0)]_{\omega \to \infty} = \frac{1}{3}[\overline{\alpha^2}]_{\omega=0}$
$$+ \frac{1}{2}[\overline{\mu\beta}]_{\omega=0}$$
$$+ \frac{1}{6}[\overline{\mu^2\alpha}]_{\omega=0} \qquad (3.55)$$

IDRI: $\qquad [\bar{\gamma}^v(-\omega; \omega, -\omega, \omega)]_{\omega \to \infty} = \frac{2}{3}[\overline{\alpha^2}]_{\omega=0}$ $\hfill (3.56)$

Similar results also exist for any diagonal tensor component $\gamma^v_{\alpha\alpha\alpha\alpha}$, as is seen in Ref. [82]. In Eqs. (3.53)–(3.56) the double-harmonic-oscillator approximation has not been invoked, though it could subsequently be used, in which case the last term in Eq. (3.55) would disappear. For ESHG the only surviving term at high optical frequencies is just the one neglected in Ref. [68] and in other investigations by these authors. The four last equations highlight how different the dispersion character of the vibrational hyperpolarizabilities is for different nonlinear optical processes, since for $\omega_i = 0$, the four values are all identical.

F. Semiempirical Approach

Though it comes late in this review, the use of experimental data to estimate vibrational hyperpolarizabilities was historically the earliest method used. It had its antecedents in the evaluation of vibrational polarizabilities from measured infrared intensities. It is, however, logical to describe semiempirical calculations in the framework of the more modern theory given in Sections A and B. Looking, for a moment, at the dynamic linear polarizability, the vibrational contribution is

$$[\mu^2] = \hbar^{-1} \sum P_{-\sigma,\, 1} \sum_k' (\mu_\alpha)_{0k}(\mu_\beta)_{k0}(\omega_k - \omega_\sigma)^{-1} \qquad (3.57)$$

The vibrational matrix elements $(\mu_\alpha)_{0k}$ are directly related to the infrared intensities, and these can thus be used to find $[\mu^2]$. Alternatively, in the double-harmonic-oscillator approximation

$$[\mu^2] \simeq [\mu^2]^{0,\, 0} = \frac{1}{2} \sum P_{-\sigma,\, 1} \sum_a (\alpha/a)(\beta/a)(\omega_a + \omega_\sigma)^{-1}(\omega_a - \omega_\sigma)^{-1} \quad (3.58)$$

as in Eq. (3.21), and the derivatives $(\alpha/a) = (\partial \mu_\alpha^e/\partial Q_a)$ can be extracted from the infrared intensity data. Vibrational polarizabilities found in this way have been collected for several dozen molecules by Bishop and Cheung [87].

For the vibrational hyperpolarizabilities β^v and γ^v, the matrix elements involve not only terms in μ^e but also α^e and β^e, and use is now also made of Raman intensity measurements, depolarization ratios, and hyper-Raman spectra. In 1981, the present author [88] used this approach, coupled with the infinite-frequency approximation, to estimate $\beta(-\omega; \omega, 0)$ for $CHCl_3$ and CHF_3; Elliott and Ward [85] applied the technique to SF_6, CH_4, and three fluorinated methanes. Shelton and co-workers have also used Raman data to determine γ^v for N_2 and O_2 [10,89] (see also Ref. [13]), CF_4 [90], SF_6 [91], and CO_2 [92]. Due to imperfections in the input data and flaws in some of the calculations, not all of these estimates were reliable. An analysis of these early semiempirical investigations has been given in a previous review [8].

Much more recently the Milan group, apparently unaware of this early work, used the same ideas to determine β^v and γ^v for a very large number of molecules [68–74]. They ignored the $[\mu\beta]$ contribution to γ^v and in all cases they only reported static values.

The use of Raman measurements also lies behind the investigation of Yaron and Silbey [34], though in a very simple model: the two-electronic-state/one-vibrational-state model. One concern, previously mentioned, which has not been properly addressed, is that in arriving at Eqs. (3.8), (3.9), and (3.12)–(3.15), it is implicitly assumed that the α^e and β^e involved are frequency independent. In fact there is an inherent but ambiguous frequency dependence present, and it may not be reflected in the quantities derived from the Raman data.

Finally, nonlinear optical experiments can themselves, in certain circumstances, be used to find the differences between various γ^v. For example, Shelton and Palubinskas [7] have experimentally found that at high optical frequencies $\bar{\gamma}^v$ (KERR) $- \bar{\gamma}^v$ (ESHG) is 289, 497, and 818 au for CH_4, CF_4, and SF_6, respectively.

G. Finite-Field Approach

This method is conceptually simple and computationally very direct. It can be set in the framework of the approach given in Section D. First, a finite field of varying strength is applied to a molecule to determine changes in the energy caused by the shift of the equilibrium geometry and then by finite differencing, rather than through analytic differentiation; β^{nr} and γ^{nr} are ascertained for the Taylor series expansion of the energy. Second, various fields are used to determine changes to the zero-point vibrational energy at the field-perturbed equilibrium geometry (in the simplest case, changes to the harmonic vibrational frequencies) and β^{curv} and γ^{curv} are found. Necessarily, only static vibrational hyperpolarizabilities are accessible. The genesis of this technique were the comments made in Ref. [8] on the calculation of Duran and co-workers [93,94] for field-perturbed infrared spectra and field-perturbed vibrational frequencies.

The Girona group has subsequently investigated a number of small molecules in this way: HF, CH_4, C_2H_4 [95]; CO_2 [96]; HF, CO, H_2O, HCHO [97]; and N_2O [98]. They have also looked at Stark vibrational effects (see Section D) using this finite-field approach for CH_4, H_2O, NH_3, HCHO, C_2H_4 [99]; HF, CO, H_2O, HCHO, CH_4 [97]; and N_2O [98].

The Cambridge group [55] has used finite-difference derivatives of the vibrational energy to test the truncations used in the Bishop–Kirtman methodology [32,37,38] (see Section B) and has demonstrated that these are acceptable. Quadratic, cubic and quartic force constants were found at the optimized geometries in the field from which harmonic frequencies and anharmonic force constants were obtained by using the program SPECTRO, which applied standard second-order perturbation theory. The vibrational energy was then found from these frequencies and constants and numerical differencing with respect to the field performed. Thus they differ from the Girona group by getting both the nuclear relaxation and curvature contributions in one step. This research was limited not only to static quantities but also to just the diagonal tensor components. The latter limitation does not occur for the same group's application of this method to Li_2 [100] and LiH [101]. The authors of Ref. [55] very honestly point out that a major drawback to the scheme is the numerical instability which comes from taking fourth-finite-difference derivatives of the vibrational energy.

An approximate scheme for obtaining dynamic γ^v and β^v values [Eqs. (3.1) and (3.2)] within the finite-field approach has been introduced by Bishop et al. [102] It is based on the assumption that the nuclear relaxation contribution is dominant and that the infinite-frequency ansatz is valid. Both of these assumptions have been shown to be satisfactory for calculations of γ^v for several molecules [66]. This means that the only significant contributions to $\beta^{vibration}$ and $\gamma^{vibration}$ [see Eqs. (3.1) and (3.2)] which are then missing are the ZPVA corrections. It is important to bear in mind that, with this method, these corrections must be calculated separately. The method simply requires determining the electronic μ^e, α^e, β^e in the presence of a uniform static electric field (F) with and without allowing the nuclei to relax to their equilibrium positions when the field is present. If we denote the equilibrium geometry as R_F or R_0 depending on whether the field is present or not, then for an electronic property P we may define

$$(\Delta P)_{R_0} = P(F, R_0) - P(0, R_0) \tag{3.59}$$

$$(\Delta P)_{R_F} = P(F, R_F) - P(0, R_0) \tag{3.60}$$

The static electronic properties can be found from Eq. (3.59) by the fit of a Taylor expansion, for example

$$(\Delta\alpha_{\alpha\beta})_{R_0} = \beta^e_{\alpha\beta\gamma}(0)F_\gamma + \tfrac{1}{2}\gamma^e_{\alpha\beta\gamma\delta}(0)F_\gamma F_\delta + \cdots \tag{3.61}$$

Similarly, Eq. (3.60) may be expanded for a given property, for example,

$$(\Delta\alpha_{\alpha\beta})_{R_F} = b_2 F_\gamma + \tfrac{1}{2}g_2 F_\gamma F_\delta + \cdots \tag{3.62}$$

It can be shown that

$$b_2 = \beta^e_{\alpha\beta\gamma}(0) + \beta^v_{\alpha\beta\gamma}(-\omega; \omega, 0)_{\omega \to \infty} \tag{3.63}$$

$$g_2 = \gamma^e_{\alpha\beta\gamma\delta}(0) + \gamma^v_{\alpha\beta\gamma\delta}(-\omega; \omega, 0, 0)_{\omega \to \infty} \tag{3.64}$$

TABLE 1
Methods for Calculating Vibrational First and Second Hyperpolarizabilities

Method	Approximation	Frequency Dependence	Illustrative References
Perturbation theory	(I) None, exact $[A]^a$	Yes	11, 103
	(II) BK truncation of $[A]^b$	Yes	104
	(III) Nuclear relaxation only^c	Yes	66
	(IV) Double-harmonic oscillator $[A]^d$	Yes	84
	(V) Double-harmonic oscillator $[A]^d$	Infinite frequency^e	78
	(VI) Double-harmonic oscillator $[A]^d$	Static	82
	(VII) Semiempirical $[A]$	Infinite frequency^e	85
	(VIII) Semiempirical $[A]^f$	Static	68
	(IX) Nuclear relaxation and curvature	Static	59
Finite-field theory	(X) Relaxation and harmonic oscillator	Static	95
	(XI) Relaxation and anharmonic oscillator	Static	55
	(XII) Relaxation only	Infinite frequency^e	102

^{a} See Eqs. (3.7) and (3.11).
^{b} See Eqs. (3.19) and (3.20).
^{c} See Eqs. (3.41) and (3.42).
^{d} See Eqs. (3.51) and (3.52).
^{e} See Eqs. (3.53)–(3.56).
^{f} $[\mu\beta]$ term neglected in $\gamma^{\text{vibrational}}$.

and from equations like these β^v and γ^v for the relevant nonlinear optical processes can be extracted. Thus we have a finite-field version of the nuclear-relaxation/infinite-frequency approximation for the vibrational hyperpolarizabilities. Use of this method has so far been limited to test cases. Recently, it has been extended to the determination of the curvature terms [102].

For quick reference, a survey of the various methods described above for obtaining $\beta^{\text{vibrational}}$ and $\gamma^{\text{vibrational}}$ is given in Table I.

IV. APPLICATIONS

In this section the applications of the various theories introduced in Section III are surveyed. In fact, since the start of this decade, considerable attention has been given to the evaluation of $\beta^{\text{vibrational}}$ and $\gamma^{\text{vibrational}}$. To a large degree, the choice of approximation used is computer driven, just as it is for electronic hyperpolarizability calculations. That is to say, the larger the molecule, the more extreme the approximation used. In order to reduce the computational burden, besides using one of the previously described approximations, there are two other strategies: (i) determine only a single or just the diagonal component(s) of $\beta^{\text{vibrational}}$ and $\gamma^{\text{vibrational}}$ and (ii) in the summations over the normal modes in Eqs. (3.8), (3.9), and (3.12)–(3.15), examine only those contributions which are deemed most important. But, no matter what, there is always more computation required than in the calculation of the electronic counterparts; examination of zero-point-vibrational averaging makes this transparent: either the electronic property (β^{e} or γ^{e}) must be found for a number of different nuclear geometries or derivatives (first and second ones in first-order perturbation theory) of the property must be determined [105–108].

It might be pointed out that centrosymmetric molecules, which contain a point of inversion, are the most tractable. These molecules have no dipole moment (μ) and no first hyperpolarizability (β). The reason for this is that these are vector quantities and under the operation of inversion will change sign, something not permissible for a molecular property when a symmetry operation is applied; therefore, they must be zero if the molecule contains a point of inversion [109]. For these systems only γ^{v} exists, and for diatomics Eq. (3.11) reduces to simply

$$\gamma^{\text{v}} = [\alpha^2] \tag{3.65}$$

In order to make the survey of the published calculations more digestible, this section is divided into three parts: small molecules, large molecules, and long-chain (polymeric-type) molecules. Necessarily, the distinction made between small and large is an arbitrary one.

A. Small Molecules

In Table II references for diatomic molecules are listed. This table excludes some of the pre-1990 calculations for H_2^+, H_2, D_2, LiH, HF, OH, and OH^+ which were reviewed in Ref. [8]. The $\Delta\gamma^{\text{ZPVA}}$ values vary quite widely: less than 1% of $\bar{\gamma}^{\text{e}}$ for N_2, Cl_2, and Br_2 and 11–12% of $\bar{\gamma}^{\text{e}}$ for H_2. Static values of β^{v} and γ^{v} as a percentage of the corresponding electronic values may be very significant; for optical frequencies the values are highly dependent on the particular nonlinear optical process being considered; in general, they remain important for $\beta(-\omega; \omega, 0)$, $\gamma(-\omega; \omega, \omega, -\omega)$, and $\gamma(-\omega; \omega, 0, 0)$, whereas for $\gamma(-2\omega; \omega, \omega, 0)$ they are important to a lesser extent. This is in line with the infinite-frequency approximation [see Eqs. (3.53)–(3.56)]. The HF molecule is a useful prototype for comparing different methods; this was done by the Cambridge group [55], who showed, as

TABLE II

Literature on Vibrational Hyperpolarizabilities for Diatomic Molecules

Molecule	Reference	Contribution[a]	Method[b]	Comment
Br_2	105	$\Delta\gamma^{ZPVA}$	—	First-order perturbation theory
	35	$\gamma^v, \Delta\gamma^{ZPVA}$	I	$[\alpha^2]$ exact, $\Delta\gamma^{ZPVA}$ numerical integration
Cl_2	35	$\gamma^v, \Delta\gamma^{ZPVA}$	I	$[\alpha^2]$ exact, $\Delta\gamma^{ZPVA}$ numerical integration
D_2	10	γ^v	I	Quantum-mechanical rotation included
	89	γ^v	I	$[\alpha^2]$ exact except for limited vibrational sum
	103	γ^v	I	Limited vibrational sum
H_2	11	$\gamma^v, \Delta\gamma^{ZPVA}$	I	Quantum-mechanical rotation included
	9	$\gamma^v, \Delta\gamma^{ZPVA}$	I	$[\alpha^2]$ exact; quantum-mechanical rotation
	10	γ^v	I	Quantum-mechanical rotation included
	11	$\gamma^v, \Delta\gamma^{ZPVA}$	I	Quantum-mechanical rotation included
	86	γ^v	V	
	13	γ^v	I	$[\alpha^2]$ exact; also rotational contribution
	103	γ^v	I	Limited vibrational sum
Li_2	100	γ^v	XI	
N_2	10	γ^v	VII	Rotation included, limited vibrational sum
	9	γ^v	VI	
	13	γ^v	VIII	Also rotational contribution
	36	$\gamma^v, \Delta\gamma^{ZPVA}$	I	$[\alpha^2]$ exact; $\Delta\gamma^{ZPVA}$ numerical integration
	89	γ^v	VII	Limited vibrational sum, optical frequency
	106	$\Delta\gamma^{ZPVA}$	—	Eq. (53) is questionable, first-order perturbation theory
	107	$\Delta\gamma^{ZPVA}$	—	Numerical integration
O_2	10	γ^v	VII	Rotation included, limited vibrational sum
	89	γ^v	VII	Limited vibrational sum, optical frequency
CO	59	β^{nr}, β^{curv}	IX	Axial term only
	97	$\beta^{nr}, \beta^{curv}, \gamma^{nr}, \gamma^{curv}$	IX	Axial term only
	108	$\beta^{ZPVA}, \gamma^{ZPVA}$	—	Rotational averaging also given, first-order perturbation theory
HF	32	β^v, γ^v	I, II	
	55	β^v, γ^v	II, XI	$\Delta\beta^{ZPVA}$ partially included in XI-type calculations
	86	β^v, γ^v	V	
	66	γ^v	II-IV	Static, $\hbar\omega = 0.072$ au, $\hbar\omega = \infty$
	97	$\beta^{nr}, \beta^{curv}, \gamma^{nr}, \gamma^{curv}$	IX	Axial term only
	118	γ^{nr}	III	Infinite frequency
LiH	101	β^v, γ^v	XI	$\Delta\beta^{ZPVA}$ partially included

previously mentioned, that finite-field calculations gave very similar results to those obtained by the Bishop–Kirtman formalism [32].

For the small polyatomic molecules (see Table III), all the different strategies outlined in Section III have been applied in one case or another. Zero-point vibrational averaging is very important in those few cases where it has been studied: around 15% for γ for CH_4, 6% for β for NH_3, and 9% for β for H_2O. The $(HF)_2$ dimer was investigated [110] in a failed attempt to account for the experimental-theoretical difference in β and γ for the monomer, the hypothesis being that the experiment might be contaminated by the dimer. Another interesting example is *trans*-butadiene [111]. Calculations based on the random-phase approximation (RPA) for γ and ignoring all vibrational effects were validated by agreement with experiment. In Ref. [111] it is shown, however, that any agreement would be fortuitous, since vibration and electron correlation may be significant factors and it is too early to recommend the RPA–γ^e technique by itself as a method for the determination of the hyperpolarizabilities of long-chain hydrocarbons. For example, $\bar{\gamma}^v$ $(-2\omega; \omega, \omega, 0)$ is 2% of the electronic component; incidentally, the static value of $\bar{\gamma}^v$ is of the same order as $\bar{\gamma}^e$.

The first vibrational hyperpolarizability of the H_2O molecule has been studied by several groups and the value obtained [60], using Eq. (3.3), is in disaccord with that of Ref. [104]. This is probably due to the neglect of certain higher order derivatives of the energy and the use of small basis sets in Ref. [60]. In Ref. [66] several molecules (CH_4, NH_3, H_2O, HF, and CO_2) are used as guinea pigs for testing various vibrational hyperpolarizability approximations.

B. Large Molecules

It is a sine qua non that the larger the molecule, the fewer the number of molecules investigated and the more drastic the approximations invoked. This is borne out by Table IV. Quite often only the $[\alpha^2]$ term of γ^v is determined. Furthermore, with the size of the molecule increasing, so does the number of normal modes and the complete summations in Eqs. (3.8), (3.9), and (3.12)–(3.15) become prohibitive. The Milan group circumvented this problem by considering only a single effective conjugation coordinate (ECC) which is the vibrational motion associated with bond length alternation. They deem this to be the major contributor to vibrational hyperpolarizabilities. However, Kirtman and Champagne [21] find that for $NH_2-(CH=CH)_3-NO_2$ an ECC-like normal mode contributes less than 30% to the total longitudinal β^v.

Further approximations are often used when calculating the required electric properties (α^e, β^e, γ^e), such as the two-state (and three-state) model,

TABLE III

Literature on Vibrational Hyperpolarizabilities for Small Polyatomic Molecules[a]

Molecule	Reference	Contribution	Method	Comment
CO_2	32	γ^v	II	Static, $\hbar\omega = 0.072$ au, $\hbar\omega = \infty$
	66	γ^v	II–IV	
	92	γ^v	VII	
	96	$\gamma^{nr}, \gamma^{curv}$	X	Axial term only
	104	γ^v	II	
H_2O	41	$\Delta\beta^{ZPVA}$	—	Variational treatment of vibrational wave functions
	60	β^{nr}, β^{curv}	IX	
	55	$\Delta\beta^{ZPVA}, \beta^v, \gamma^v$	XI	
	66	γ^v	II–IV	Static, $\hbar\omega = 0.072$ au, $\hbar\omega = \infty$
	86	β^v, γ^v	V	
	97	$\beta^{nr}, \beta^{curv}, \gamma^{nr}, \gamma^{curv}$	X	Axial term only
	104	$\Delta\beta^{ZPVA}, \beta^v, \gamma^v$	II	
N_2O	98	$\beta^{nr}, \beta^{curv}, \gamma^{nr}, \gamma^{curv}$	X	Axial term only
NH_3	40	$\Delta\beta^{ZPVA}$	—	Effective β for an approximate vibrational Hamiltonian
	66	γ^v	II–IV	Static, $\hbar\omega = 0.072$ au, $\hbar\omega = \infty$
	104	$\Delta\beta^{ZPVA}, \beta^v, \gamma^v$	II	
HCHO	97	$\beta^{nr}, \beta^{curv}, \gamma^{nr}, \gamma^{curv}$	X	Axial term only
$(HF)_2$	110	$\Delta\beta^{ZPVA}, \beta^v, \gamma^v$	II	
CH_4	6	γ^v	II	
	7	$(\gamma^K - \gamma^E)$ (vibration)	—	Experimental differences (KERR–ESHG)
	54	$\Delta\gamma^{ZPVA}$	—	
	66	γ^v	II–IV	Static, $\hbar\omega = 0.072$ au, $\hbar\omega = \infty$
	85	γ^v	VII	
	86	β^v, γ^v	V	
	95	$\gamma^{nr}, \gamma^{curv}$	X	Axial term only
	118	γ^{nr}	III	Infinite frequency
CF_4	7	$(\gamma^K - \gamma^E)$ (vibration)	—	Experimental differences (KERR–ESHG)
	85	γ^v	VII	
	90	γ^v	VII	
	118	γ^{nr}	III	Infinite frequency

CHF$_3$	85	γ^v	VII	
	88	β^v	VII	
CHCl$_3$	85	γ^v	VII	
	88	β^v	VII	
CH$_2$F$_2$	85	γ^v	VII	
C$_2$H$_4$	95	$\gamma^{nr}, \gamma^{curv}$	X	Diagonal terms only
SF$_6$	7	$(\gamma^K - \gamma^E)$ (vibration)	—	Experimental differences (KERR–ESHG)
	85	γ^v	VII	
	91	γ^v	VII	
	118	γ^{nr}	III	Infinite frequency
C$_2$H$_6$	77	γ^v	VI	Axial term only
C$_4$H$_6$	24	γ^v	IV	The authors quote B. Kirtman (UCSB); see also Ref. [111]
	111	γ^v	II	
C$_5$H$_5$N	60	β^{nr}, β^{curv}	IX	

[a] See footnotes to Table II.

TABLE IV

Literature on Vibrational Hyperpolarizabilities for Large Polyatomic Molecules[a]

Molecule	Reference	Contribution	Method	Comment
p-Nitroaniline	72	β^{nr}	VI	Comparison with
	112	β^{v}	VI	semiempirical
	79	β^{v}, γ^{v}	V	
Polyenovanillins	71	β^{nr}, γ^{nr}	VIII	Only one component; one mode
1,3,5-Triamino-2,4,6-trinitro benzene	72	β^{nr}	VI	One mode
4-Amino-4'-nitro1,1'-biphenyl	72	β^{nr}	VI	
Octane	68	γ^{nr}	VIII	$[\mu\beta]^{0, 0}$ neglected
Octatetraene	68	γ^{nr}	VIII	$[\mu\beta]^{0, 0}$ neglected
Oligo-p-phenylenes	74	γ^{nr}	VIII	$[\mu\beta]^{0, 0}$ neglected
Oligorylenes	74	γ^{nr}	VIII	$[\mu\beta]^{0,0}$ neglected
Oligoacenes	74	γ^{nr}	VIII	$[\mu\beta]^{0, 0}$ neglected
Mono- and di-substituted benzenes	83	β^{v}	VI	Only longitudinal component
Trisilane	84	γ^{v}	IV	
Pentasilane	84	γ^{v}	IV	

[a] See footnotes to Table II.

where only the ground state and one (two) electronic excited state(s) are included in the sum-over-states formulation of these properties. In many cases, as well, only the diagonal or longitudinal components of the β^{v} and γ^{v} tensors are determined.

Included in Table IV are trisilane and pentasilane [84]. These molecules gave the first indication that γ^{v} may be important in quasi-linear conjugated systems which are σ-conjugated rather than π-conjugated. Kirtman and Hasan [84] found that, contrary to received wisdom, $|\gamma_L^{v}(-\omega; \omega, \omega, -\omega)| > |\gamma_L^{v}(0)|$ and $|\gamma_L^{v}(-2\omega; \omega, \omega, 0)| > |\gamma_L^{v}(-\omega; \omega, 0, 0)|$, where the subscript L indicates the longitudinal component. This arose from the delicate balance between the $[\alpha^2]^{0, 0}$ and $[\mu\beta]^{0, 0}$ terms, which are of opposite sign and each of which changes substantially with the given nonlinear optical process. This discovery is a warning not to neglect $[\mu\beta]$ terms in an ad hoc manner.

C. Long-Chain Molecules

Long-chain molecules are the most likely candidates for practical nonlinear optical materials. This makes the vibrational hyperpolarizability of these species particularly relevant. Most of the investigations in this domain have been carried out by Kirtman and co-workers. Since the review by Kirtman and Champagne [21] devotes considerable space to this topic, little needs be said here. In Table V the appropriate references and methods are dis-

TABLE V
Literature on Vibrational Hyperpolarizabilities for Long-Chain Molecules[a]

Molecule	Reference	Contribution	Method
Polyynes	80	γ^v	V
Polyactelynes	78	γ^v	V
	81	γ^v	V
Polydiacetylenes	82	γ^v	IV
Polyethylenes	78	γ^v	V
Polybutatrienes	82	γ^v	IV
Polysilanes	78	γ^v	V
Push–pull polyenes	112	β^v	VI

[a] See footnotes to Table II.

played. In all cases the double-harmonic-oscillator approximation is utilized; to go any further one would need knowledge of the anharmonic force constants, something not easily come by for these molecules.

One aspect not dwelt upon in the aforesaid review is the extrapolation procedure to be used to obtain values for the infinite-chain limit. It is implicitly assumed in Refs. [80] and [82], for example, that the form of the dependence of γ^v on chain length is the same as for γ^e. There seems no a priori physical justification for this. For the electronic linear polarizability (α^e) several extrapolation techniques have been tested [113] on $C_{2n}H_2$, and it would be interesting to make a similar study for β^v and γ^v.

V.   RELATION BETWEEN ELECTRONIC AND VIBRATIONAL HYPERPOLARIZABILITIES?

Does a relation, either quantitative or qualitative, exist between the vibrational and electronic hyperpolarizabilities of a given species? This is the question posed in this section. At first sight, this appears to be a strange question, since one would not expect the ability of the electrons to relax in an electric field to be in any way connected with the ability of the nuclei to relax in the same field. And, since we are considering hyperpolarizabilities, it is a second- or third-order ability. However, the question must be addressed because it has been concluded, in several articles [70,72,112] by the Milan group, that the answer is yes. This leads, in the opinion of these authors [72], to an alternative way to measure nonlinear optical properties. The route would be to calculate the nuclear relaxation terms (β^{nr} and γ^{nr}) and then use them as a measure of β^e and γ^e. Since calculating ab initio vibrational hyperpolarizabilities is more computationally demanding than the electronic hyperpolarizabilities themselves, the saving of effort would only occur if β^{nr} and γ^{nr} were evaluated by semiempirical means (in the manner of Bishop [88], Elliott and Ward [85], and Shelton and co-workers [89–91]) from infrared and Raman measurements. This would then, as a corollary, map infrared and Raman studies on the nonlinear optical field, leading to a rejuvenation of the former [72].

Clearly, such a far-reaching conjecture requires close scrutiny. This can be done in two ways: (a) by looking at any theoretical underpinnings or (b) by looking at the numerical data. Taking up the first point, a proof, flawed in the opinion of some [114], has been offered by Castiglioni et al. [112] that β^e and β^{nr}, in the double-harmonic-oscillator approximation (i.e., $[\mu\alpha]^{0,\,0}$), are the same within a factor of 2. Interestingly, no theoretical justification was attempted in Ref. [112] for the second hyperpolarizability, but see Ref. [115]. The proof was based on a number of quite drastic approximations: (i) that the electronic properties can be accounted for by a two-state model, where the electronic ground state and one excited state are described by mixtures of a valence bond (VB) state and a charge transfer (CT) state; (ii) that in the summation over the vibrational modes only the quasi-normal mode described by the effective conjugation coordinate is important; (iii) that the derivative with respect to this normal coordinate (Q) of the electronic polarizability (α^e) can be taken as if the numerator in the two-state (sum-over-states) form of α^e were independent of Q; (iv) that the shapes of the potential curves for the ground and single electronic excited state are identical (same force constants); (v) that all electrical and mechanical anharmonicities can be ignored (i.e., the $[\mu\alpha]^{0,\,0}$ approximation); (vi) that only one component of $\bar{\beta}$ (i.e., β_{zzz}) need be considered; and (vii) that dispersion, or frequency dependence, can be ignored.

In the original article [112] some of these points, particularly (iii), were obscure and were only clarified by a later comment [114]. In a response [116] to that comment, Castiglioni et al. justified approximation (iii) on the grounds of a single calculation for $(CN)_2C(CH)_5N(CH_3)_2$, where the measured Raman intensity for the most intense band is 6.7×10^{-7} cm^4 g^{-1} and that calculated with approximation (iii) is 1.37×10^{-6} cm^4 g^{-1}. On this point it should be noted that the ab initio calculations of β^v, which were the inspiration for the initial conjecture, do not make this approximation and thus there is an inconsistency between the initial observation and the theory.

Kim et al. [115] have constructed a very similar model but without approximation (iii) and which includes the second as well as the first hyperpolarizability. The ratios β^v/β^e and γ^e/γ^v are then expressed in terms of t (the charge transfer integral), k (the single force constant), Q_{CT}^0 and Q_{VB}^0 (equilibrium positions for the CT and VB states), V (the energy between these states), and Δ (the gap between the ground and excited state). These formulas, in themselves, do not show any deep insight into a possible relation between electronic and vibrational hyperpolarizabilities, but calculations for $(CH_3)_2N-(CH=CH)_3-CHO$ of β_{zzz}^v and β_{zzz}^e, and γ_{zzzz}^v and γ_{zzzz}^e, versus the parameter f, which describes the CT character in the electronic ground state, do. In both the β and γ cases the vibrational values mimic the electronic ones, and hence in the two-state VB–CT model, the two types of hyperpolarizability are related by the bond length alternation parameter f. This confirms the qualitative reasoning given in Ref. [112]. This writer finds this pragmatic result, within the well-known limitations of the model, more convincing than the mathematical treatment in Ref. [112]. Simply put [21], bond length alternation is the major factor, in these systems, in determining electronic hyperpolarizabilities, and ECC is the vibrational motion associated with bond length alternation, and this motion is often the major contributor to vibrational hyperpolarizabilities; thus both electronic and vibrational hyperpolarizabilities are related to bond length alternation. Paradoxically, the original ab initio results which led to the conjecture were not based on the VB–CT model.

Now let us consider other types of calculation and measurement of β^e, β^v, γ^e, and γ^v. Many of the comparisons go far beyond molecules which can legitimately be described as push–pull polyenes (see Table I of Ref. [112]) and often do so on the basis of log–log plots; thus the comparison is only one of orders of magnitude and not of exact or near equality. Clearly, as well, there are many molecules where there is absolutely no similarity between γ^e and γ^v at all (think of H_2^+). But even if we consider, for the moment, only push–pull polyenes, there are obvious problems when we go beyond the crude VB–CT approximation, which, though it may have helped in our qualitative understanding, is not an established predictive tool. For example, Kirtman and Champagne [21] have studied the ratio of $\beta_L^v(0)$ to $\beta_L^e(0)$ for (A) $NO_2-(CH=CH)_N-NH_2$, (B) $NO_2-(C\equiv C)_N-NH_2$, and (C) $NO_2-(Th)_{N/2}-NH_2$. The ratios for $N = 2$ are 2.20, 1.21, and 1.68 for A, B, and C, respectively; for $N = 4$ they are, respectively, 2.11, 1.13, and 2.88. It is apparent that there is not only a chain length dependence for the ratio, quite striking in the case of the thiophenes (C) but also a dependence on the linker type.

If we try to extend the relationship conjecture to polydiacetylene (PDA) and polybutatriene (PBT) [82], we note that Perpète et al. found the $\Delta[\alpha^2]_L^{0,\,0}/\Delta\gamma_L^e(0)$ ratio, where Δ indicates the per-unit-cell quantity, converges with chain length to 0.857 (PDA) and to 3.34 (PBT). The latter number exceeding the upper limit of 2 given by Castiglioni et al. [112]. For the polyynes [80], $C_{2n}H_2$, the $\Delta[\alpha^2]_L^{0,\,0}/\Delta\gamma_L^e(0)$ ratio drops from 20.4 to 0.93 as n goes from 1 to 8. Champagne [83] has studied a number of disubstituted benzenes and calculates static values of β^e and β^v corresponding to the longitudinal axis. The work includes one of the prototypes,

$NH_2-C_6H_4-NO_2$, quoted in Ref. [112]. Similar electronic and vibrational contributions are found, but the analysis is very different from that in Ref. [112]. Instead Champagne has found that variations in $\beta_L^e(0)$ are almost solely accounted for by mesomeric effects, whereas those in $\beta_L^v(0)$ require inductive effects as well.

It is the writer's opinion, in answer to the question posed at the beginning of this section, that β^v and γ^v are a means of estimating (crudely) β^e and γ^e for push–pull polyenes in those cases where both the electronic and vibrational properties are adequately described by a VB–CT scenario and there is a single dominant mode of vibration. It would be necessary, if any practical predictive use of this were to be entertained, that the dispersion character of the quantities discussed not be an issue.

One approach to this subject, which has yet to be explored and might be fruitful, is to look at the relation which exists between derivatives of μ^e, α^e, and β^e with respect to the normal coordinates (which occur in the vibrational hyperpolarizabilities) and the next higher electronic (hyper)polarizability densities [39]. The connecting thread is that electrons respond to changes in the nuclear Coulomb field (changes due to an infinitesimal shift in nuclear position) through the same nonlocal (hyper)polarizability densities that characterize their response to external electric fields.

VI. CONCLUSIONS

Since 1990, the time of the last review [8], nearly a hundred papers have been published that in one way or another touch on vibrational hyperpolarizabilities (and their role in nonlinear optics), and one can now consider the field as an established one. As is witnessed by the preceding sections, the general methods of computation are all linked through perturbation theory. One of the biggest stumbling blocks is the efficient evaluation of the required derivatives (with respect to the normal coordinates) of the electric properties μ^e, α^e, β^e. Programs are available at the Hartree–Fock level but, in general, not beyond that. Furthermore, derivative programs which incorporate time dependence do not exist. To get around these problems a number of approximations, faute de mieux, have been introduced. In the writer's opinion not enough has been done to validate these approximations, though doubtless, with experience, the pitfalls, if any, will become apparent. Finite-field techniques are less computationally intensive (certainly as far as programming is concerned), though sometimes numerically unstable, but they do nothing to get around the time dependence obstacle: they can only be used to obtain static quantities which have, in principle, no experimental relevance. At the present time only the Bishop–Kirtman formulation allows for the accurate evaluation of the frequency-dependent vibrational hyperpolarizabilities β^v and γ^v. Other methods rely on the infinite-frequency approximation.

Future development is required in considering how these properties might be obtained for the all-important condensed phase; such work is cited in Ref. [21] as being in progress. Development is also anticipated on the quantum-mechanical rotational contributions to hyperpolarizabilities and, as in the conclusion to the last review, on the investigation of non-Born–Oppenheimer effects.

This review has been concerned only with electric fields, but the questions raised also come into play when dealing with magnetic fields and such phenomena as the Cotton–Mouton effect [117].

ACKNOWLEDGMENTS

The author thanks the Natural Sciences and Engineering Research Council of Canada for continuous financial support. He also thanks all present and past collaborators.

REFERENCES

1. C. A. Coulson, A. Macoll, and L. E. Sutton, *Trans. Faraday Soc.*, **48**, 106 (1952).

2. L. Ebert, *Z. Phys. Chem.*, **113**, 1 (1924).

3. D. M. Bishop and S. A. Solunac, *Phys. Rev. Lett.*, **55**, 1986 and 2627 (1985).

4. R. Tammer, K. Löblein, K. H. Peting, and W. Hüttner, *Chem. Phys.*, **168**, 151 (1992).

5. D. M. Bishop and D. W. De Kee, *J. Chem. Phys.*, **104**, 9876 (1996); *J. Chem. Phys.*, **105**, 8247 (1996).

6. D. M. Bishop and J. Pipin, *J. Chem. Phys.*, **103**, 4980 (1995).

7. D. P. Shelton and J. J. Palubinskas, *J. Chem. Phys.*, **104**, 2482 (1996).

8. D. M. Bishop, *Rev. Mod. Phys.*, **62**, 343 (1990).

9. D. M. Bishop and B. Lam, *Chem. Phys. Lett.*, **134**, 283 (1987); *J. Chem. Phys.*, **89**, 1571 (1988).

10. D. P. Shelton, *Phys. Rev. A*, **42**, 2578 (1990).

11. D. M. Bishop, J. Pipin, and S. M. Cybulski, *Phys. Rev. A*, **43**, 4845 (1991).

12. D. P. Shelton and B. Rugar, *Chem. Phys. Lett.*, **201**, 364 (1993) and references therein.

13. S. C. Read, A. D. May, and G. D. Sheldon, *Can. J. Phys.*, **75**, 211 (1997).

14. D. P. Shelton and J. E. Rice, *Chem. Rev.*, **94**, 3 (1994).

15. J. L. Brédas, C. Adant, P. Tackx, A. Persoons, and B. M. Pierce, *Chem. Rev.*, **94**, 243 (1994).

16. A. A. Hasanein, *Adv. Chem. Phys.*, **85**, 415 (1993).

17. Y. Luo, H. Agren, P. Jørgensen, and K. V. Mikkelsen, *Adv. Quant. Chem.*, **26**, 165 (1995).

18. S. P. Karna and A. T. Yeates, Eds., *Nonlinear Optical Materials*, ACS Symp. Ser. 628, 1996. Especially Chapter 2 by R. J. Bartlett and H. Sekino and Chapter 3 by B. Kirtman.

19. Special issue on molecular nonlinear optics, M. A. Ratner, Ed., *Int. J. Quant. Chem.*, **43**, 5 (1992).

20. D. M. Bishop, *Adv. Quant. Chem.*, **25**, 1 (1994).

21. B. Kirtman and B. Champagne, *Int. Rev. Phys. Chem.*, **16**, 389 (1997).

22. A. Willetts, J. E. Rice, D. M. Burland, and D. P. Shelton, *J. Chem. Phys.*, **97**, 7590 (1992).

23. D. A. Kleinman, *Phys. Rev.*, **126**, 1977 (1962).

24. H. Sekino and R. J. Bartlett, *Chem. Phys. Lett.*, **234**, 87 (1995).

25. H. Sekino and R. J. Bartlett, *J. Chem. Phys.*, **94**, 3665 (1991).

26. C. Daniel and M. Dupuis, *Chem. Phys. Lett.*, **171**, 209 (1990).

27. D. M. Bishop, *J. Chem. Phys.*, **100**, 6535 (1994).

28. M. Born and K. Huang, *Dynamical Theory of Crystal Lattices*, Oxford University Press, New York, 1956.

29. W. Cencek and W. Kutzelnigg, *Chem. Phys. Lett.*, **266**, 383 (1997).

30. D. M. Bishop, B. Kirtman, and B. Champagne, *J. Chem. Phys.*, **107**, 5780 (1997).

31. D. M. Bishop, L. M. Cheung, and A. D. Buckingham, *Mol. Phys.*, **41**, 1225 (1980).

32. D. M. Bishop and B. Kirtman, *J. Chem. Phys.*, **95**, 2646 (1991).

33. G. P. Das, *J. Chem. Phys.*, **101**, 4474 (1994).

34. D. Yaron and R. Silbey, *J. Chem. Phys.*, **95**, 563 (1991).

35. E. F. Archibong and A. J. Thakkar, *Chem. Phys. Lett.*, **201**, 485 (1993).

36. E. F. Archibong and A. J. Thakkar, *J. Chem. Phys.*, **100**, 7471 (1994).

37. B. Kirtman and D. M. Bishop, *Chem. Phys. Lett.*, **175**, 601 (1990).

38. D. M. Bishop and B. Kirtman, *J. Chem. Phys.*, **97**, 5255 (1992). For a recent extension to $[\mu^2]^{0,2}$ and $[\mu^4]^{0,2}$ terms, see D. M. Bishop, J. M. Luis, and B. Kirtman, *J. Chem. Phys.* (in press).

39. E. L. Tisko, X. Li, and K. L. C. Hunt, *J. Chem. Phys.*, **103**, 6873 (1995) and references therein.

40. V. Spirko, Y. Luo, H. Agren, and P. Jørgensen, *J. Chem. Phys.*, **99**, 9815 (1993).

41. Y. Luo, H. Agren, O. Vahtras, P. Jørgensen, V. Spirko, and H. Hettema, *J. Chem. Phys.*, **98**, 7159 (1993).

42. W. C. Ermler and C. W. Kern, *J. Chem. Phys.*, **55**, 4851 (1971).

43. L. L. Sprandel and C. W. Kern, *Mol. Phys.*, **24**, 1383 (1972).

44. B. J. Krohn, W. C. Ermler, and C. W. Kern, *J. Chem. Phys.*, **60**, 22 (1974).

45. G. Riley, W. T. Raynes, and P. W. Fowler, *Mol. Phys.*, **38**, 877 (1979).

46. W. T. Raynes, P. Lazzeretti, and R. Zanasi, *Mol. Phys.*, **64**, 1061 (1988).

47. P. W. Fowler, *Mol. Phys.*, **43**, 591 (1981); a misprint is corrected in Ref. 50.

48. P. W. Fowler, *Mol. Phys.*, **51**, 1423 (1984); a misprint is corrected in Ref. 51.

49. A. J. Russell and M. A. Spackman, *Mol. Phys.*, **84**, 1239 (1995).

50. A. J. Russell and M. A. Spackman, *Mol. Phys.*, **88**, 1109 (1996).

51. A. J. Russell and M. A. Spackman, *Mol. Phys.*, **90**, 251 (1997).

52. H.-J. Werner and W. Meyer, *Mol. Phys.*, **31**, 855 (1976); C. W. Kern and R. L. Matcha, *J. Chem. Phys.*, **49**, 2081 (1968).

53. C. Schlier, *Fortschr. Phys.*, **9**, 455 (1961).

54. D. M. Bishop and S. P. A. Sauer, *J. Chem. Phys.*, **107**, 8502 (1997).

55. M. J. Cohen, A. Willetts, R. D. Amos, and N. C. Handy, *J. Chem. Phys.*, **100**, 4467 (1994).

56. P. Rozyczko and R. J. Bartlett, *J. Chem. Phys.*, **107**, 10823 (1997); see also H. Sekino and R. J. Bartlett, *J. Chem. Phys.*, **84**, 2726 (1986).

57. D. M. Bishop and J. Pipin, *J. Chem. Phys.*, **98**, 522 (1993); *J. Chem. Phys.*, **98**, 4003 (1993).

58. J. Martí and D. M. Bishop, *J. Chem. Phys.*, **99**, 3860 (1993); note that there is a misprint in Eq. (19): a factor of 4 should precede the $a_{21}^2 a_{20}^2$ term.

59. J. M. Luis, J. Martí, M. Duran, and J. L. Andrés, *J. Chem. Phys.*, **102**, 7573 (1995).

60. J. M. Luis, M. Duran, and J. L. Andrés, *J. Chem. Phys.*, **107**, 1501 (1997).

61. D. M. Bishop, *J. Chem. Phys.*, **98**, 3179 (1993); erratum, *J. Chem. Phys.*, **99**, 4875 (1993).

62. D. M. Bishop, unpublished results; see also Ref. 66.

63. J. E. Gready, G. B. Bacskay, and N. S. Hush, *Chem. Phys.*, **24**, 333 (1977); see also N. S. Hush and M. L. Williams, *J. Mol. Spectrosc.*, **50**, 349 (1974).

64. N. S. Hush and J. R. Reimers, *J. Phys. Chem.*, **99**, 15798 (1995).

65. J. R. Reimers, J. Zeng, and N. S. Hush, *J. Phys. Chem.*, **100**, 1498 (1996).

66. D. M. Bishop and E. K. Dalskov, *J. Chem. Phys.*, **104**, 1004 (1996).

67. C. Castiglioni, M. Gassoni, M. Del Zoppo, and G. Zerbi, *Sol. State Commun.*, **82**, 13 (1992).

68. M. Del Zoppo, C. Castiglioni, G. Zerbi, M. Rui, and M. Gussoni, *Synth. Met.*, **51**, 135 (1992).

69. M. Del Zoppo, C. Castiglioni, M. Veronelli, and G. Zerbi, *Synth. Met.*, **55**, 3919 (1993).

70. C. Castiglioni, M. Del Zoppo, and G. Zerbi, *J. Raman Spectrosc.*, **24**, 485 (1993).

71. P. Zuliani, M. Del Zoppo, C. Castiglioni, G. Zerbi, C. Andraud, T. Brotin, and A. Collet, *J. Phys. Chem.*, **99**, 16242 (1995).

72. M. Del Zoppo, C. Castiglioni, and G. Zerbi, *Nonlinear Opt.*, **9**, 73 (1995).

73. C. Castiglioni, M. Del Zoppo, P. Zuliani, and G. Zerbi, *Synth. Met.*, **74**, 171 (1995).

74. M. Rumi, G. Zerbi, K. Müllen, G. Müller, and M. Rehahn, *J. Chem. Phys.*, **106**, 24 (1997).

75. D. M. Bishop, *Synth. Met.*, **68**, 293 (1995).

76. M. Del Zoppo, C. Castiglioni, and G. Zerbi, *Synth. Met.*, **68**, 295 (1995).

77. G. P. Das, A. T. Yeates, and D. Dudis, *Chem. Phys. Lett.*, **212**, 671 (1993).

78. B. Kirtman, B. Champagne, and J.-M. André, *J. Chem. Phys.*, **104**, 4125 (1996).

79. B. Champagne, *Chem. Phys. Lett.*, **261**, 57 (1996).

80. E. A. Perpète, B. Champagne, J.-M. André, and B. Kirtman, *J. Molec. Struct.*, (THEOCHEM), **425**, 115 (1998).

81. B. Champagne, E. A. Perpète, J.-M. André, and B. Kirtman, *Synth. Met.*, **85**, 1047 (1997).

82. E. A. Perpète, B. Champagne, and B. Kirtman, *J. Chem. Phys.*, **107**, 2463 (1997).

83. B. Champagne, *Int. J. Quant. Chem.*, **65**, 689 (1997).

84. B. Kirtman and M. Hasan, *J. Chem. Phys.*, **96**, 470 (1992).

85. D. S. Elliott and J. F. Ward, *Mol. Phys.*, **51**, 45 (1984).

86. T. W. Rowlands, Ph.D. Thesis, University of Cambridge, England (1989).

87. D. M. Bishop and L. M. Cheung, *J. Phys. Chem. Ref. Data*, **11**, 119 (1982).

88. D. M. Bishop, *Mol. Phys.*, **42**, 1219 (1981).

89. D. P. Shelton, *Mol. Phys.*, **60**, 65 (1987); correction in *Phys. Rev. A*, **36**, 3461 (1987).

90. Z. Lu and D. P. Shelton, *J. Chem. Phys.*, **87**, 1967 (1987); some errors corrected in Ref. 91.

91. D. P. Shelton and L. Ulivi, *J. Chem. Phys.*, **89**, 149 (1988).

92. D. P. Shelton, *J. Chem. Phys.*, **85**, 4234 (1986).

93. M. Duran, J. L. Andrés, A. Lledós, and J. Bertrán, *J. Chem. Phys.*, **90**, 328 (1989).

94. J. L. Andrés, J. Martí, M. Duran, A. Lledós, and J. Bertrán, *J. Chem. Phys.*, **95**, 3521 (1991).

95. J. Martí, J. L. Andrés, J. Bertrán, and M. Duran, *Mol. Phys.*, **80**, 625 (1993).

96. J. L. Andrés, J. Bertrán, M. Duran, and J. Martí, *J. Phys. Chem.*, **98**, 2803 (1994).

97. J. M. Luis, J. Martí, M. Duran, and J. L. Andrés, *Chem. Phys.*, **217**, 29 (1997).

98. J. L. Andrés, J. Bertrán, M. Duran, and J. Martí, *Int. J. Quant. Chem.*, **52**, 9 (1994).

99. J. Martí, A. Lledós, J. Bertrán, and M. Duran, *J. Comp. Chem.*, **13**, 821 (1992).

100. M. G. Papadopoulos, A. Willetts, N. C. Handy, and A. D. Buckingham, *Mol. Phys.*, **85**, 1193 (1995).

101. M. G. Papadopoulos, A. Willetts, N. C. Handy, and A. E. Underhill, *Mol. Phys.*, **88**, 1063 (1996).

102. D. M. Bishop, M. Hasan, and B. Kirtman, *J. Chem. Phys.*, **103**, 4157 (1995). For a recent extension to curvature terms, see B. Kirtman, J. M. Luis, and D. M. Bishop, *J. Chem. Phys.* (in press).

103. V. Mizrahi and D. P. Shelton, *Phys. Rev. A*, **32**, 3454 (1985).

104. D. M. Bishop, B. Kirtman, H. A. Kurtz, and J. E. Rice, *J. Chem. Phys.*, **98**, 8024 (1993).

105. G. Maroulis and C. Makris, *Mol. Phys.*, **91**, 333 (1997).

106. G. Maroulis and A. J. Thakkar, *J. Chem. Phys.*, **88**, 7623 (1988); erratum *J. Chem. Phys.*, **89**, 6558 (1988).

107. Y. Luo, O. Vahtras, H. Agren, and P. Jørgensen, *Chem. Phys. Lett.*, **205**, 555 (1993).

108. G. Maroulis, *J. Phys. Chem.*, **100**, 13466 (1996).

109. D. M. Bishop, *Group Theory and Chemistry*, Dover, New York, 1993, p. 19.

110. D. M. Bishop, J. Pipin, and B. Kirtman, *J. Chem. Phys.*, **102**, 6778 (1995).

111. B. Kirtman, J. L. Toto, C. Breneman, C. P. de Melo, and D. M. Bishop, *J. Chem. Phys.*, **108**, 4355 (1998).

112. C. Castiglioni, M. Del Zoppo, and G. Zerbi, *Phys. Rev. B*, **53**, 13319 (1996).

113. E. K. Dalskov, J. Oddershede, and D. M. Bishop, *J. Chem. Phys.*, **108**, 2152 (1998).

114. D. M. Bishop and B. Kirtman, *Phys. Rev. B*, **56**, 2273 (1997).

115. H.-S. Kim, M. Cho and S.-J. Jeon, *J. Chem. Phys.*, **107**, 1936 (1997). There is a misprint in Eq. (6): a factor of 3 is missing from the numerator. Note that these authors use a power series rather than a Taylor series expansion, so that their β and γ values are $\frac{1}{2}$ and $\frac{1}{6}$, respectively, of those of Bishop and Kirtman. Because of this a factor of 8 should have been used, instead of 4, in their Eq. (11). See also J. Y. Lee and K. S. Kim, *J. Chem. Phys.*, **107**, 6515 (1997).

116. C. Castiglioni, M. Del Zoppo, and G. Zerbi, *Phys. Rev. B*, **56**, 2275 (1997).

117. C. Rizzo, A. Rizzo, and D. M. Bishop *Int. Rev. Phys. Chem.*, **16**, 81 (1997).

118. J. M. Luis, J. Martí, M. Duran, J. L. Andrés, and B. Kirtman, *J. Chem. Phys.* **108**, 4123 (1998).

LOCAL MODE VIBRATIONS IN POLYATOMIC MOLECULES

LAURI HALONEN

Laboratory of Physical Chemistry, P.O. Box 55 (A. I. Virtasen aukio 1), FIN-00014 University of Helsinki, Finland

CONTENTS

Advances in Chemical Physics, Volume 104, Edited by I. Prigogine and Stuart A. Rice.
ISBN 0-471-29338-5   © 1998 John Wiley & Sons, Inc.

I. INTRODUCTION

Overtone spectroscopy has experienced a tremendous revival during the last 20 years [1–5]. Before this time the sensitivity of the experimental techniques available was such that it was often impossible to study overtone transitions as they are usually very weak, particularly in the case of high overtones. Therefore, most of the work in the past was done on fundamentals. Another crucial factor was the limited value of vibration–rotation theory 20 years ago. Much of it was based on the normal coordinate concept. The fact that this coordinate is rectilinear by definition already indicates that it might not be the best choice for large-amplitude displacements, which often occur in overtone spectroscopy. Due to the lack of computational power, the eigenvalues of the traditional models were obtained by perturbation theory or by other related methods at that time. This did not encourage the tackling of overtone problems with the existing theory. It was natural to think that the approximate methods available had little value for highly excited vibrational states. Finally, one reason for not studying overtones was the generally accepted assumption that due to the large increase of density of states as energy increases, overtone spectroscopy would become an extremely difficult topic. It was a surprise to observe that to a certain extent there are unexpected simplifications.

The development of modern laser methods, where tunable monochromatic sources could cover large portions of the near-infrared and the visible regions, opened a new world in overtone spectroscopy [6]. The high resolution offered by these devices allowed observation of the overtone bands in detail. As an example, Fig. 1 shows a low-resolution photoacoustic overtone spectrum of C_2H_2 in the region 10,900–14,300 cm^{-1}. Figure 2 shows some of the bands recorded using high resolution. These spectra were obtained by placing the sample cell inside a Ti–sapphire ring laser cavity [7]. In this way the sensitivity was increased drastically when compared with traditional absorption experiments. Lasers, combined with novel methods such as double-resonance experiments and molecular beams, made it possible to resolve congested spectra. It is not surprising that a large amount of high-quality overtone data became available. This naturally stimulated developments on the theory side. Modern computational power made it possible to go beyond traditional perturbation theory solutions. First variational calculations were performed for triatomic molecules or for molecular systems where only part of the degrees of freedom was included. In the case of the overtone studies this was fortunate because it became unnecessary to deal with all degrees of freedom to understand much of the overtone spectral structure, and it became possible to extend the calculations to high overtones. Consequently, the results obtained were not

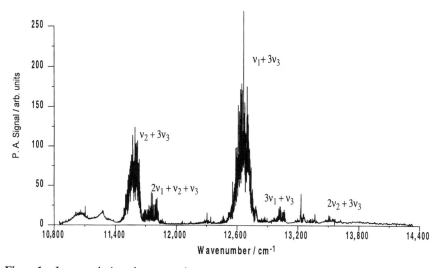

Figure 1. Low-resolution photoacoustic overtone spectrum of C_2H_2. The usual normal mode labeling scheme is such that ν_1, ν_2, and ν_3 are the symmetric CH stretch, CC stretch, and antisymmetric CH stretch, respectively. Note the highest intensity of the pure stretching vibrational band $\nu_1 + 3\nu_3$. The vertical axis is the photoacoustic signal in arbitrary units. (From Ref. [7] with permission.)

hidden inside complicated models.

It was noticed some time ago that with low resolution the overtone spectra of molecules such as benzene are simple [8]. Bands at slightly decreasing intervals (about 3000 cm^{-1}) were much stronger than other bands nearby. This was interpreted as a sign of independent CH bond oscillators. This indicates a departure from the traditional normal-mode-based vibrational picture where during a normal vibration all atoms of the molecule move in phase. The concept of local modes is based on uncoupled bond oscillators. The surprising thing at the beginning was the observation that the anharmonic bond oscillators seemed to become decoupled as the stretching energy increased. Thus, the local mode concept seemed to offer an improved way of describing highly excited stretching states of anharmonic vibrations such as hydrogen stretches. However, there remained some problems with this concept. The idea of exciting single oscillators seemed to contradict with ideas on symmetry, and it took some time until the meaning of this was fully understood.

The local mode and the normal mode models are almost of equal age [9,10]. Therefore, it is somewhat surprising that only in recent years spectroscopists have started to accept local mode interpretations of experimental spectra of polyatomic molecules. The original local mode models

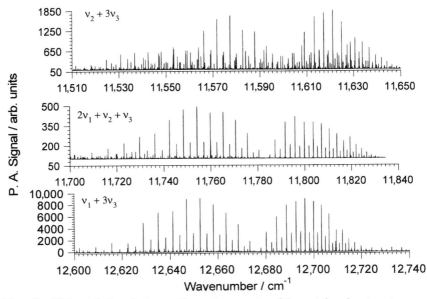

Figure 2. High-resolution photoacoustic overtone spectra of the $v_2 + 3v_3$, $2v_1 + v_2 + v_3$, and $v_1 + 3v_3$ bands in C_2H_2. For the explanation of the labels see the caption of Fig. 1. (From Ref. [7] with permission.)

included just the stretching vibrations, and much of the development concentrated on this aspect of the theory. It is fair to say that modern work has explained the relation between the normal and local mode concepts.

The restriction to stretching vibrations was a drawback in the original local mode approach. In order to have the theory in a useful form, it is necessary to be able to include bending vibrations in the model. Of course, at least in principle, it is possible to use modern methods to form exact vibrational Hamiltonians that include all vibrational degrees of freedom. However, these models are useful only for small molecules and for low-lying states. A practical vibrational Hamiltonian is obtained by including the most important effects and disregarding, for example, some of the nonresonance interactions. In many hydrogen-containing molecules anharmonic resonances (Fermi resonances) between CH stretches and bends, for example, must be included. Using these guidelines, local mode models were extended in a simple way to include all vibrational degrees of freedom [5].

Simple stretching vibrational local mode models showed that the vibrational energy level structure of high overtones possesses unusual patterns, such as close degeneracies in some symmetrical molecules, when compared

with the traditional theory. Later experimental work confirmed these find-
ings. It was a natural extension to consider the rotational motion in these
unusual vibrational states.

This review deals with local mode models, choosing mainly XH_2-, XH_3-,
and XH_4-type molecules as examples. Emphasis is put on bent triatomic
molecules. In addition to stretching motion, bending vibrations are also
included. Fermi resonance interactions between stretches and bends are
examined. Full variational methods to obtain eigenvalues of vibrational
Hamiltonians are discussed. The derivation of exact vibrational Hamiltonians is presented. Finally, rotational motion is included in the local mode
models.

II. SIMPLE LOCAL- AND NORMAL-MODE MODELS

A. Bent Symmetrical Triatomic Molecules

The models presented are aimed at explaining modern experimental
vibration–rotation spectra of polyatomic molecules. A large part of this
chapter deals with small symmetrical molecules. For this reason, experimental vibrational spectra of water (H_2O) and sulfur dioxide (SO_2) have
been chosen as examples. Table I contains the observed stretching vibrational term values for both molecules [11,12]. In the customary vibration–
rotation theory these are called overtone and combination levels, but for
the sake of simplicity they are called overtone levels in this chapter. The
left-hand columns in Table I provide assignments of the vibrational states
in two different labeling schemes. The customary normal mode labels v_1
and v_3 are the symmetric and antisymmetric stretching vibrational normal
mode quantum numbers, respectively. The local mode quantum numbers
v_{r_1} and v_{r_2} describe the excitations of the two bond oscillators. The symbols
$\pm$ are symmetry labels, that is $+$ states are symmetric and $-$ states are
antisymmetric, with respect to permutation of identical nuclei. Table I also
contains splittings of some of the close-lying levels within different overtone
manifolds, which are defined as sets of levels with $v = v_1 + v_3 = v_{r_1} + v_{r_2} =$
const. In water, a striking feature is the rapid decrease of the splittings
between the lowest members in each of the overtone manifolds as the total
stretching quantum number v increases. At high v the states become degenerate. The same behavior occurs in the splitting between second lowest
members of each overtone manifold (and so on for the higher members),
although the splittings decrease more slowly than in the lowest pairs. The
stretching overtone levels of sulfur dioxide behave in a different manner.
The corresponding splittings hardly decrease as v increases. These two molecules provide examples of different vibrational behavior. The stretching

TABLE I

Experimental Stretching Vibrational Energy Levels and Energy Level Splittings of Water and Sulfur Dioxide[a]

Normal v_1v_3	Local $v_{r_1}v_{r_2} \pm$	H_2O (cm^{-1})		$^{32}SO_2$ (cm^{-1})	
		E	ΔE	E	ΔE
10	10+	3657.05	98.88	1151.71	210.35
01	10−	3755.93		1362.06	
20	20+	7201.54	48.28	2295.81	204.06
11	20−	7249.82		2499.87	
02	11+	7445.07		2715.46	
30	30+	10599.69	13.67	3431.19	198.42
21	30−	10613.36		3629.61	
12	21+	10868.88	163.53	3837.06	
03	21−	11032.41		4054.26	
40	40+	13828.28	2.66		
31	40−	13830.94		4751.23	
22	31+	14221.16	97.65		
13	31−	14318.81		5165.64	
04	22+	14536.87			
50	50+	16898.40	0.44		
41	50−	16898.84			
32	41+	17458.35	37.18		
23	41−	17495.53			
14	32+	17748.07			

[a] Experimental data are taken from Refs. [11] (H_2O) and [12] (SO_2) where references to original experimental work are found.

vibrational energy level structure of SO_2, with almost evenly spaced levels within each overtone manifold, is typical for a normal mode molecule; that is, the stretching vibrations are best described in terms of normal modes. The high overtones of water with close degeneracies are typical for a local mode molecules; that is, the stretching vibrations are best described in terms of independent bond oscillators [3,13].

The starting point is a model where only stretching vibrations are included in the well-bent XY_2 molecules such as H_2O. The Hamiltonian is expressed in terms of internal bond displacement coordinates r_1 and r_2 and in terms of their conjugate momentum operators $p_{r_1} = -i\hbar \, \partial/\partial r_1$ and $p_{r_2} = -i\hbar \, \partial/\partial r_2$. $i = \sqrt{-1}$ and $\hbar = h/2\pi$, where h is Planck's constant. With this operator form of the quantum-mechanical momenta, the vibrational volume element for integration is equal to unity (see Section IV). The curvilinear coordinate describing valence angle displacement is constrained to its equilibrium value. Consequently, the momentum conjugate to the valence angle displacement coordinate disappears. Thus, the bending and the stretching degrees of freedom are decoupled. Morse oscillator Hamiltonians [14] are employed to describe the bond oscillators which are coupled by bilinear kinetic and potential energy terms. In fact, the potential coupling is bilinear in the Morse variable y_i ($i = 1, 2$) and is defined below. By expanding the coupling term as a Taylor series and retaining the first term, the coupling becomes bilinear in the variable r_i. The Hamiltonian takes the form

$$H = T + V = H_1^{(0)} + H_2^{(0)} + g_{rr'}^{(e)} p_{r_1} p_{r_2} + f_{rr'} a_r^{-2} y_1 y_2 \qquad (2.1)$$

where T and V are kinetic and potential energy operators, respectively. In this chapter, the superscripts are included in parentheses in order to distinguish them from exponents. The Morse oscillator Hamiltonian $H_i^{(0)}$ ($i = 1, 2$) is given as

$$H_i^{(0)} = \tfrac{1}{2} g_{rr} p_{r_i}^2 + D_e y_i^2 \qquad (2.2)$$

Its eigenvalues in wavenumber units are

$$E_i = \omega_r(v_{r_i} + \tfrac{1}{2}) + x_{rr}(v_{r_i} + \tfrac{1}{2})^2 \qquad (2.3)$$

where $\omega_r = \hbar(2a_r^2 D_e g_{rr})^{1/2}/hc_0 = \hbar(f_{rr} g_{rr})^{1/2}/hc_0$ is the harmonic wavenumber and $x_{rr} = -a_r^2 \hbar^2 g_{rr}/2hc_0$ which is often denoted $-\omega x$, is the bond anharmonicity parameter; $y_i = 1 - \exp(-a_r r_i)$ is the Morse variable; and c_0 is the speed of light in a vacuum. The Morse dissociation energy D_e, the Morse steepness parameter a_r, the harmonic force constant

$$f_{rr} = \left(\frac{\partial^2 V}{\partial r_1^2}\right)_e = \left(\frac{\partial^2 V}{\partial r_2^2}\right)_e = 2a_r^2 D_e$$

and the coupling coefficient

$$f_{rr'} = \left(\frac{\partial^2 V}{\partial r_1 \partial r_2}\right)_e$$

are potential energy parameters; $g_{rr} = g_{rr}^{(e)} = (1/m_X) + (1/m_Y)$ and $g_{rr'}^{(e)} = (\cos \alpha_e)/m_X$ are kinetic energy coefficients; m_X and m_Y are masses of the X and Y atoms, respectively; and α_e is the equilibrium valence angle. The index e indicates that the Hamiltonian coefficients are evaluated at the equilibrium geometry. The potential energy coupling term in $f_{rr'}$ is expressed in terms of the Morse variables y_1 and y_2 in order to ensure realistic asymptotic behav-

ior at large displacements from the equilibrium configuration [15]. The
coupling coefficient takes its usual form $f_{rr'} r_1 r_2$ if the Morse variables y_1
and y_2 are expanded as Taylor series and the first term is retained in both
expansions.

The eigenvalues of the Hamiltonian in Eq. (2.1) are best obtained variationally. The local mode Hamiltonian provides a starting point for a
simpler model, the harmonically coupled anharmonic oscillator (HCAO)
model [3,13,16], which is discussed next.

The key idea in the HCAO model is to approximate the matrix elements
of the coupling terms in $g_{rr}^{(e)}$ and in $f_{rr'}$ by harmonic oscillator formulas,
which for the momentum operator and for the displacement coordinate
operator are given in Table II [17–19]. A bond product harmonic oscillator
eigenfunction basis $|v_{r_1} v_{r_2}\rangle = |v_{r_1}\rangle |v_{r_2}\rangle$ is employed. These functions are
chosen to be consistent with the harmonic wavenumber ω_r, which appears
in Eq. (2.3). The notation used with the basis functions of this chapter is
such that quantum numbers associated with equivalent oscillators are
written together. A comma separates other quantum numbers, which also
include a vibrational quantum number and its associated vibrational
angular momentum quantum number. Due to the selection rules of the
given coupling, only the states with $\Delta v = 0, \pm 2$, where $v = v_{r_1} + v_{r_2}$, are
coupled. The energy level separation between $\Delta v = \pm 2$ states is large, of the
order $2\omega_r$. The effects of the $\Delta v = \pm 2$ coupling can be taken into account
by second-order perturbation theory. On the other hand, the $\Delta v = 0$ coupling connects close-lying states, and approximate treatments such as nondegenerate perturbation theory do not produce accurate results. It is
necessary to set up and diagonalize block diagonal Hamiltonian matrices in
v in order to take this coupling properly into account. In this model v
remains a good quantum number, that is, states with different v values are
not coupled in the model where the second-order perturbation theory is
used to take nonresonance couplings ($\Delta v = \pm 2$) into account.

The diagonal matrix elements in the HCAO model are a sum of two
Morse oscillator eigenvalues as given for a single oscillator in Eq. (2.3). The
nonzero diagonal and off-diagonal matrix elements are

$$\langle v_{r_1} v_{r_2} | \frac{H}{hc_0} | v_{r_1} v_{r_2} \rangle = \omega'_r (v_{r_1} + v_{r_2} + 1) + x_{rr}[(v_{r_1} + \tfrac{1}{2})^2 + (v_{r_2} + \tfrac{1}{2})^2]$$

$$\langle (v_{r_1} + 1)(v_{r_2} - 1) | \frac{H}{hc_0} | v_{r_1} v_{r_2} \rangle = \lambda_r [(v_{r_1} + 1)v_{r_2}]^{1/2}$$

$$\langle (v_{r_1} - 1)(v_{r_2} + 1) | \frac{H}{hc_0} | v_{r_1} v_{r_2} \rangle = \lambda_r [v_{r_1}(v_{r_2} + 1)]^{1/2} \tag{2.4}$$

TABLE II

Nonzero Coordinate and Momentum Operator Matrix Elements for the One- and Two-Dimensional Harmonic Oscillators[a]

$$\langle v+1|q_s|v\rangle = \langle v+1|q_s^+|v\rangle = \frac{1}{\sqrt{2}}(v+1)^{1/2}$$

$$\langle v-1|q_s|v\rangle = \langle v-1|q_s^-|v\rangle = \frac{1}{\sqrt{2}}v^{1/2}$$

$$\langle v+1|p_s|v\rangle = \frac{i}{\sqrt{2}}(v+1)^{1/2} \qquad \langle v-1|p_s|v\rangle = \frac{-i}{\sqrt{2}}v^{1/2}$$

$$\langle v+1,l+1|q_{t+}|v,l\rangle = \langle v+1,l+1|q_{t+}^+|v,l\rangle = \frac{1}{\sqrt{2}}(v+l+2)^{1/2}$$

$$\langle v-1,l+1|q_{t+}|v,l\rangle = \langle v-1,l+1|q_{t+}^-|v,l\rangle = \frac{1}{\sqrt{2}}(v-l)^{1/2}$$

$$\langle v+1,l-1|q_{t-}|v,l\rangle = \langle v+1,l-1|q_{t-}^+|v,l\rangle = \frac{1}{\sqrt{2}}(v-l+2)^{1/2}$$

$$\langle v-1,l-1|q_{t-}|v,l\rangle = \langle v-1,l-1|q_{t-}^-|v,l\rangle = \frac{1}{\sqrt{2}}(v+l)^{1/2}$$

$$\langle v+1,l+1|p_{t+}|v,l\rangle = \frac{i}{\sqrt{2}}(v+l+2)^{1/2} \qquad \langle v-1,l+1|p_{t+}|v,l\rangle = \frac{-i}{\sqrt{2}}(v-l)^{1/2}$$

$$\langle v+1,l-1|p_{t-}|v,l\rangle = \frac{i}{\sqrt{2}}(v-l+2)^{1/2} \qquad \langle v-1,l-1|p_{t-}|v,l\rangle = \frac{-i}{\sqrt{2}}(v+l)^{1/2}$$

[a] All nonzero matrix elements for the operators in question have been given. Here, q_s and $q_t = (q_{ta}, q_{tb})$ are dimensionless normal coordinates, and p_s and $p_t = (p_{ta}, p_{tb})$ are their conjugate momentum operators; $q_{t\pm} = q_{ta} \pm iq_{tb}$ and $p_{t\pm} = p_{ta} \pm ip_{tb}$; and $q_s^\pm = \frac{1}{2}(q_s \mp ip_s)$, $q_{t+}^\pm = \frac{1}{2}(q_{t+} \mp ip_{t+})$, and $q_{t-}^\pm = \frac{1}{2}(q_{t-} \mp ip_{t-})$. In the case of diatomic molecules, the dimensionless normal coordinate q and its conjugate momentum p are related to the internal displacement coordinate r and its conjugate momentum operator p_r as $q = \alpha^{1/2}r$ and $p = \hbar^{-1}\alpha^{-1/2}p_r$. Also,

$$\alpha = \frac{2\pi c_0 \omega_r}{hg_{rr}} = \frac{(f_{rr}g_{rr}^{-1})^{1/2}}{\hbar} = \frac{(2a_r^2 D_e g_{rr}^{-1})^{1/2}}{\hbar}$$

where the last equality holds for the dimensionless normal coordinate of a Morse oscillator. The parameter ω_r is the harmonic wavenumber, g_{rr} is the inverse of the reduced mass of the oscillator, and f_{rr} is the harmonic force constant. The matrix elements of r and p_r are obtained by applying the scaling given. All other matrix elements of positive-integer powers of dimensionless normal coordinates (internal displacement coordinates) and their momentum operators and their products can be obtained from these results by using the completeness relations

$$1 = \sum_{v=0}^{\infty} |v\rangle\langle v| \quad \text{and} \quad 1 = \sum_{v,l} |v,l\rangle\langle v,l|$$

for the one- and two-dimensional case, respectively.

where

$$\omega_r' = \omega_r - \tfrac{1}{8}\omega_r\left(\frac{g_{rr}^{(e)}}{g_{rr}} - \frac{f_{rr'}}{f_{rr}}\right)^2 \tag{2.5}$$

and

$$\lambda_r = \tfrac{1}{2}\omega_r\left(\frac{g_{rr}^{(e)}}{g_{rr}} + \frac{f_{rr'}}{f_{rr}}\right) \tag{2.6}$$

Second-order perturbation theory is used to remove the $\Delta v = \pm 2$ couplings. The second term in ω_r' [Eq. (2.5)] usually makes a small contribution. In water, for example, it would be about $0.004\ \mathrm{cm}^{-1}$. The effects of both the kinetic energy coupling (the term in $g_{rr'}^{(e)}$) and the potential energy coupling (the term in $f_{rr'}$) can be described by a single parameter, λ_r, as is obvious from Eqs. (2.4) and (2.6). The HCAO model contains only three parameters: the harmonic wavenumber ω_r', the bond anharmonicity parameter x_{rr}, and the coupling coefficient λ_r.

The Hamiltonian matrices can be further factorized by employing symmetrized basis functions

$$|v_{r_1}v_{r_2}\pm\rangle = \frac{1}{\sqrt{2}}(|v_{r_1}v_{r_2}\rangle \pm |v_{r_2}v_{r_1}\rangle) \qquad v_{r_1} > v_{r_2}$$
$$|v_{r_1}v_{r_2}+\rangle = |v_{r_1}v_{r_1}\rangle \qquad v_{r_1} = v_{r_2} \tag{2.7}$$

The plus and minus symbols on the left-hand sides can be replaced by the usual C_{2v} point group symmetry labels $A_1(=\mathrm{plus})$ and $B_1(=\mathrm{minus})$ if desired. When the zero-point energy is subtracted from all diagonal elements, the resulting Hamiltonian matrices up to $v = 5$ are in wavenumber units:

$$H^\pm(v=1) = \omega_r' + 2x_{rr} \pm \lambda_r$$

$$H^+(v=2) = \begin{pmatrix} 2\omega_r' + 6x_{rr} & 2\lambda_r \\ 2\lambda_r & 2\omega_r' + 4x_{rr} \end{pmatrix}$$

$$H^-(v=2) = 2\omega_r' + 6x_{rr}$$

$$H^\pm(v=3) = \begin{pmatrix} 3\omega_r' + 12x_{rr} & \sqrt{3}\,\lambda_r \\ \sqrt{3}\,\lambda_r & 3\omega_r' + 8x_{rr} \pm 2\lambda_r \end{pmatrix}$$

$$H^+(v=4) = \begin{pmatrix} 4\omega_r' + 20x_{rr} & 2\lambda_r & 0 \\ 2\lambda_r & 4\omega_r' + 14x_{rr} & \sqrt{12}\,\lambda_r \\ 0 & \sqrt{12}\,\lambda_r & 4\omega_r' + 12x_{rr} \end{pmatrix}$$

$$H^-(v=4) = \begin{pmatrix} 4\omega_r' + 20x_{rr} & 2\lambda_r \\ 2\lambda_r & 4\omega_r' + 14x_{rr} \end{pmatrix}$$

$$H^\pm(v=5) = \begin{pmatrix} 5\omega_r' + 30x_{rr} & \sqrt{5}\,\lambda_r & 0 \\ \sqrt{5}\,\lambda_r & 5\omega_r' + 22x_{rr} & \sqrt{8}\,\lambda_r \\ 0 & \sqrt{8}\,\lambda_r & 5\omega_r' + 18x_{rr} \pm 3\lambda_r \end{pmatrix} \quad (2.8)$$

The three parameters ω_r', x_{rr}, and λ_r can be obtained from experimental spectra with matrices given or by optimizing them in a least-squares calculation with experimental vibrational term values as data. Child and Lawton have demonstrated that the model works well for H_2O, SO_2, C_2H_2, and C_2D_2 [13].

The HCAO model differs in spirit from customary theories, which are based on normal coordinates [17]. For the sake of clarity, the customary model is first given in terms of matrix elements. Conventionally, anharmonic expansions including both harmonic and anharmonic terms are employed [9]. In the case of stretching vibrations in bent triatomic molecules, this expansion in wavenumber units is

$$E = \omega_1(v_1 + \tfrac{1}{2}) + \omega_3(v_3 + \tfrac{1}{2}) + x_{11}(v_1 + \tfrac{1}{2})^2$$
$$+ x_{33}(v_3 + \tfrac{1}{2})^2 + x_{13}(v_1 + \tfrac{1}{2})(v_3 + \tfrac{1}{2}) + \cdots \quad (2.9)$$

where ω_1 and ω_3 are symmetric and antisymmetric stretching harmonic wavenumbers, respectively, and x parameters describe stretching anharmonicities. In 1940 Darling and Dennison found that this expansion does not describe well all stretching overtone and combination states [20]. The model is improved by including quartic anharmonic resonance terms, so-called Darling–Dennison resonance terms, which can be defined by their matrix elements as [21]

$$\langle v_1 + 2, v_3 - 2 | \frac{H}{hc_0} | v_1, v_3 \rangle = \tfrac{1}{4}K_{11;33}[(v_1 + 1)(v_1 + 2)v_3(v_3 - 1)]^{1/2}$$

$$\langle v_1 - 2, v_3 + 2 | \frac{H}{hc_0} | v_1, v_3 \rangle = \tfrac{1}{4}K_{11;33}[v_1(v_1 - 1)(v_3 + 1)(v_3 + 2)]^{1/2}$$

$$(2.10)$$

Here, $|v_1, v_3\rangle = |v_1\rangle|v_3\rangle$ is a product of harmonic oscillator eigenfunctions expressed in terms of normal coordinates. The product functions are chosen to be consistent with the normal mode harmonic wavenumbers ω_1 and ω_3. The parameter $K_{11;33}$ describes the strength of this coupling. It is a function of quadratic, cubic, and quartic force constants [for a theoretical expression obtained by perturbation theory see Ref. [21] and Eq. (2.32) later]. In water, the Darling–Dennison resonance coupling exists, for example, between $2v_1$ and $2v_3$, between $3v_1$ and $v_1 + 2v_3$, and between $2v_1 + v_3$ and $3v_3$. The notation is such that v_1 and v_3 denote the symmetric and the antisymmetric stretching fundamentals, respectively, and the normal mode labels refer to interacting states. Note that as in the HCAO model, $v = v_1 + v_3$ remains a good quantum label; that is, only states with the same v can be coupled.

The customary model including Darling–Dennison resonances was given in Eqs. (2.9) and (2.10) in matrix form. It can also be given in operator form as follows [3]:

$$\frac{H}{hc_0} = \tfrac{1}{2}\omega_1(p_1^2 + q_1^2) + \tfrac{1}{2}\omega_3(p_3^2 + q_3^2) + \tfrac{1}{4}x_{11}(p_1^2 + q_1^2)^2 + \tfrac{1}{4}x_{33}(p_3^2 + q_3^2)^2$$

$$+ \tfrac{1}{4}x_{13}(p_1^2 + q_1^2)(p_3^2 + q_3^2) + K_{11;33}[(q_1^+)^2(q_3^-)^2 + (q_1^-)^2(q_3^+)^2]$$

$$(2.11)$$

where q_1 and q_3 are the dimensionless normal coordinates associated with the symmetric and antisymmetric stretch, respectively. Their conjugate momentum operators are $p_1 = -i\,\partial/\partial q_1$ and $p_3 = -i\,\partial/\partial q_3$. Dimensionless normal coordinates q_s are related to usual normal coordinates Q_s as $q_s = (2\pi c_0\omega_s/\hbar)^{1/2}Q_s$, where $s = 1,\ 3$ [19,22,23]. Also, $q_s^\pm = \tfrac{1}{2}(q_s \mp ip_s)$. Using product harmonic oscillator basis functions $|v_1, v_3\rangle = |v_1\rangle|v_3\rangle$, which are expressed in terms of q_1 and q_3 and which are consistent with the harmonic wavenumbers ω_1 and ω_3, the matrix elements given in Eqs. (2.9) and (2.10) can be obtained. The appropriate harmonic oscillator matrix elements can be obtained from the results given in Table II. Note that if, for example, the coupling term in $K_{11;33}$ had been written $K_{11;33}\,q_1^2 q_3^2$, then it would have the undesirable property of possessing diagonal matrix elements in addition to off-diagonal matrix elements. Theoretical justification of the Hamiltonian is given later. The Hamiltonian is presented in the operator form in order to emphasize the quartic nature of the vibrationally off-diagonal Darling–Dennison resonance term in $K_{11;33}$. This should be contrasted with the local mode model where the off-diagonal coupling is provided by harmonic bilinear kinetic and potential energy terms.

The customary theory with Darling–Dennison resonance terms included and the HCAO model used for stretching vibrational states of well-bent triatomic molecules look different. However, the two matrix representations as given in Eqs. (2.4) and in Eqs. (2.9) and (2.10) produce identical eigenvalues if the following relations, x–K relations, are applied between vibrational parameters [21,24]:

$$\omega_1 = \omega_r' + \lambda_r$$
$$\omega_3 = \omega_r' - \lambda_r \tag{2.12}$$
$$x_{11} = x_{33} = \tfrac{1}{4}x_{13} = \tfrac{1}{4}K_{11;33} = \tfrac{1}{2}x_{rr}$$

When the zero-point energy has been subtracted from all diagonal elements, the Hamiltonian matrices (in reciprocal centimeters) in the Darling–Dennison resonance model up to $v = v_1 + v_3 = 5$ become

$$H^+(v = 1) = \omega_1 + 2x_{11} + \tfrac{1}{2}x_{13} = \omega_r' + 2x_{rr} + \lambda_r$$

$$H^-(v = 1) = \omega_3 + 2x_{33} + \tfrac{1}{2}x_{13} = \omega_r' + 2x_{rr} - \lambda_r$$

$$H^+(v = 2) = \begin{pmatrix} 2\omega_1 + 6x_{11} + x_{13} & \tfrac{1}{2}K_{11;33} \\ \tfrac{1}{2}K_{11;33} & 2\omega_3 + 6x_{33} + x_{13} \end{pmatrix}$$
$$= \begin{pmatrix} 2\omega_r' + 5x_{rr} + 2\lambda_r & x_{rr} \\ x_{rr} & 2\omega_r' + 5x_{rr} - 2\lambda_r \end{pmatrix}$$

$$H^-(v = 2) = \omega_1 + \omega_3 + 2x_{11} + 2x_{33} + 2x_{13} = 2\omega_r' + 6x_{rr}$$

$$H^+(v = 3) = \begin{pmatrix} 3\omega_1 + 12x_{11} + \tfrac{3}{2}x_{13} & \tfrac{1}{2}\sqrt{3}\,K_{11;33} \\ \tfrac{1}{2}\sqrt{3}\,K_{11;33} & \omega_1 + 2\omega_3 + 2x_{11} + 6x_{33} + \tfrac{7}{2}x_{13} \end{pmatrix}$$
$$= \begin{pmatrix} 3\omega_r' + 9x_{rr} + 3\lambda_r & \sqrt{3}\,x_{rr} \\ \sqrt{3}\,x_{rr} & 3\omega_r' + 11x_{rr} - \lambda_r \end{pmatrix}$$

$$H^-(v = 3) = \begin{pmatrix} 2\omega_1 + \omega_3 + 6x_{11} + 2x_{33} + \tfrac{7}{2}x_{13} & \tfrac{1}{2}\sqrt{3}\,K_{11;33} \\ \tfrac{1}{2}\sqrt{3}\,K_{11;33} & 3\omega_3 + 12x_{33} + \tfrac{3}{2}x_{13} \end{pmatrix}$$
$$= \begin{pmatrix} 3\omega_r' + 11x_{rr} + \lambda_r & \sqrt{3}\,x_{rr} \\ \sqrt{3}\,x_{rr} & 3\omega_r' + 9x_{rr} - 3\lambda_r \end{pmatrix}$$

$$H^+(v = 4) = \begin{pmatrix} 4\omega_1 + 20x_{11} + 2x_{13} & \tfrac{1}{2}\sqrt{6}\,K_{11;33} & 0 \\ \tfrac{1}{2}\sqrt{6}\,K_{11;33} & 2\omega_1 + 2\omega_3 + 6x_{11} + 6x_{33} + 6x_{13} & \tfrac{1}{2}\sqrt{6}\,K_{11;33} \\ 0 & \tfrac{1}{2}\sqrt{6}\,K_{11;33} & 4\omega_3 + 20x_{33} + 2x_{13} \end{pmatrix}$$
$$= \begin{pmatrix} 4\omega_r' + 14x_{rr} + 4\lambda_r & \sqrt{6}\,x_{rr} & 0 \\ \sqrt{6}\,x_{rr} & 4\omega_r' + 18x_{rr} & \sqrt{6}\,x_{rr} \\ 0 & \sqrt{6}\,x_{rr} & 4\omega_r' + 14x_{rr} - 4\lambda_r \end{pmatrix}$$

$$H^-(v=4) = \begin{pmatrix} 3\omega_1 + \omega_3 + 12x_{11} + 2x_{33} + 5x_{13} & \frac{3}{2}K_{11;33} \\ \frac{3}{2}K_{11;33} & \omega_1 + 3\omega_3 + 2x_{11} + 12x_{33} + 5x_{13} \end{pmatrix}$$

$$= \begin{pmatrix} 4\omega_r' + 17x_{rr} + 2\lambda_r & 3x_{rr} \\ 3x_{rr} & 4\omega_r' + 17x_{rr} - 2\lambda_r \end{pmatrix}$$

$$H^+(v=5) = \begin{pmatrix} 5\omega_1 + 30x_{11} + \frac{5}{2}x_{13} & \frac{1}{2}\sqrt{10}\,K_{11;33} & 0 \\ \frac{1}{2}\sqrt{10}\,K_{11;33} & 3\omega_1 + 2\omega_3 + 12x_{11} + 6x_{33} + \frac{17}{2}x_{13} & \frac{3}{2}\sqrt{2}\,K_{11;33} \\ 0 & \frac{3}{2}\sqrt{2}\,K_{11;33} & \omega_1 + 4\omega_3 + 2x_{11} + 20x_{33} + \frac{13}{2}x_{13} \end{pmatrix}$$

$$= \begin{pmatrix} 5\omega_r' + 20x_{rr} + 5\lambda_r & \sqrt{10}\,x_{rr} & 0 \\ \sqrt{10}\,x_{rr} & 5\omega_r' + 26x_{rr} + \lambda_r & 3\sqrt{2}\,x_{rr} \\ 0 & 3\sqrt{2}\,x_{rr} & 5\omega_r' + 24x_r - 3\lambda_r \end{pmatrix}$$

$$H^-(v=5) = \begin{pmatrix} 4\omega_1 + \omega_3 + 20x_{11} + 2x_{33} + \frac{13}{2}x_{13} & \frac{3}{2}\sqrt{2}\,K_{11;33} & 0 \\ \frac{3}{2}\sqrt{2}\,K_{11;33} & 2\omega_1 + 3\omega_3 + 6x_{11} + 12x_{33} + \frac{17}{2}x_{13} & \frac{1}{2}\sqrt{10}\,K_{11;33} \\ 0 & \frac{1}{2}\sqrt{10}\,K_{11;33} & 5\omega_3 + 30x_{33} + \frac{5}{2}x_{13} \end{pmatrix}$$

$$= \begin{pmatrix} 5\omega_r' + 24x_{rr} + 3\lambda_r & 3\sqrt{2}\,x_{rr} & 0 \\ 3\sqrt{2}\,x_{rr} & 5\omega_r' + 26x_{rr} - \lambda_r & \sqrt{10}\,x_{rr} \\ 0 & \sqrt{10}\,x_{rr} & 5\omega_r' + 20x_{rr} - 5\lambda_r \end{pmatrix} \quad (2.13)$$

The matrices in Eqs. (2.8) and (2.13) are different. It is not immediately obvious that they produce identical eigenvalues when the x–K relations are obeyed. Lehmann was the first to prove this [24]. The equivalence of the two models is easy to demonstrate by using creation and annihilation operators. Lehmann did this for the first time in the case of the stretching vibrations of benzene [25], and later Della Valle extended this work [26]. The local mode creation and annihilation operators a_1^+, a_1^-, a_2^+, and a_2^- are defined by the equations

$$a_1^+ | v_{r_1} v_{r_2} \rangle = (v_{r_1} + 1)^{1/2} | (v_{r_1} + 1) v_{r_2} \rangle$$
$$a_2^+ | v_{r_1} v_{r_2} \rangle = (v_{r_2} + 1)^{1/2} | v_{r_1} (v_{r_2} + 1) \rangle$$
$$a_1^- | v_{r_1} v_{r_2} \rangle = v_{r_1}^{1/2} | (v_{r_1} - 1) v_{r_2} \rangle$$
$$a_2^- | v_{r_1} v_{r_2} \rangle = v_{r_2}^{1/2} | v_{r_1} (v_{r_2} - 1) \rangle$$

$$(2.14)$$

where $| v_{r_1} v_{r_2} \rangle = | v_{r_1} \rangle | v_{r_2} \rangle$ is a local mode product basis function. These operators obey the usual commutation rules $[a_i^-, a_j^+] = \delta_{ij}$ and $[a_i^+, a_j^+] = [a_i^-, a_j^-] = 0$. The number operators $v_{r_1} = a_1^+ a_1^-$ and $v_{r_2} = a_2^+ a_2^-$, where v_{r_1} and v_{r_2} are vibrational populations, form a complete set of commuting operators. Their basis is the basis of the local mode Hamiltonian defined through the matrix elements in Eqs. (2.4) [27]. It is obtained from the ground state $| 00 \rangle$ as [27]

$$| v_{r_1} v_{r_2} \rangle = \frac{1}{\sqrt{v_{r_1}! v_{r_2}!}} (a_1^+)^{v_{r_1}} (a_2^+)^{v_{r_2}} | 00 \rangle \tag{2.15}$$

The local mode Hamiltonian defined through the matrix elements in Eq. (2.4) is expressed in terms of the creation and annihilation operators as

$$\frac{H}{hc_0} = \omega_r'(a_1^+ a_1^- + a_2^+ a_2^- + 1)$$
$$+ x_{rr}[a_1^+ a_1^- (a_1^+ a_1^- + 1) + a_2^+ a_2^- (a_2^+ a_2^- + 1) + \tfrac{1}{2}]$$
$$+ \lambda_r(a_1^+ a_2^- + a_2^+ a_1^-) \tag{2.16}$$

The transformation between the local and normal mode representations is found by considering the couplings between the upper states of harmonic stretching fundamentals in both cases. Unlike in the local mode case, the harmonic coupling is absent in the normal mode representation. This is part of the normal coordinate definition. In the present application of two identical bond oscillators, the unitary transformation between the two representations is

$$\begin{pmatrix} \omega_1 & 0 \\ 0 & \omega_3 \end{pmatrix} = \mathbf{D}^\dagger \mathbf{N} \mathbf{D} = \mathbf{D}^{-1} \mathbf{N} \mathbf{D} = \begin{pmatrix} \dfrac{1}{\sqrt{2}} & \dfrac{1}{\sqrt{2}} \\ \dfrac{1}{\sqrt{2}} & -\dfrac{1}{\sqrt{2}} \end{pmatrix} \begin{pmatrix} \omega_r' & \lambda_r \\ \lambda_r & \omega_r' \end{pmatrix} \begin{pmatrix} \dfrac{1}{\sqrt{2}} & \dfrac{1}{\sqrt{2}} \\ \dfrac{1}{\sqrt{2}} & -\dfrac{1}{\sqrt{2}} \end{pmatrix}$$

$$(2.17)$$

where N is the Hamiltonian matrix in the HCAO model for stretching fundamentals when the basis functions used are in the unsymmetrized form ($|v_{r_1}v_{r_2}\rangle = |10\rangle$ and $|v_{r_1}v_{r_2}\rangle = |01\rangle$). The normal mode creation and annihilation operators A_s^+ and A_s^- can be expressed as linear combinations of the local mode creation and annihilation operators a_i^+ and a_i^- in a general case as [26]

$$A_s^+ = \sum_i D_{is} a_i^+ \qquad A_s^- = \sum_i D_{is}^* a_i^- \qquad (2.18)$$

These equations can be inverted easily because D is a unitary matrix. For the present application, the normal mode creation and annihilation operators A_1^+, A_1^-, A_3^+ and A_3^- are given as the linear combinations

$$A_1^+ = \frac{1}{\sqrt{2}}(a_1^+ + a_2^+) \qquad A_1^- = \frac{1}{\sqrt{2}}(a_1^- + a_2^-)$$

$$A_3^+ = \frac{1}{\sqrt{2}}(a_1^+ - a_2^+) \qquad A_3^- = \frac{1}{\sqrt{2}}(a_1^- - a_2^-) \qquad (2.19)$$

These definitions, together with commutation relations between the local mode creation and annihilation operators, ensure that the normal mode operators satisfy the usual commutation relations $[A_s^-, A_{s'}^+] = \delta_{ss'}$ and $[A_s^+, A_{s'}^+] = [A_s^-, A_{s'}^-] = 0$. Therefore, it is justified to interpret the normal mode operators as creation and annihilation operators. The number operators $v_1 = A_1^+ A_1^-$ and $v_3 = A_3^+ A_3^-$ also form a complete set of commuting operators, The basis $|v_1, v_3\rangle = |v_1\rangle|v_3\rangle$ of these number operators is called the normal mode basis. The operator equations take the forms

$$A_1^+ |v_1, v_3\rangle = (v_1 + 1)^{1/2} |v_1 + 1, v_3\rangle \qquad A_1^- |v_1, v_3\rangle = v_1^{1/2} |v_1 - 1, v_3\rangle$$

$$A_3^+ |v_1, v_3\rangle = (v_3 + 1)^{1/2} |v_1, v_3 + 1\rangle \qquad A_3^- |v_1, v_3\rangle = v_3^{1/2} |v_1, v_3 - 1\rangle$$

$$(2.20)$$

The normal mode basis functions are obtained from the ground state $|00\rangle = |0, 0\rangle$ as

$$|v_1, v_3\rangle = \frac{1}{\sqrt{v_1! v_3!}} (A_1^+)^{v_1}(A_3^+)^{v_3} |0,0\rangle \qquad (2.21)$$

Equations (2.15) and (2.21) provide a link in relating both the normal and the local mode basis sets with each other. The Darling–Dennison resonance Hamiltonian defined by its matrix elements in Eqs. (2.9) and (2.10) can be written in terms of these symmetrized creation and annihilation operators as

$$
\frac{H}{hc_0} = \omega_1(A_1^+ A_1^- + \tfrac{1}{2}) + \omega_3(A_3^+ A_3^- + \tfrac{1}{2})
$$

$$
+ x_{11}[A_1^+ A_1^-(A_1^+ A_1^- + 1) + \tfrac{1}{4}]
$$

$$
+ x_{33}[A_3^+ A_3^-(A_3^+ A_3^- + 1) + \tfrac{1}{4}]
$$

$$
+ x_{13}(A_1^+ A_1^- A_3^+ A_3^- + \tfrac{1}{2}(A_1^+ A_1^- + A_3^+ A_3^- + \tfrac{1}{2})
$$

$$
+ \tfrac{1}{4}K_{11;\,33}(A_1^+ A_1^+ A_3^- A_3^- + A_3^+ A_3^+ A_1^- A_1^-) \qquad (2.22)
$$

Using Eq. (2.16) and x–K relations, it can be proved that the Hamiltonians in Eqs. (2.16) and (2.22) represent the same problem in different representations. The eigenvalues are identical when the x–K relations are applied (note that there is an insignificant difference in the zero-point energy [21]).

There exists an approximate but physically revealing way to relate the local mode and the normal mode Hamiltonians [21]. It is based on perturbation theory, which is often used in the conventional vibration–rotation theory. The starting point is the local mode Hamiltonian as given in Eq. (2.1). The Morse potential energy functions are expanded up to the fourth order and the potential energy coupling term in $f_{rr'}$ up to the second order. This yields the result

$$
H = \tfrac{1}{2}g_{rr}(p_{r_1}^2 + p_{r_2}^2) + g_{rr'}^{(e)} p_{r_1} p_{r_2} + \tfrac{1}{2} f_{rr}(r_1^2 + r_2^2)
$$

$$
+ \tfrac{1}{6} f_{rrr}(r_1^3 + r_2^3) + \tfrac{1}{24} f_{rrrr}(r_1^4 + r_2^4) + f_{rr'} r_1 r_2 \qquad (2.23)
$$

Note that only bilinear potential energy coupling terms in variables r_1 and r_2 have been included in this procedure. This is justified because the effects of the higher order potential energy terms are small when compared with the terms included in the model. The Hamiltonian in Eq. (2.23) is transformed to the symmetry coordinate representation, where these coordinates are defined in the standard way as

$$
S_1 = \frac{1}{\sqrt{2}} (r_1 + r_2) \qquad S_3 = \frac{1}{\sqrt{2}} (r_1 - r_2) \qquad (2.24)
$$

The Hamiltonian is expressed in terms of the S_j coordinates and their conjugate momenta $P_{S_j} = -i\hbar \, \partial/\partial S_j \, (j = 1, 3)$ as

$$H = \tfrac{1}{2}G_{11}^{(e)}P_{S_1}^2 + \tfrac{1}{2}G_{33}^{(e)}P_{S_3}^2 + \tfrac{1}{2}F_{11}S_1^2 + \tfrac{1}{2}F_{33}S_3^2$$
$$+ \tfrac{1}{6}F_{111}S_1^3 + \tfrac{1}{2}F_{133}S_1S_3^2 + \tfrac{1}{24}F_{1111}S_1^4$$
$$+ \tfrac{1}{4}F_{1133}S_1^2S_3^2 + \tfrac{1}{24}F_{3333}S_3^4 \tag{2.25}$$

where

$$G_{11}^{(e)} = g_{rr} + g_{rr'}^{(e)} \qquad G_{33}^{(e)} = g_{rr} - g_{rr'}^{(e)}$$
$$F_{11} = f_{rr} + f_{rr'} = 2a_r^2 D_e + f_{rr'}$$
$$F_{33} = f_{rr} - f_{rr'} = 2a_r^2 D_e - f_{rr'}$$
$$F_{111} = F_{133} = \frac{1}{\sqrt{2}} f_{rrr} = -\frac{1}{\sqrt{2}} 6a_r^3 D_e \tag{2.26}$$
$$F_{1111} = F_{1133} = F_{3333} = \tfrac{1}{2} f_{rrrr} = 7a_r^4 D_e$$

In the present model, the dimensionless normal coordinates q_1 and q_3 are obtained by appropriate scaling of the symmetry coordinates S_1 and S_3 as

$$q_i = \left(\frac{2\pi c_0 \omega_i}{\hbar G_{ii}^{(e)}}\right)^{1/2} S_i = \alpha_i^{1/2} S_i = \left(\frac{F_{ii}}{G_{ii}^{(e)}\hbar^2}\right)^{1/4} S_i \tag{2.27}$$

because the bending mode is completely decoupled from the symmetric stretch and because the quadratic cross term between S_1 and S_3 disappears due to symmetry (the operator S_1S_3 is not totally symmetric as S_1 is symmetric and S_3 is antisymmetric with respect to the permutation of r_1 and r_2). The Hamiltonian is written in terms of q_1 and q_3 and their conjugate momenta $p_1 = -i \, \partial/\partial q_1$ and $p_3 = -i \, \partial/\partial q_3$ in wavenumber units as

$$\frac{H}{hc_0} = \tfrac{1}{2}\omega_1(p_1^2 + q_1^2) + \tfrac{1}{2}\omega_3(p_3^2 + q_3^2) + \tfrac{1}{6}\phi_{111}q_1^3 + \tfrac{1}{2}\phi_{133}q_1q_3^2$$
$$+ \tfrac{1}{24}\phi_{1111}q_1^4 + \tfrac{1}{4}\phi_{1133}q_1^2q_3^2 + \tfrac{1}{24}\phi_{3333}q_3^4 \tag{2.28}$$

where

$$\omega_1 = \frac{F_{11}\alpha_1^{-1}}{hc_0} = \frac{(G_{11}F_{11})^{1/2}}{2\pi c_0} \qquad \omega_3 = \frac{F_{33}\alpha_3^{-1}}{hc_0} = \frac{(G_{33}F_{33})^{1/2}}{2\pi c_0}$$

$$\phi_{111} = \frac{F_{111}\alpha_1^{-3/2}}{hc_0} \qquad \phi_{133} = \frac{F_{133}\alpha_1^{-1/2}\alpha_3^{-1}}{hc_0} \qquad (2.29)$$

$$\phi_{1111} = \frac{F_{1111}\alpha_1^{-2}}{hc_0} \qquad \phi_{1133} = \frac{F_{1133}\alpha_1^{-1}\alpha_3^{-1}}{hc_0} \qquad \phi_{3333} = \frac{F_{3333}\alpha_3^{-2}}{hc_0}$$

The eigenvalues of the Hamiltonian given in Eq. (2.28) can be obtained variationally using harmonic oscillator eigenfunctions as basis functions. The Hamiltonian matrices (one of A_1 and the other of B_1 symmetry) obtained in this way are not in a block diagonal form in $v = v_1 + v_3$. The normal mode Hamiltonian with Darling–Dennison resonance terms included is derived from Eq. (2.28) by applying perturbation theory to the anharmonic terms. In order to obtain a block diagonal Hamiltonian in $v = v_1 + v_3$, second-order perturbation theory is used for the cubic terms and first-order perturbation theory for the quartic terms. Theoretical expressions for the normal mode vibrational anharmonicity parameters x_{11}, x_{33}, and x_{13} and for the Darling–Dennison resonance interaction constant $K_{11;33}$ are obtained with this procedure. This gives a theoretical justification for the normal Hamiltonian given in Eq. (2.11). A further approximation is made by setting $\omega = \omega_1 = \omega_3$ and $\alpha = \alpha_1 = \alpha_3$. This leads to relations $\phi_{111} = \phi_{133}$ and $\phi_{1111} = \phi_{1133} = \phi_{3333}$. This approximation corresponds to the neglect of some small off-diagonal matrix elements connecting different values of v [21]. The final results are most easily obtainable by using the general formulas both for x and for K parameters [21–23]. In bent XY_2 molecules, the following theoretical expressions have been derived:

$$x_{ss} = \frac{1}{16}\phi_{ssss} - \frac{1}{16}\sum_{s'}\phi_{sss'}^2\frac{8\omega_s^2 - 3\omega_{s'}^2}{\omega_s(4\omega_s^2 - \omega_{s'}^2)} \qquad s = 1, 3 \qquad (2.30)$$

$$x_{13} = \tfrac{1}{4}\phi_{1133} - \frac{\tfrac{1}{4}\phi_{111}\phi_{133}}{\omega_1} - \frac{\tfrac{1}{4}\phi_{112}\phi_{233}}{\omega_2}$$

$$- \tfrac{1}{2}\phi_{133}^2\frac{\omega_3}{4\omega_3^2 - \omega_1^2} + C_e(\zeta_{13}^{(c)})^2\left(\frac{\omega_1}{\omega_3} + \frac{\omega_3}{\omega_1}\right) \qquad (2.31)$$

and

$$K_{11;33} = \tfrac{1}{4}\phi_{1133} + \frac{\tfrac{1}{6}\phi_{111}\phi_{133}}{\omega_1 + \omega_3}$$

$$+ \tfrac{1}{4}\phi_{112}\phi_{233} \frac{\omega_2}{(\omega_1+\omega_3)^2 - \omega_2^2} - \frac{\phi_{133}^2}{2\omega_3} - 4C_e(\zeta_{13}^{(c)})^2 \quad (2.32)$$

where the summation index s' goes over the three normal modes, C_e is the smallest of the three equilibrium rotational constants, $\zeta_{13}^{(c)}$ is the Coriolis coupling coefficient between symmetric and antisymmetric stretching modes around the c axis (this axis, which is perpendicular to the molecular plane, corresponds to the largest principal moments of inertia), and the subindex 2 refers to the bending normal vibration. In the present model, the cubic force constants ϕ_{112} and ϕ_{233} disappear because the bending and stretching modes are assumed to be decoupled. It is a good approximation to set $\zeta_{13}^{(c)} = 0$, which implies that there is no Coriolis coupling between the rotational states of the symmetric and antisymmetric stretching states [28]. After these simplifications, the following relations are obtained:

$$x_{11} = x_{33} = \tfrac{1}{4}x_{13} = \tfrac{1}{4}K_{11;33} = \tfrac{1}{16}\phi_{1111} - \frac{5\phi_{111}^2}{48\omega_r'} \quad (2.33)$$

Thus, the previously derived x–K relations are recovered.

It is remarkable that the Darling–Dennison resonance Hamiltonian is obtained from the local mode Hamiltonian after the approximations made. The Morse potential energy functions have been expanded only up to the fourth order and perturbation theory has been used. In diatomic molecules, a similar kind of approximate solution based on a Taylor series expansion and second-order perturbation theory produces the Morse oscillator energy level formula exactly. In this light, it seems likely that the anharmonic vibrational energy levels of polyatomic molecules calculated from the perturbation theory treatment represent the exact solution up to higher energies than the truncated polynomial expansion of the Morse potential would suggest [21].

The HCAO model and the Darling–Dennison resonance model with x–K relations applied provide a simple and physically clear way to investigate the structure of stretching vibrational states in water-type well-bent triatomic molecules. These two equivalent models contain only three independent parameters: the harmonic wavenumber ω_r', the bond anharmonicity parameter x_{rr}, and the coupling coefficient λ_r. The harmonic wavenumber ω_r' affects just the separation of different overtone manifolds.

The two other parameters, x_{rr} and λ_r, determine the energy level structure within different overtone manifolds [3,13]. In the vibrational local mode limit there is no vibrational coupling between the bond oscillators. This can be achieved by setting $\lambda_r = 0$. The energy level structure in this limit is obtained from Eqs. (2.4) with $\lambda_r = 0$. This situation is shown in the left-hand side of Fig. 3. This figure also shows which states are coupled by the off-diagonal terms in λ_r. It is seen that if the $|v_{r_1} v_{r_2}\rangle$ state is labeled as $[v_{r_1} v_{r_2}]$ then the [10] and [01] states are coupled in the first order (they are directly coupled by matrix elements in λ_r), [20] and [02] are coupled in the second order ([20] is coupled to [11], which is coupled to [02]), [30] and [03] are coupled in the third order, [40] and [04] are coupled in the fourth order, and so on. Near the local mode limit ($|\lambda_r| \ll |x_{rr}|$), splittings between the lowest members of each overtone manifold rapidly decrease in moving to higher overtone manifolds. The same conclusion holds for the second, third, and so on, lowest pairs in each overtone manifold. This can be interpreted physically by observing that the bond anharmonicity terms become dominant at large vibrational quantum numbers when compared with interbond coupling terms in λ_r. Due to the anharmonicity term, the energy difference between the interacting states is proportional to the vibrational quantum number v_r, whereas the coupling term is proportional to $\sqrt{v_r}$. At large energies, the bond oscillators become effectively decoupled from each other, and the vibrational energies of the states near the local mode limit can be described well with the uncoupled Morse oscillator energy level formula. This argument holds particularly well for the so-called local mode pair of states $[v_r 0+]$ and $[v_r 0-]$, which in the near-local-mode molecules get quickly decoupled from the other states as the vibrational quantum number v_r increases. This is the reason why the high overtone spectra of many molecules such as different hydrocarbons [1] and benzene [8] have been described successfully by an expression derived from the Morse energy level formula.

The normal mode limit can be achieved by setting $x_{rr} = 0$ in Eqs. (2.4). The levels of each overtone manifold possess a separation $2\lambda_r$ between successive levels. Near this limit, the coupling term in λ_r dominates over the bond anharmonicity term in x_{rr}. In this case, close local mode degeneracies do not occur.

The positions of the vibrational levels between the two limiting cases are obtained by diagonalizing the Hamiltonian matrices in either of the models. A convenient way to investigate vibrational energy level structures is to construct correlation diagrams by plotting the reduced eigenvalues [13]

$$\varepsilon = \frac{E - \bar{E}}{(\lambda_r^2 + x_{rr}^2)^{1/2}} \tag{2.34}$$

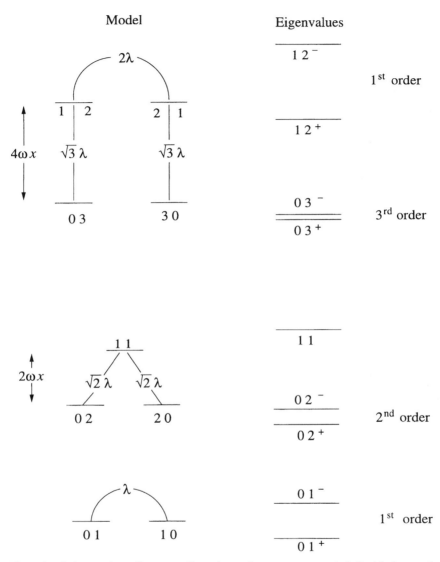

Figure 3. Anharmonic oscillator coupling schemes for $v = v_{r_1} + v_{r_2} = 1, 2, 3$ with the zeroth-order states labeled by $|v_{r_1}v_{r_2}\rangle$. Those levels arising from the coupling are approximately $|v_{r_1}v_{r_2}\pm\rangle = (1/\sqrt{2})(|v_{r_1}v_{r_2}\rangle \pm |v_{r_2}v_{r_1}\rangle)$ if $v_{r_1} \neq v_{r_2}$ and $|v_{r_1}v_{r_2}\rangle = |v_{r_1}v_{r_1}\rangle$ if $v_{r_1} = v_{r_2}$; $\omega x = -x_{rr}$ and $\lambda = \lambda_r$. (From Ref. [3] (Child and Halonen) with permission.)

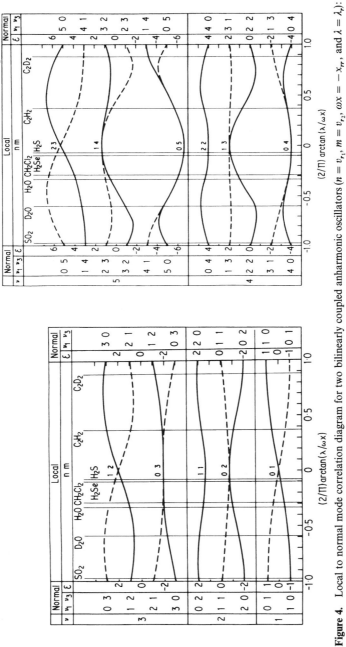

Figure 4. Local to normal mode correlation diagram for two bilinearly coupled anharmonic oscillators ($n = v_{r_1}$, $m = v_{r_2}$, $\omega x = -x_{rr}$, and $\lambda = \lambda_r$):
(a) $v = v_{r_1} + v_{r_2} = 1$, 2, 3 and (b) $v = v_{r_1} + v_{r_2} = 4$, 5. (From Ref. [3] (Child and Halonen) with permission.)

TABLE III

Harmonically Coupled Anharmonic Oscillator Model Parameters for Water, Acetylene, and Sulfur Dioxide[a]

Parameter	H_2O	C_2H_2	C_2D_2	SO_2
ω_r', cm^{-1}	3848.71 (420)	3428.462 (780)	2617.65 (150)	1270.16 (160)
x_{rr}, cm^{-1}	−77.914 (980)	−50.487 (160)	−23.447 (310)	−7.169 (440)
λ_r, cm^{-1}	−43.67 (270)	45.770 (630)	135.123 (700)	−103.875 (430)

[a] The uncertainties given in parentheses are one standard errors in the least significant digit.

as a function of the local mode parameter $\xi = -x_{rr}/\lambda_r$. Here, E is the energy of the vibrational level in question, and $\bar{E}$ is the mean energy of a particular overtone manifold. This kind of diagram is shown in Fig. 4a and b, where the normal mode limits $(|\lambda_r| \gg |x_{rr}|)$ are on the left- and right-hand sides and the local mode limit $(|\lambda_r| \ll |x_{rr}|)$ is in the center [3]. A striking result seen in Fig. 4 is the large difference in the energy level patterns of the two limits: The local mode limit is characterized by unusual degeneracies according to the standard vibration–rotation theory, and the normal mode limit is characterised by equally spaced levels within each overtone manifold.

The local mode parameter ξ is an experimentally determinable quantity, and its value is known for many bent triatomic molecules. The positions of H_2O, H_2S, H_2Se, D_2O, CH_2Cl_2, SO_2, C_2H_2, and C_2D_2 are shown in Fig. 4. The H_2S and H_2Se molecules are close to the vibrational local model limit. The stretching vibrational overtone and combination states are best described by local mode labels $[v_{r_1}v_{r_2} \pm]$. On the other hand, SO_2 and C_2D_2, are normal mode molecules in their vibrational behavior. Consequently, the customary normal mode labels (v_1v_3) remain a good representation up to relatively high energies. At low stretching energies, the intermediate cases, H_2O, D_2O, D_2S, and C_2H_2, can be described well by both labeling schemes. At high energies, the close local mode degeneracies start to dominate in the energy level structure and the local mode labels are physically more meaningful than the customary normal mode labels.

The fact that the simple vibrational model presented reproduces experimental stretching vibrational term values accurately is demonstrated by applying the model to H_2O, C_2H_2, C_2D_2, and SO_2. The three model parameters ω_r', x_{rr}, and λ_r, have been optimized by the nonlinear least-squares method using vibrational term values as data [11,12,29–33]. The parameters are given in Table III and the fits are shown in Table IV. The model works well for both acetylene isotopic species and sulfur dioxide,

TABLE IV
Experimental and Model Vibrational Term Values (cm^{-1}) for Water, Sulfur Dioxide, and Acetylene[a]

Normal $v_1 v_3$	Local $v_{r_1} v_{r_2} \pm$	H$_2$O		^{32}SO$_2$	
		v_{obs}	v_{calc}	v_{obs}	v_{calc}
10	10+	3657.05	3649.21	1151.71	1151.95
01	10−	3755.93	3736.56	1362.06	1359.70
20	20+	7201.54	7190.81	2295.81	2296.60
11	20−	7249.82	7229.94	2499.87	2497.31
02	11+	7445.07	7424.90	2715.46	2712.35
30	30+	10599.69	10588.05	3431.19	3433.95
21	30−	10613.36	10597.32	3629.61	3627.39
12	21+	10868.88	10858.61	3837.06	3835.88
03	21−	11032.41	11024.03	4054.26	4057.95
40	40+	13828.28	13819.58	—	4563.98
31	40−	13830.94	13820.79	4751.23	4749.91
22	31+	14221.16	14225.86	—	4951.71
13	31−	14318.81	14319.85	5165.64	5167.63
04	22+	14536.87	14555.09	—	5396.47
50	50+	16898.40	16890.77	—	5686.67
41	50−	16898.84	16890.88	—	5864.83
32	41+	17458.35	17479.83	—	6059.77
23	41−	17495.53	17512.04	—	6269.47
14	32+	17748.07	17775.12	—	6492.52
05	32−	—	18004.83	—	6727.93

Normal $v_1 v_3$	Local $v_{r_1} v_{r_2} \pm$	C$_2$H$_2$		C$_2$D$_2$	
		v_{obs}	v_{calc}	v_{obs}	v_{calc}
10	10+	3372.85	3373.26	2705.16	2705.88
01	10−	3294.84[b]	3281.72	2439.24	2435.63
02	20+	6502.32	6499.95	—	4846.80
11	20−	6556.48	6554.00	5097.20	5094.62
20	11+	6709.02	6709.03	—	5389.32
12	30+	9663.35[c]	9659.50	—	7457.11
03	30−	9639.85	9638.15	7230.04	7233.23
30	21+	—	9993.08	—	8050.09
21	21−	9835.16	9831.35	7734.01	7733.47
04	40+	12671.55[c]	12673.69	—	9594.53

whereas for water the fit obtained is less pleasing. In order to obtain better results for water, it is necessary to relax the x–K relations and it is necessary to include effects of the bending vibration [34]. Vibrational cubic interaction called Fermi resonance between stretching and bending vibrations becomes significant in high overtones of water. This aspect will be discussed later.

TABLE IV—*Continued*

Normal v_1v_3	Local $v_{r_1}v_{r_2} \pm$	C$_2$H$_2$		C$_2$D$_2$	
		v_{obs}	v_{calc}	v_{obs}	v_{calc}
13	40−	12675.68	12678.60	9794.08	9792.74
22	31+	—	12915.82	—	10050.68
31	31−	13033.29	13032.55	10347.92	10351.25
40	22+	—	13229.65	—	10688.01
14	50+	—	15602.02	—	12100.68
05	50−	15600.16	15601.29	11905.37[d]	11930.06
32	41+	—	16012.13	—	12625.13
23	41−	15948.48	15949.73	12344.54	12340.58
50	32+	—	16416.02	—	13302.99
41	32−	16203.45	16204.53	—	12947.42
06	60+	—	18425.34	—	14238.82
15	60−	18430.07	18425.42	14377.8	14379.79
24	51+	—	18899.24	—	14602.95
33	51−	18915.39	18920.00	—	14871.52
42	42+	—	19145.10	—	15179.41
51	42−	—	19318.22	—	15521.65
60	33+	—	19553.05	—	15894.95

[a] Experimental data are taken from Refs. [11] (H$_2$O), [12] (SO$_2$), [29,30] (C$_2$H$_2$), and [32] (C$_2$D$_2$), where references to original experimental work are found. In Refs. [6], Jungner and Halonen provide data for the [30+] state and Barnes et al. for the [40+] state of C$_2$H$_2$. See also Ref. [33] where vibrational term values are available for H$_2$O above 18000 cm^{-1}, and Ref. [31], where vibrational term values are available for C$_2$H$_2$ above 20000 cm^{-1}. For SO$_2$, low-resolution fluorescence data (as given in Refs. [12]) were not included in the fit. All data included in the fits were given unit weights.

[b] Not included in the fit. Perturbed value due to an anharmonic resonance between v_3 and $v_2 + v_4^1 + v_5^1$.

[c] Recent observation. Not included in the fit.

[d] Not included in the fit.

As the two approaches, the HCAO model and the customary normal mode model, give identical eigenvalues when x–K relations are applied, there is perhaps little to choose between the two models. However, an energy level fit, that is, the agreement between observed and calculated levels alone, does not provide the whole picture. For example, in the neighborhood of the local mode limit, the customary model produces eigenvalues and eigenvalues which possess no obvious connection with the zeroth-order normal mode picture. A similar comment, of course, applies to the local mode model near the normal mode limit. At least in the neighborhood of the two limits, it is physically more suited to describe molecular vibrations with a model whose diagonal terms provide the best zeroth-order picture. The clear physical interpretation of the local mode model, that is, coupled anharmonic bond oscillators, favors its use over a wide range. The normal mode model with the quartic anharmonic coupling terms (Darling–Dennison resonance terms) appears somewhat unfamiliar to nonexperts in the field of spectroscopy. On the other hand, it is more straightforward to include the rotational motion in the normal-mode-based approach because the whole machinery of the conventional vibration–rotation theory is readily usable. However, this situation might be changing [35,36].

At the end of the discussion of these simple models, it is appropriate to say something about the history of overtone vibrations. The idea to treat some of the vibrations in polyatomic molecules as local modes is old. Ellis, Mecke, and their co-workers observed as early as about 80 years ago that hydrogen stretching vibrational overtone spectra of H_2O, C_2H_2, and C_6H_6, for example, can be described well with the Morse-oscillator-type eigenvalue expression [10]. At the same time, the customary normal mode theory started to develop. In 1940, Darling and Dennison observed that in water it is necessary to include a quartic anharmonic coupling term, later called the Darling–Dennison resonance term, to account for the overtone stretching states [20]. Herzberg, in his book *Infrared and Raman Spectroscopy*, adopted the conventional normal mode approach to describe molecular vibrations, although the earlier work by Mecke using a simple local mode approach, for example, on acetylene is mentioned (see page 289 in Herzberg's book [9]). Later Wilson, Decius, and Cross, in their book *Molecular Vibrations*, which became a landmark in the theory of molecular vibrations, used just normal modes without mentioning the work by Ellis, Mecke, and their co-workers [17]. Considering this background, it is not surprising that the early work on local modes was disregarded for many years.

The Darling–Dennison resonance term was later included in the analyses of hydrogen stretching overtone spectra of C_2H_2 (Herzberg [9] first suggested that this resonance should be included in the analysis of acetylene overtone spectra), H_2S, and H_2Se [37–39]. In spite of this success, it is astonishing that these ideas were not extended until recently to more complicated molecules such as ammonia and methane [21], ethylene (C_2H_4) and propadiene (C_3H_4) [40], and benzene [25].

The local mode model was rediscovered about 30 years ago in a series of papers by Siebrand, Henry, and their co-workers [1,8,41]. These papers launched an intensive experimental and theoretical work in this area. Of later significant developments, the work by Wallace in vibrational models based on Morse oscillators [42], the work by Watson, Henry, and Ross [43] on a model which already contains some characteristics of the HCAO model, and the variational quantum-mechanical calculations of excited states in water by Lawton and Child [44] should be mentioned. The breakthrough was the introduction of the HCAO model by Child [13] and later by Mortensen et al. [16]. These papers made the local mode concept a real alternative to normal mode models. This development was followed by a paper by Lehmann [24] which shows that the local and normal mode approaches give identical eigenvalues when x–K relations are applied, as discussed in this chapter. The possibility that simple relations between diagonal anharmonicity constants might exist is already apparent in some earlier work by Henry and co-workers [1], but these relations were published explicitly for the first time in the case of XY_4 tetrahedral molecules [45] and acetylene [46].

B.   Extension to Larger Molecules

The models described for the stretching vibrational states of bent XY_2 molecules can be extended to other types of molecules. There is a technical difficulty in generating symmetrized local mode basis functions corresponding to Eqs. (2.7). A simple way to solve the problem is discussed below. If the stretching vibrational local mode Hamiltonians are expressed in terms of unsymmetrized coordinates, then the resulting Hamiltonian matrices can be symmetry factorized by employing symmetrized local mode basis functions. This procedure, if done by hand, is tedious for molecules with high symmetry. If the Hamiltonian is expressed in terms of normal coordinates or in terms of symmetrized internal coordinates, it might also be a good idea to symmetrize basis sets when Cartesian representations of doubly and triply degenerate vibrations are used. This aspect will be discussed briefly later in the context of stretch–bend vibrational models in tetrahedral molecules (see Section III.C).

The symmetrization of the local mode basis of larger molecules is conveniently done with a method first presented for XY_6 octahedral molecules [47] and later used for tetrahedral molecules [48–50]. Tetrahedral XY_4 molecules are taken as an example. Because the symmetric stretch v_1 and the antisymmetric stretch v_3 span A_1 and F_2 species, respectively, the symmetrized A_1 and F_2 basis functions

$$|1000A_1\rangle = \tfrac{1}{2}(|1_1\rangle + |1_2\rangle + |1_3\rangle + |1_4\rangle)$$
$$|1000F_{2x}\rangle = \tfrac{1}{2}(|1_1\rangle - |1_2\rangle + |1_3\rangle - |1_4\rangle)$$
$$|1000F_{2y}\rangle = \tfrac{1}{2}(|1_1\rangle - |1_2\rangle - |1_3\rangle + |1_4\rangle) \tag{2.35}$$
$$|1000F_{2z}\rangle = \tfrac{1}{2}(|1_1\rangle + |1_2\rangle - |1_3\rangle - |1_4\rangle)$$

provide the starting point. The symbol $|a_i\rangle$ denotes the ath excitation of the ith bond. These linear combinations are similar to the definition of symmetrized internal stretching coordinates [see Eqs. (3.50) in Section III]. All symmetrized basis functions required are obtained by applying symmetrized promotion operators as follows. The totally symmetric promotion operator, for example, is

$$O(A_1) = \sigma_1 + \sigma_2 + \sigma_3 + \sigma_4 \qquad (2.36)$$

where the shift operator σ_i is defined by the equation

$$\sigma_i | a_i \rangle = \sqrt{a + 1} | (a + 1)_i \rangle \qquad (2.37)$$

Note that in the earlier work the square root factor was not included. This is not an error in the present application because the same square root term appears as a common factor in the linear combination representation of a basis function. However, it should not necessarily be omitted when the present method is extended to the symmetrization of other types of basis functions such as Cartesian representations of doubly and triply degenerate vibrations. Using the method described, for example, the result

$$\begin{aligned}
O(A_1) | 1000F_{2z} \rangle &= \tfrac{1}{2}(2 | 1_1 1_2 \rangle - 2 | 1_3 1_4 \rangle + \sqrt{2} | 2_1 \rangle \\
&\quad + \sqrt{2} | 2_2 \rangle - \sqrt{2} | 2_3 \rangle - \sqrt{2} | 2_4 \rangle) \\
&= \sqrt{2}(| 1100F_{2z} \rangle + | 2000F_{2z} \rangle) \qquad (2.38)
\end{aligned}$$

is obtained. A new type of symmetrized basis function

$$| 1100F_{2z} \rangle = \frac{1}{\sqrt{2}} (| 1_1 1_2 \rangle - | 1_3 1_4 \rangle) \qquad (2.39)$$

has been generated. The function $| 2000F_{2z} \rangle$ is already of the form given in Eqs. (2.35).

In order to obtain all possible symmetry combinations, symmetrized promotion operators, which span different symmetry species, can be defined, and appropriate vector coupling coefficients from Ref. [51] can be used. For example, in order to generate the $| 1100E_a \rangle$ and $| 1100E_b \rangle$ pair of symmetrized functions, the following symmetrized promotion operators can be employed:

$$\begin{aligned}
O(F_{2x}) &= \sigma_1 - \sigma_2 + \sigma_3 - \sigma_4 \\
O(F_{2y}) &= \sigma_1 - \sigma_2 - \sigma_3 + \sigma_4 \qquad (2.40) \\
O(F_{2z}) &= \sigma_1 + \sigma_2 - \sigma_3 - \sigma_4
\end{aligned}$$

TABLE V
Vector Coupling Coefficients

$F_2 \times F_2$		E	
		a	b
x	x	$\dfrac{1}{\sqrt{6}}$	$-\dfrac{1}{\sqrt{2}}$
y	y	$\dfrac{1}{\sqrt{6}}$	$\dfrac{1}{\sqrt{2}}$
z	z	$-\dfrac{2}{\sqrt{6}}$	—

$E \times E$		E	
		a	b
a	a	$-\dfrac{1}{\sqrt{2}}$	—
b	b	$\dfrac{1}{\sqrt{2}}$	—
a	b	—	$\dfrac{1}{\sqrt{2}}$
b	a	—	$\dfrac{1}{\sqrt{2}}$

$E \times F_2$		F_2		
		x	y	z
a	x	$-\frac{1}{2}$	—	—
a	y	—	$-\frac{1}{2}$	—
a	z	—	—	1
b	x	$\frac{1}{2}\sqrt{3}$	—	—
b	y	—	$-\frac{1}{2}\sqrt{3}$	—
b	z	—	—	—

with the vector coupling coefficients given in Table V. Note again the analogy between Eqs. (2.40) and (3.50). The result for $|1100E_b\rangle$ is

$$-\frac{1}{\sqrt{2}}\, O(F_{2x})\,|\,1000F_{2x}\rangle + \frac{1}{\sqrt{2}}\, O(F_{2y})\,|\,1000F_{2y}\rangle$$

$$= \frac{1}{\sqrt{2}}\left(-4\,|\,1_1 1_3\rangle + 4\,|\,1_1 1_4\rangle + 4\,|\,1_2 1_3\rangle - 4\,|\,1_2 1_4\rangle\right) \quad (2.41)$$

which after normalization takes the form

$$|1100E_b\rangle = \tfrac{1}{2}(-|1_1 1_3\rangle + |1_1 1_4\rangle + |1_2 1_3\rangle - |1_2 1_4\rangle) \quad (2.42)$$

The corresponding symmetrized basis function $|1100E_a\rangle$ is

$$|1100E_a\rangle = \frac{1}{2\sqrt{3}}(-2|1_1 1_2\rangle + |1_1 1_3\rangle + |1_1 1_4\rangle$$

$$+ |1_2 1_3\rangle + |1_2 1_4\rangle - 2|1_3 1_4\rangle) \quad (2.43)$$

An additional complication arises, for example, in the case of $|2100F_2\rangle$-type functions, which contain two sets of F_2 functions. The Gram–Schmidt orthogonalization is required to form orthogonal linear combinations. Reference [49] gives a more thorough description of this method in tetrahedral XY_4 molecules. All different types of symmetrized basis functions of XY_4 tetrahedral molecules are easily obtainable by hand, but in XY_6 octahedra it is advisable to generate them by computers.

The HCAO model is easily extended to other molecular symmetries following the procedure described in C_{2v} triatomic molecules. The resulting Hamiltonian matrices can be symmetry factorized using symmetrized basis functions obtained as described above. These extensions of the original model have been published for C_{3v} [52,53], T_d [45], and O_h [47] point group molecules [XH_3- (and XH_3Y-), XH_4-, and XY_6-type molecules, respectively] with vibrational energy level correlation diagrams between local and normal mode limits. These diagrams are reproduced in Fig. 5 for three identical coupled anharmonic oscillators (C_{3v}), and in Ref. [45] these are given for four equivalent coupled anharmonic oscillators (T_d). It is of interest to note that SiH_4 (and also both GeH_4 and SnH_4, which are almost at the same position as SiH_4) [50,54] is close to the local mode limit. The small interbond coupling parameter λ_r in SiH_4 is the result of an almost equal cancellation of nonzero potential and kinetic energy couplings [45]. In the other tetrahedral molecules (CH_4, CD_4, and SiD_4), this kind of cancellation is less complete and the molecules lie about halfway between the two limits. In light of this, SiH_3D is close to the local mode limit (see Fig. 5) but $SiHD_3$, CHD_3, and CH_3D are about halfway between the two limits. A simple extension of the original HCAO model has been applied to overtone spectra of CH_2CF_2 and CH_2CCl_2 [55].

The customary normal mode model with vibrationally off-diagonal Darling–Dennison resonance coupling has been extended to the hydrogen stretching vibrations in ammonia [21], methane [21], ethylene (C_2H_4) [40], propadiene (C_3H_4) [40], and benzene [25] type molecules. When compared

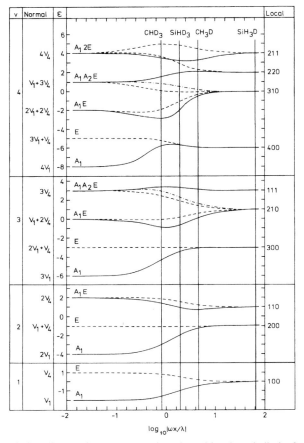

Figure 5. Correlation diagram between normal mode and local mode limits for the C_{3v} point group: $v = v_{r_1} + v_{r_2} + v_{r_3}$ is the total stretching quantum number; $\omega x = -x_{rr}$ and $\lambda = \lambda_r$. (From Ref. [52] with permission.)

with water-type molecules, new kinds of Darling–Dennison resonance terms appear as given in Table VI for these molecules. Table VII contains x–K relations in these cases.

The normal mode labeling of vibrational states is well explained in the book *Infrared and Raman Spectra* by Herzberg [9]. The local mode labeling, on the other hand, is less well established but the local mode quantum labels together with symmetry labels are often given first for locallike vibrations. The quantum labels for normal-type vibrations are given next. For example, in symmetrical isotopic species of water (H_2O, D_2O, and T_2O),

TABLE VI
Matrix Elements in the Normal Mode Basis[a]

1. Asymmetric and symmetric top molecules[b]:

$$\langle v\,|\,H/hc_0\,|\,v\rangle = \sum_s \omega_s(v_s + \tfrac{1}{2}d_s) + \sum_s x_{ss}(v_s + \tfrac{1}{2}d_s)^2 + \sum_{s<s'} x_{ss'}(v_s + \tfrac{1}{2}d_s)(v_{s'} + \tfrac{1}{2}d_{s'})$$
$$+ \sum_t g_{tt}\,l_t^2 + \sum_{t<t'} g_{tt'}\,l_t\,l_{t'}$$

(a) v_s and $v_{s'}$ both nondegenerate:

$$\langle v_s + 2, v_{s'} - 2\,|\,H/hc_0\,|\,v_s, v_{s'}\rangle = \tfrac{1}{4}K_{ss;\,s's'}[(v_s + 1)(v_s + 2)v_{s'}(v_{s'} - 1)]^{1/2}$$

(b) v_s, $v_{s'}$, and $v_{s''}$ are nondegenerate:

$$\langle v_s + 1, v_{s'} - 1, v_{s''}\,|\,H/hc_0\,|\,v_s, v_{s'}, v_{s''}\rangle$$
$$= \tfrac{1}{4}[(K_{ss;\,ss'}(v_s + 1) + K_{ss';\,s's'}v_{s'} + 2K_{ss'';\,s's''}(v_{s''} + \tfrac{1}{2})][(v_s + 1)v_{s'}]^{1/2}$$
$$\langle v_s + 2, v_{s'} - 1, v_{s''} - 1\,|\,H/hc_0\,|\,v_s, v_{s'}, v_{s''}\rangle = \tfrac{1}{4}K_{ss;\,s's'}[(v_s + 1)(v_s + 2)v_{s'}v_{s''}]^{1/2}$$

(c) v_s, $v_{s'}$, v_t, and $v_{t'}$ nondegenerate:

$$\langle v_s + 1, v_{s'} + 1, v_t - 1, v_{t'} - 1\,|\,H/hc_0\,|\,v_s, v_{s'}, v_t, v_{t'}\rangle = \tfrac{1}{4}K_{ss';\,tt'}[(v_s + 1)(v_{s'} + 1)v_t v_{t'}]^{1/2}$$

(d) v_s nondegenerate and v_t doubly degenerate:

$$\langle v_s + 2, v_t - 2, l_t\,|\,H/hc_0\,|\,v_s, v_t, l_t\rangle = \tfrac{1}{4}K_{ss;\,tt}[(v_s + 1)(v_s + 2)(v_t^2 - l_t^2)]^{1/2}$$
$$\langle v_s + 1, v_t - 1, l_t \pm 3\,|\,H/hc_0\,|\,v_s, v_t, l_t\rangle = \tfrac{3}{8}K_{st;\,tt}[(v_s + 1)(v_t \pm l_t + 2)(v_t \mp l_t - 2)(v_t \mp l_t)]^{1/2}$$

(e) v_s and $v_{s'}$ nondegenerate, v_t doubly degenerate:

$$\langle v_s + 1, v_{s'} + 1, v_t - 2, l_t \pm 2\,|\,H/hc_0\,|\,v_s, v_{s'}, v_t, l_t\rangle$$
$$= \tfrac{1}{4}K_{ss';\,tt}[(v_s + 1)(v_{s'} + 1)(v_t \mp l_t)(v_t \mp l_t - 2)]^{1/2}$$
$$\langle v_s + 1, v_{s'} - 1, v_t, l_t \pm 2\,|\,H/hc_0\,|\,v_s, v_{s'}, v_t, l_t\rangle = \tfrac{1}{4}K_{ss';\,tt}\{(v_s + 1)v_{s'}[(v_t + 1)^2 - (l_t \pm 1)^2]\}^{1/2}$$

(f) v_t doubly degenerate:

$$\langle v_t, l_t + 4\,|\,H/hc_0\,|\,v_t, l_t\rangle = \tfrac{3}{4}K_{tt;\,tt}\{[(v_t + 1)^2 - (l_t + 1)^2][(v_t + 1)^2 - (l_t + 3)^2]\}^{1/2}$$

2. XY_4 spherical top molecules[c]:

(g) Stretching vibrational matrix elements:

$$\langle v_1, v_{3x} v_{3y} v_{3z}\,|\,H/hc_0\,|\,v_1, v_{3x} v_{3y} v_{3z}\rangle$$
$$= \omega_1(v_1 + \tfrac{1}{2}) + \omega_3(v_3 + \tfrac{3}{2})$$
$$+ x_{11}(v_1 + \tfrac{1}{2})^2 + x_{13}(v_1 + \tfrac{1}{2})(v_3 + \tfrac{3}{2}) + x_{33}(v_3 + \tfrac{3}{2})^2$$
$$+ 2(G_{33} + 2T_{33})(v_3 + v_{3y}v_{3z} + v_{3z}v_{3x} + v_{3x}v_{3y})$$
$$+ T_{33}[4(v_{3x}^2 + v_{3y}^2 + v_{3z}^2) - 12(v_{3y}v_{3z} + v_{3z}v_{3x} + v_{3x}v_{3y}) - 8v_3]$$

$$\langle v_{3x}(v_{3y} + 2)(v_{3z} - 2)\,|\,H/hc_0\,|\,v_{3x}v_{3y}v_{3z}\rangle$$
$$= -(G_{33} + 2T_{33})[(v_{3y} + 1)(v_{3y} + 2)v_{3z}(v_{3z} - 1)]^{1/2}$$

$$\langle v_1 + 2, (v_{3x} - 2)v_{3y}v_{3z}\,|\,H/hc_0\,|\,v_1, v_{3x}v_{3y}v_{3z}\rangle$$
$$= \tfrac{1}{4}K_{11;\,33}[(v_1 + 1)(v_1 + 2)v_{3x}(v_{3x} - 1)]^{1/2}$$

$$\langle v_1 + 1, (v_{3x} + 1)(v_{3y} - 1)(v_{3z} - 1)\,|\,H/hc_0\,|\,v_1, v_{3x}v_{3y}v_{3z}\rangle$$
$$= \tfrac{1}{4}K_{13;\,33}[(v_1 + 1)(v_{3x} + 1)v_{3y}v_{3z}]^{1/2}$$

(h) Bending vibrational matrix elements:

$$\langle v_{2a}v_{2b}, v_{4x}v_{4y}v_{4z}\,|\,H/hc_0\,|\,v_{2a}v_{2b}, v_{4x}v_{4y}v_{4z}\rangle$$
$$= \omega_2(v_2 + 1) + \omega_4(v_4 + \tfrac{3}{2})$$
$$+ x_{22}(v_2 + 1)^2 + x_{24}(v_2 + 1)(v_4 + \tfrac{3}{2}) + x_{44}(v_4 + \tfrac{3}{2})^2 + 2G_{22}[(v_{2a} + \tfrac{1}{2})(v_{2b} + \tfrac{1}{2}) - \tfrac{1}{4}]$$
$$+ 2(G_{44} + 2T_{44})(v_4 + v_{4y}v_{4z} + v_{4z}v_{4x} + v_{4x}v_{4y})$$
$$+ T_{44}[4(v_{4x}^2 + v_{4y}^2 + v_{4z}^2) - 12(v_{4y}v_{4z} + v_{4z}v_{4x} + v_{4x}v_{4y}) - 8v_4]$$
$$+ 4T_{24}(v_{2a} - v_{2b})(v_4 - 3v_{4z})$$

$$\langle (v_{2a} + 2)(v_{2b} - 2)\,|\,H/hc_0\,|\,v_{2a}v_{2b}\rangle$$
$$= -G_{22}[(v_{2a} + 1)(v_{2a} + 2)v_{2b}(v_{2b} - 1)]^{1/2}$$

$$\langle v_{4x}(v_{4y} + 2)(v_{4z} - 2)\,|\,H/hc_0\,|\,v_{4x}v_{4y}v_{4z}\rangle$$
$$= -(G_{44} + 2T_{44})[(v_{4y} + 1)(v_{4y} + 2)v_{4z}(v_{4z} - 1)]^{1/2}$$

$$\langle (v_{2a} + 1)(v_{2b} - 1), v_{4x}v_{4y}v_{4z}\,|\,H/hc_0\,|\,v_{2a}v_{2b}, v_{4x}v_{4y}v_{4z}\rangle$$
$$= 4\sqrt{3}\,T_{24}[(v_{2a} + 1)v_{2b}]^{1/2}(v_{4x} - v_{4y})$$

$$\langle (v_{2a} + 1)v_{2b}, (v_{4x} + 1)(v_{4y} - 1)(v_{4z} - 1)\,|\,H/hc_0\,|\,v_{2a}v_{2b}, v_{4x}v_{4y}v_{4z}\rangle$$
$$= -\tfrac{1}{2}K_{24;\,44}[(v_{2a} + 1)(v_{4x} + 1)v_{4y}v_{4z}]^{1/2}$$

$$\langle (v_{2a} + 1)v_{2b}, (v_{4x} - 1)(v_{4y} + 1)(v_{4z} - 1)\,|\,H/hc_0\,|\,v_{2a}v_{2b}, v_{4x}v_{4y}v_{4z}\rangle$$
$$= -\tfrac{1}{2}K_{24;\,44}[(v_{2a} + 1)v_{4x}(v_{4y} + 1)v_{4z}]^{1/2}$$

$$\langle (v_{2a} + 1)v_{2b}, (v_{4x} - 1)(v_{4y} - 1)(v_{4z} + 1)\,|\,H/hc_0\,|\,v_{2a}v_{2b}, v_{4x}v_{4y}v_{4z}\rangle$$
$$= K_{24;\,44}[(v_{2a} + 1)v_{4x}v_{4y}(v_{4z} + 1)]^{1/2}$$

TABLE VI—*Continued*

$\langle v_{2a}(v_{2b}+1),(v_{4x}+1)(v_{4y}-1)(v_{4z}-1)|H/hc_0|v_{2a}v_{2b},v_{4x}v_{4y}v_{4z}\rangle$
$$=\frac{\sqrt{3}}{2}K_{24;\,44}[(v_{2b}+1)(v_{4x}+1)v_{4y}v_{4z}]^{1/2}$$

$\langle v_{2a}(v_{2b}+1),(v_{4x}-1)(v_{4y}+1)(v_{4z}-1)|H/hc_0|v_{2a}v_{2b},v_{4x}v_{4y}v_{4z}\rangle$
$$=-\frac{\sqrt{3}}{2}K_{24;\,44}[(v_{2b}+1)v_{4x}(v_{4y}+1)v_{4z}]^{1/2}$$

$\langle v_{2a}(v_{2b}+1),(v_{4x}-1)(v_{4y}-1)(v_{4z}+1)|H/hc_0|v_{2a}v_{2b},v_{4x}v_{4y}v_{4z}\rangle=0$

$K_{22;\,44A_1}$:

$\langle(v_{2a}+2)v_{2b},(v_{4x}-2)v_{4y}v_{4z}|H/hc_0|v_{2a}v_{2b},v_{4x}v_{4y}v_{4z}\rangle$
$$=\tfrac{1}{4}K_{22;\,44A_1}[(v_{2a}+1)(v_{2a}+2)v_{4x}(v_{4x}-1)]^{1/2}$$

$\langle(v_{2a}+2)v_{2b},v_{4x}(v_{4y}-2)v_{4z}|H/hc_0|v_{2a}v_{2b},v_{4x}v_{4y}v_{4z}\rangle$
$$=\tfrac{1}{4}K_{22;\,44A_1}[(v_{2a}+1)(v_{2a}+2)v_{4y}(v_{4y}-1)]^{1/2}$$

$\langle(v_{2a}+2)v_{2b},v_{4x}v_{4y}(v_{4z}-2)|H/hc_0|v_{2a}v_{2b},v_{4x}v_{4y}v_{4z}\rangle$
$$=\tfrac{1}{4}K_{22;\,44A_1}[(v_{2a}+1)(v_{2a}+2)v_{4z}(v_{4z}-1)]^{1/2}$$

$\langle v_{2a}(v_{2b}+2),(v_{4x}-2)v_{4y}v_{4z}|H/hc_0|v_{2a}v_{2b},v_{4x}v_{4y}v_{4z}\rangle$
$$=\tfrac{1}{4}K_{22;\,44A_1}[(v_{2b}+1)(v_{2b}+2)v_{4x}(v_{4x}-1)]^{1/2}$$

$\langle v_{2a}(v_{2b}+2),v_{4x}(v_{4y}-2)v_{4z}|H/hc_0|v_{2a}v_{2b},v_{4x}v_{4y}v_{4z}\rangle$
$$=\tfrac{1}{4}K_{22;\,44A_1}[(v_{2b}+1)(v_{2b}+2)v_{4y}(v_{4y}-1)]^{1/2}$$

$\langle v_{2a}(v_{2b}+2),v_{4x}v_{4y}(v_{4z}-2)|H/hc_0|v_{2a}v_{2b},v_{4x}v_{4y}v_{4z}\rangle$
$$=\tfrac{1}{4}K_{22;\,44A_1}[(v_{2b}+1)(v_{2b}+2)v_{4z}(v_{4z}-1)]^{1/2}$$

$K_{22;\,44E}$:

$\langle(v_{2a}+2)v_{2b},(v_{4x}-2)v_{4y}v_{4z}|H/hc_0|v_{2a}v_{2b},v_{4x}v_{4y}v_{4z}\rangle$
$$=\tfrac{1}{4}K_{22;\,44E}[(v_{2a}+1)(v_{2a}+2)v_{4x}(v_{4x}-1)]^{1/2}$$

$\langle(v_{2a}+2)v_{2b},v_{4x}(v_{4y}-2)v_{4z}|H/hc_0|v_{2a}v_{2b},v_{4x}v_{4y}v_{4z}\rangle$
$$=\tfrac{1}{4}K_{22;\,44E}[(v_{2a}+1)(v_{2a}+2)v_{4y}(v_{4y}-1)]^{1/2}$$

$\langle(v_{2a}+2)v_{2b},v_{4x}v_{4y}(v_{4z}-2)|H/hc_0|v_{2a}v_{2b},v_{4x}v_{4y}v_{4z}\rangle$
$$=-\tfrac{1}{2}K_{22;\,44E}[(v_{2a}+1)(v_{2a}+2)v_{4z}(v_{4z}-1)]^{1/2}$$

$\langle v_{2a}(v_{2b}+2),(v_{4x}-2)v_{4y}v_{4z}|H/hc_0|v_{2a}v_{2b},v_{4x}v_{4y}v_{4z}\rangle$
$$=-\tfrac{1}{4}K_{22;\,44E}[(v_{2b}+1)(v_{2b}+2)v_{4x}(v_{4x}-1)]^{1/2}$$

$\langle v_{2a}(v_{2b}+2),v_{4x}(v_{4y}-2)v_{4z}|H/hc_0|v_{2a}v_{2b},v_{4x}v_{4y}v_{4z}\rangle$
$$=-\tfrac{1}{4}K_{22;\,44E}[(v_{2b}+1)(v_{2b}+2)v_{4y}(v_{4y}-1)]^{1/2}$$

$\langle v_{2a}(v_{2b}+2),v_{4x}v_{4y}(v_{4z}-2)|H/hc_0|v_{2a}v_{2b},v_{4x}v_{4y}v_{4z}\rangle$
$$=\tfrac{1}{2}K_{22;\,44E}[(v_{2b}+1)(v_{2b}+2)v_{4z}(v_{4z}-1)]^{1/2}$$

$\langle(v_{2a}+1)(v_{2b}+1),(v_{4x}-2)v_{4y}v_{4z}|H/hc_0|v_{2a}v_{2b},v_{4x}v_{4y}v_{4z}\rangle$
$$=\frac{\sqrt{3}}{2}K_{22;\,44E}[(v_{2a}+1)(v_{2b}+1)v_{4x}(v_{4x}-1)]^{1/2}$$

$\langle(v_{2a}+1)(v_{2b}+1),v_{4x}(v_{4y}-2)v_{4z}|H/hc_0|v_{2a}v_{2b},v_{4x}v_{4y}v_{4z}\rangle$
$$=-\frac{\sqrt{3}}{2}K_{22;\,44E}[(v_{2a}+1)(v_{2b}+1)v_{4y}(v_{4y}-1)]^{1/2}$$

$\langle(v_{2a}+1)(v_{2b}+1),v_{4x}v_{4y}(v_{4z}-2)|H/hc_0|v_{2a}v_{2b},v_{4x}v_{4y}v_{4z}\rangle=0$

---

[a] In the case of off-diagonal matrix elements, symmetry restrictions are imposed, for example in 1c the product of the symmetries of r, s, t, and u must be totally symmetric. Also note that, for example, the matrix element given in 1f disappears in the C_{3v} point group. The label l_t is the vibrational angular momentum quantum number for the doubly degenerate modes. It takes the values v_t, $v_t - 2$, ..., $-v_t$, where v_t is the vibrational quantum number. The basis set notation is such that, for example, $|v_s, v_{s'}, v_t, l_t\rangle = |v_s\rangle|v_{s'}\rangle|v_t, l_t\rangle$, where $|v_s\rangle$ and $|v_{s'}\rangle$ are one-dimensional harmonic oscillator eigenfunctions and $|v_t, l_t\rangle$ is a two-dimensional harmonic oscillator eigenfunction, which are all expressed in the normal coordinate representation.

[b] In the case of diagonal matrix elements, t spans degenerate modes, and these modes are counted only once. Indices s and s' span all modes, d_s and $d_{s'}$ being the corresponding degeneracies. Degenerate modes are counted only once.

[c] More off-diagonal matrix elements are obtained by Hermitian conjugation in all cases and by permuting x, y, and z in the case G_{33}, T_{33}, G_{44}, T_{44}, $K_{11;\,33}$, and $K_{13;\,33}$; $v_2 = v_{2a} + v_{2b}$, $v_3 = v_{3x} + v_{3y} + v_{3z}$, and $v_4 = v_{4x} + v_{4y} + v_{4z}$.

the appropriate local mode labels would be $[v_{r_1}v_{r_2} \pm]v_2$ (or $[v_{r_1}v_{r_2} A_1]v_2$ and $[v_{r_1}v_{r_2} B_1]v_2$, which correspond to $+$ and $-$ labels, respectively), where v_2 is the normal mode quantum number for the bend. The labels $+$ and $-$ describe the symmetry of the eigenfunctions with respect to permutation of hydrogen atoms. For symmetrical isotopic species of acetylene (C_2H_2, C_2D_2) the local mode label is $[v_{r_1}v_{r_2} \pm]v_2 v_4^{l_4}v_5^{l_5}$ (or $[v_{r_1}v_{r_2} \Sigma_g^+]v_2 v_4^{l_4}v_5^{l_5}$ and $[v_{r_1}v_{r_2} \Sigma_u^+]v_2 v_4^{l_4}v_5^{l_5}$, which correspond to $+$ and $-$ labels, respectively). The labels v_2, v_4, and v_5 are the usual normal mode vibrational quantum numbers for the C≡C stretch and for the symmetric and antisymmetric bending vibrations, respectively. Here, l_i ($= -v_i$, $-v_i + 2$, ..., v_i), where $i = 4, 5$, is the vibrational angular momentum quantum number for the ith doubly degenerate bending vibration. When just the stretching vibrations are studied, it is feasible to leave out the bending quantum numbers to obtain the label $[v_{r_1}v_{r_2} \pm]v_2$ for acetylene. For near-local-mode molecules, the local mode quantum numbers of the hydrogen stretching states provide

TABLE VII

The x–K Relations for Hydrogen Stretching Vibrations in Some Small Symmetrical Molecules[a]

1. H_2O-type molecules (C_{2v}):
$$\omega_1(A_1) = \omega_r' + \lambda_r, \quad \omega_3(B_1) = \omega_r' - \lambda_r$$
$$x_{11} = x_{33} = \tfrac{1}{4}x_{13} = \tfrac{1}{4}K_{11;\,33} = \tfrac{1}{2}x_{rr}$$

2. NH_3-type molecules (C_{3v}):
$$\omega_1(A_1) = \omega_r' + 2\lambda_r, \quad \omega_3(E) = \omega_r' - \lambda_r$$
$$x_{11} = \tfrac{2}{3}x_{33} = \tfrac{1}{4}x_{13} = -2g_{33} = \tfrac{1}{4}K_{11;\,33} = \frac{3}{4\sqrt{2}}K_{13;\,33} = \tfrac{1}{3}x_{rr}$$

3. CH_4-type molecules (T_d):
$$\omega_1(A_1) = \omega_r' + 3\lambda_r, \quad \omega_3(F_2) = \omega_r' - \lambda_r$$
$$x_{11} = \tfrac{5}{9}x_{33} = \tfrac{1}{4}x_{13} = -\tfrac{3}{5}G_{33} = -5T_{33} = \tfrac{1}{4}K_{11;\,33} = \tfrac{1}{16}K_{13;\,33} = \tfrac{1}{4}x_{rr}$$

4. $H_2C{=}CH_2$-type molecules (D_{2h})[b]:
$$\omega_1(A_g) = \omega_r' + \lambda_a + \lambda_c + \lambda_t, \quad \omega_5(B_{1g}) = \omega_r' - \lambda_a - \lambda_c + \lambda_t$$
$$\omega_9(B_{2u}) = \omega_r' - \lambda_a + \lambda_c - \lambda_t, \quad \omega_{11}(B_{3u}) = \omega_r' + \lambda_a - \lambda_c - \lambda_t$$
$$x_{ss} = \tfrac{1}{4}x_{ss'} = \tfrac{1}{4}K_{ss;\,s's'} = \tfrac{1}{16}K_{15;\,9,\,11} = \tfrac{1}{16}K_{19;\,5,\,11} = \tfrac{1}{4}x_{rr}$$
$$s, s' = 1, 5, 9, 11 \text{ and } s < s'$$

5. $H_2C{=}C{=}CH_2$-type molecules (D_{2d})[c]:
$$\omega_1(A_1) = \omega_r' + \lambda_a + 2\lambda_b, \quad \omega_5(B_2) = \omega_r' + \lambda_a - 2\lambda_b, \quad \omega_8(E) = \omega_r' - \lambda_a$$
$$x_{11} = x_{55} = \tfrac{1}{4}x_{15} = \tfrac{1}{4}x_{18} = \tfrac{1}{4}x_{58} = \tfrac{2}{3}x_{88} = -2g_{88}$$
$$= \tfrac{1}{4}K_{11;\,55} = \tfrac{1}{4}K_{11;\,88} = \tfrac{1}{4}K_{55;\,88} = \tfrac{1}{4}K_{15;\,88} = 3K_{88;\,88} = \tfrac{1}{4}x_{rr}$$

[a] Definitions of the conventional spectroscopic parameters (ω, x, g, G, T, and K) are given in Table VI.

[b] λ_a represents the interaction between the two CH bonds in the same methylene group; λ_c and λ_t represent cis and trans interactions, respectively, between the CH bonds at opposite ends of the molecule.

[c] λ_a represents the interaction between the two CH bonds in the same methylene group and λ_b between the CH bonds at opposite ends of the molecule.

physically the best picture, as is obvious from the normal mode–local mode correlation diagrams presented in Figs. 4 and 5. Finally, local mode quantum labels have also been used to describe the bending states in XH_3-type hydrides [56,57]. This will be discussed later in this chapter.

C. Local Modes in the Absence of Symmetry

So far only polyatomic molecules with two or more equivalent bond oscillators have been considered. These symmetrical systems show the most dramatic change in the overtone spectral structure with unusual vibrational degeneracies, according to the standard vibration–rotation theory as the local mode limit is approached. A similar kind of approach can also be applied to molecules with nonequivalent bond oscillators. The HCAO model has been extended with success to the acetylene isotopic species $H^{12}C^{13}CH$ where the symmetry is only slightly broken [58]. An application of this approach to $^{12}C_2HD$ failed because there is a strong bilinear harmonic kinetic energy coupling between the close-lying CC and CD stretching fundamental states ($v_{CC} = 1853.8$ cm^{-1} and $v_{CD} = 2583.6$ cm^{-1}) [59]. Note that because $v_{CH} = 3335.6$ cm^{-1} the harmonic coupling between the CH and CC oscillators can be taken into account by second-order perturbation theory. The important kinetic energy coupling term [see Eq. (2.1)] is

$$H = g_{rr'}^{(e)} p_r p_{r'} = \frac{\cos \alpha_e}{m_C} p_r p_{r'} = -\frac{1}{m_C} p_r p_{r'} \qquad (2.44)$$

where p_r and $p_{r'}$ are momenta conjugate to CD and CC bond-stretching coordinates, respectively, m_C is the mass of carbon atom, and α_e is the equilibrium valence angle. It is significant for linear molecules due to the large magnitude of the $g_{rr'}^{(e)}$ coefficient. This should be contrasted with near-local-mode molecules such as H_2S, H_2Se, and H_2Te where the bilinear kinetic energy coefficients are small in magnitude due to almost perpendicular valence bonds. In $^{12}C_2HD$, it is impossible to use an effective Hamiltonian for a single overtone manifold defined by fixed $v_{stretch} = v_{CH} + v_{CD}$ and fixed v_{CC}. Here, v_{CH}, v_{CD}, and v_{CC} are the CH, CD, CC stretching quantum numbers, respectively, in the internal coordinate representation. The inter-manifold couplings are important and second-order perturbation theory is not an appropriate way to take into account interactions between different manifolds (see later the section which discusses limitations of the simple vibrational models). In order to improve the model, it would be necessary to couple all states with the same total stretching quantum number $v_{total} = v_{CH} + v_{CD} + v_{CC}$. Coupling between the CH and the CD stretching motions might be insignificant because there is only small potential energy coupling present and because the CH and CD stretching fundamentals are not close. However, this kind of coupling can become important at high stretching energies (see below the discussion for $CHDCl_2$).

The harmonic interbond-type kinetic and potential energy couplings described above have also been found significant at higher CH and CD overtones of $CHDCl_2$ and CHD_2Cl [60]. The CH and CD stretching-fundamentals are well separated, but at higher overtones the levels differing by ± 1 in the CH stretching quantum number and by ∓ 1 in the CD stretching quantum number are close due to the large CH anharmonicity. For example, in $CHDCl_2$, the bilinear coupling operators in $g_{rr'}^{(e)}$ and $f_{rr'}$ are given in Eq. (2.1) with indices 1 and 2 referring to CH and CD bonds, respectively. In the HCAO model given in Eqs. (2.4), the interbond bilinear coupling parameter is

$$\lambda_{rr'} = \frac{1}{2}(\omega_r \omega_{r'})^{1/2}\left[\frac{g_{rr'}^{(e)}}{(g_{rr} g_{r'r'})^{1/2}} + \frac{f_{rr'}}{f_{rr}}\right] \qquad (2.45)$$

where indices r and r' refer to the CH and the CD stretching oscillators, respectively. Note that it has been assumed that $f_{rr} = f_{r'r'}$. In the methanol overtone spectrum obtained in a molecular beam by infrared laser-assisted photofragment spectroscopy, the same kind of coupling has been observed between $5v_{OH}$ and $4v_{OH} + v_{CH}$ states, where v_{CH} is the unique CH stretching fundamental [61].

The customary normal mode theory based models with quartic vibrational Darling–Dennison resonances included have also been extended to molecules with unequivalent bond oscillators [62,63]. The appropriate resonance matrix elements are given in Eqs. (1b) and (1c) in Table VI. Theoretical expressions for the resonance coefficients in terms of quartic and cubic dimensionless normal coordinate force constants have been derived in Refs. [62] and [63]. A model which includes these new interactions has been applied to the CH (CD) and CN stretching states in HCN (DCN) with success [62]. The diagonal terms are the usual anharmonic x expansions [with parameters ω_1 and ω_2, x_{11}, x_{22}, and x_{12}; the index 1 is for the CH(CD) stretch and the index 2 for the CN stretch]. There are altogether three different kinds of off-diagonal terms which must be included. The terms in $K_{12;22}$ and $K_{11;12}$ connect $|v_1, v_2\rangle$ states with $|v_1 \pm 1, v_2 \mp 1\rangle$ states and the terms in $K_{11;22}$ connect $|v_1, v_2\rangle$ states with $|v_1 \pm 2, v_2 \mp 2\rangle$ states. The total stretching quantum number $v = v_1 + v_2$ remains a good quantum label in this model. The eigenvalues are obtained by diagonalizing a $(v + 1)$-dimensional Hamiltonian matrix for each v.

Unitary transformations between normal and local mode representations

in the case of unequivalent bond oscillators have also been considered
[26,63,64]. In Ref. [63], the theory has been applied to CHD_2Cl. In the
above-mentioned methanol case assuming the staggered configuration, the
required transformation matrix is obtained from the matrix equation [64]

$$\begin{pmatrix} \omega_1(A') & 0 & 0 & 0 \\ 0 & \omega_2(A') & 0 & 0 \\ 0 & 0 & \omega_3(A') & 0 \\ 0 & 0 & 0 & \omega_9(A'') \end{pmatrix} = \mathbf{D}^\dagger \mathbf{N} \mathbf{D}$$

$$= \mathbf{D}^\dagger \begin{pmatrix} \omega_{OH} & \lambda_{Rr} & \lambda_{Rr'} & \lambda_{Rr'} \\ \lambda_{Rr} & \omega_{CH} & \lambda_{Rr'} & \lambda_{rr'} \\ \lambda_{Rr'} & \lambda_{Rr'} & \omega_{CH'} & \lambda_{r'r'} \\ \lambda_{Rr'} & \lambda_{rr'} & \lambda_{r'r'} & \omega_{CH'} \end{pmatrix} \mathbf{D} \quad (2.46)$$

Here, ω_1, ω_2, ω_3, and ω_9 are harmonic normal mode wavenumbers for the
OH, symmetric CH, symmetric CH', and antisymmetric CH' stretch, respec-
tively, where the prime is used to label the two equivalent CH bonds; ω_{OH},
ω_{CH}, and $\omega_{CH'}$ are the local mode harmonic wavenumbers for the OH
stretch, the unique CH stretch, and the two nonunique CH stretches,
respectively; $\lambda_{r'r'}$ and $\lambda_{rr'}$ are given by expressions similar to Eqs. (2.6) and
(2.45); and λ_{Rr} and $\lambda_{Rr'}$ describe harmonic coupling between the OH and
unique and nonunique CH stretching oscillators, respectively. There is no
direct kinetic energy coupling between OH and CH oscillators, but the
oscillators are coupled by direct harmonic terms and indirectly by both
kinetic and potential energy terms via some other modes. Indirect couplings
can be modeled by second-order Van Vleck perturbation theory [65]. An
ab initio harmonic force field [66] has been used to calculate the various
contributions. A much larger coupling occurs between the OH and the
unique CH vibration than between the OH vibration and the other two CH
vibrations $(|\lambda_{Rr}| \gg |\lambda_{Rr'}|)$. Direct potential energy coupling makes the
largest contribution to the coupling parameter λ_{Rr}. Assuming that the two
interacting states are degenerate, the calculated parameter -12.1 cm^{-1}
gives rise to splitting of 54 cm^{-1}, which agrees well with the experimental
observation of about 50 cm^{-1}. Unlike in the case of XH_2 molecules, the $\mathbf{D}$
matrix elements are best obtained by diagonalizing the $\mathbf{N}$ matrix numerically. The normal mode diagonal and off-diagonal anharmonicity parameters can be obtained from the local mode anharmonicity parameters x_{OH},
x_{CH}, and $x_{CH'}$ as follows $(i = 1, 2, 3, 4$ below, corresponding to OH, CH,

CH′, and CH′, respectively [26]):

$$X_{st;\,uv} = \sum_{i=1}^{4} x_{ii} D_{is}^* D_{it}^* D_{iu} D_{iv} \tag{2.47}$$

where

$$K_{ss;\,tt} = x_{st} = 4X_{ss;\,tt} = 4X_{st;\,st} \qquad x_{ss} = X_{ss;\,ss}$$

$$K_{ss;\,st} = 8X_{ss;\,st} \qquad K_{st;\,uu} = 8X_{st;\,uu} \qquad K_{st;\,su} = 8X_{st;\,su} \tag{2.48}$$

The normal mode indices s, t, u, and v differ from each other. The equations given above can be used to calculate the appropriate normal mode diagonal and off-diagonal parameters from the local mode parameters (and from the internal coordinate surface equally well if this is preferred). In the normal mode picture, the coupling between the OH and the unique CH stretching vibrations is caused by Darling–Dennison resonances as follows (see Table VI for the matrix elements) [67]:

$$\langle nv_1 | \frac{H_{DD}}{hc_0} | (n-1)v_1 + v_2 \rangle$$

$$= \tfrac{1}{4}(K_{11;\,12}\,n + K_{12;\,22} + K_{13;\,23} + K_{19;\,39})\sqrt{n} \tag{2.49}$$

where $n = 1, 2, 3, \ldots$. In the present observation $n = 5$. The resonance parameters are calculated to be $K_{11;\,12} = 11.8$ cm^{-1}, $K_{12;\,22} = -6.1$ cm^{-1}, $K_{13;\,23} = -1.5$ cm^{-1}, and $K_{19;\,29} = -0.6$ cm^{-1}. The application of the results given in Ref. [26] leads to a different formula because the unitary transformation from the local to the normal mode representation is made in Ref. [26] from harmonically coupled anharmonic oscillators to uncoupled anharmonic oscillators. Mathematically, this approach is equivalent to the one discussed here (the eigenvalues obtained, e.g., for methanol are the same) but the resonance coefficients as given in Ref. [26] are not related to the potential energy surface via traditional perturbation theory formulas [62,63]. Of course, the unitary transformation from the local to the normal mode picture provides a physical interpretation for the coefficients.

D. Limitations of Simple Stretching Vibrational Models

It is worth pointing out the limitations of the HCAO model and the customary model with Darling–Dennison resonance coupling before proceeding to more complicated modifications of these simple models. There are three major approximations: the decoupling of different manifolds, the neglect of some of the vibrations, and the use of harmonic oscillator formulas. These are discussed below.

In nonresonance cases, the intermanifold couplings can be taken into account by perturbation theory. This has already been done in deriving standard theoretical expressions for spectroscopic parameters in the conventional normal-mode-based theory. In the same way, the effects of the intermanifold coupling due to the bilinear harmonic terms in XH_2-like molecules are taken into account by second-order perturbation theory. This leads to the modified harmonic wavenumber as given in Eq. (2.5). Although it seems that in nonresonance cases the perturbation theory solution is adequate in calculating energies, it should be realized that small nonresonance interactions can have large effects on very weak overtone transition intensities. It might be that it is often better in accurate intensity calculations to take these intermanifold couplings into account explicitly by treating the problem variationally. This aspect is discussed in Section IV.

The effects of the neglected vibrations can be modeled by perturbation theory or adiabatic approximations in the HCAO model if the interactions between locallike modes and other modes are small. These ideas are discussed in the next section.

Interactions between locallike stretching modes and other modes may destroy the x–K relations to a certain extent. These relations are not exactly valid for any real molecule, but they are surprisingly well obeyed for many symmetrical molecules containing hydrogen atoms, as is indicated by the data given in Table VIII. This table gives stretching vibrational x, g, and K parameters for H_2O, H_2S, C_2H_2, NH_3, and CH_4 calculated with the x–K relations and assuming reasonable values for the Morse anharmonicity (see Table VIII for details). In Table VIII, the same parameters obtained experimentally [37–39,68–71] and/or by ab initio quantum-chemical methods [72–75] are given. In the standard normal-mode-based models, the nonresonance interactions with other vibrations can be taken into account by relaxing the x–K relations; that is, x, g, and K parameters are treated as independent free parameters in fitting experimental vibrational term value data. This leads to nonequivalence of the modified normal-mode-based model (six independent parameters in XH_2) and to the HCAO model (three independent parameters in XH_2). The HCAO model can be modified to retain the equivalence by adding higher order terms to it [34]. One way of deriving these terms is to start from the normal mode Hamiltonian expressed in terms of symmetrized creation and annihilation operators as given in Eq. (2.22). With the help of Eqs. (2.19), this Hamiltonian can be written in terms of local mode creation and annihilation operators. This procedure gives an equivalent local mode Hamiltonian given by its matrix

TABLE VIII

Stretching Anharmonicity Parameters (cm^{-1}) for Water, Hydrogen Sulfide, Acetylene, Ammonia, and Methane[a]

Molecule	Parameter	Normal Species			Deuterated Species	
		Calculated	Observed	Ab initio	Calculated	Observed
H_2O	x_{11}	-42.0	-42.6	-40.1	-22.3	-22.6
	x_{13}	-168.0	-165.8	-157.2	-89.0	-87.2
	x_{33}	-42.0	-47.6	-45.4	-22.3	-26.2
	$K_{11;33}$	-168.0	-155.0	-159.1	-89.0	-81.2
H_2S	x_{11}	-24.0	-25.1	-24.5	-12.4	
	x_{13}	-96.0	-94.7	-96.7	-49.5	
	x_{33}	-24.0	-24.0	-24.6	-12.4	
	$K_{11;33}$	-96.0	-91.0	-95.4	-49.5	
C_2H_2	x_{11}	-24.0	-24.1	—	-12.9	-12.5
	x_{13}	-96.0	-99.0	—	-51.8	-47.3
	x_{33}	-24.0	-25.7	—	-12.9	-12.9
	$K_{11;33}$	-96.0	-98.0	—	-51.8	-47.7
NH_3	x_{11}	-24.0	-25.8	-27.0	-12.8	
	x_{13}	-96.0	-93.0	-101.3	-51.3	
	x_{33}	-36.0	-44.2	-45.2	-19.2	
	g_{33}	12.0	12.3	14.5	6.4	
	$K_{11;33}$	-96.0	-98.9	-89.7	-51.3	
	$K_{13;33}$	-45.3	-46.6	-42.0	-24.2	
CH_4	x_{11}	-15.0	-11.0	-12.5	-8.1	
	x_{13}	-60.0	-50.3	-52.5	-32.3	
	x_{33}	-27.0	-29.2	-28.3	-14.6	
	G_{33}	9.0	12.1	11.5	4.9	
	T_{33}	3.0	3.4	3.3	1.6	
	$K_{11;33}$	-60.0	—	-55.2	-32.3	
	$K_{13;33}$	-240.0	—	-222.8	-129.4	

[a] Calculated values are obtained by applying x–K relations given in Table VII and by assuming that $x_{rr}(OH) = -84$ cm^{-1}, $x_{rr}(SH) = -48$ cm^{-1}, $x_{rr}(CH) = -48$ cm^{-1} (acetylenic), $x_{rr}(NH) = -72$ cm^{-1}, and $x_{rr}(CH) = -60$ cm^{-1} (methane). For the deuterated species, the values of x_{rr} are calculated using the expression $x_{rr}(XD) = x_{rr}(XH)g_{rr}(XD)/g_{rr}(XH)$. The observed values are from Refs. [37–39] and [68–71] and the ab initio values are from Refs. [72–75] (in NH_3 and CH_4 the ab initio K coefficients are taken from Ref. [72] and the other vibrational parameters from Refs. [74] and [75]). In methane, the observed values are calculated using the full experimental anharmonic force field [71].

elements as

$$\langle v_{r_1} v_{r_2} | \frac{H}{hc_0} | v_{r_1} v_{r_2} \rangle = \omega_r'(v_{r_1} + v_{r_2} + 1) + x_{rr}[(v_{r_1} + \tfrac{1}{2})^2$$

$$+ (v_{r_2} + \tfrac{1}{2})^2] + x_{rr'}(v_{r_1} + \tfrac{1}{2})(v_{r_2} + \tfrac{1}{2})$$

$$\langle (v_{r_1} - 1)(v_{r_2} + 1) | \frac{H}{hc_0} | v_{r_1} v_{r_2} \rangle = [\lambda_r + \lambda_{rr}(v_{r_1} + v_{r_2} + 1)][v_{r_1}(v_{r_2} + 1)]^{1/2}$$

$$\langle (v_{r_1} + 1)(v_{r_2} - 1) | \frac{H}{hc_0} | v_{r_1} v_{r_2} \rangle = [\lambda_r + \lambda_{rr}(v_{r_1} + v_{r_2} + 1)][(v_{r_1} + 1)v_{r_2}]^{1/2}$$

$$\langle (v_{r_1} - 2)(v_{r_2} + 2) | \frac{H}{hc_0} | v_{r_1} v_{r_2} \rangle = \delta_r[v_{r_1}(v_{r_1} - 1)(v_{r_2} + 1)(v_{r_2} + 2)]^{1/2}$$

$$\langle (v_{r_1} + 2)(v_{r_2} - 2) | \frac{H}{hc_0} | v_{r_1} v_{r_2} \rangle = \delta_r[(v_{r_1} + 1)(v_{r_1} + 2)v_{r_2}(v_{r_2} - 1)]^{1/2} \qquad (2.50)$$

where the coefficients $x_{rr'}$, λ_{rr}, and δ_r are small in magnitude for near-local-mode molecules and they disappear when x–K relations are exactly obeyed. The other coefficients (ω_r', x_{rr}, and λ_r) are from the HCAO model. This derivation of the extended HCAO model contains the drawback that it does not provide a physical interpretation of the new parameters. The same kind of extension of the anharmonic oscillator model to ammonia, methane, ethylene, allene, and benzene has also been worked out recently [63,76].

In the case of XH_2 molecules, the HCAO model has been extended to an anharmonically coupled oscillator model which retains the simple block diagonal Hamiltonian matrix structure [77]. This model starts from the model stretching vibrational Hamiltonian given in Eq. (2.1), but unlike in the HCAO model, the bilinear potential energy coupling term $f_{rr'} y_1 y_2$ is not expanded as a Taylor series. Following an earlier work [78], the stretching vibrational Hamiltonian is expressed in terms of a dimensionless coordinate $x_j = \sqrt{(\omega_r'/2 | x_{rr}|)} y_j$ and the conjugate momentum $p_{x_j} = -i(\partial/\partial x_j)$. The advantage of this representation lies in the zeroth-order model which is the sum of two uncoupled harmonic oscillators. Therefore, matrix elements of the full Hamiltonian can be evaluated using harmonic oscillator matrix elements. Using second- and third-order Van Vleck perturbation theory [65], it has been shown that the Hamiltonian matrix takes the same form as that given in Eqs. (2.50). The $x_{rr'}$, λ_{rr}, and δ_r coefficients in Eqs. (2.50) can be identified as

$$x_{rr'} = -2x_{rr}\frac{f_{rr'}}{f_{rr}} \qquad \lambda_{rr} = \frac{3}{4}x_{rr}\left(\frac{f_{rr'}}{f_{rr}} + \frac{g_{rr'}^{(e)}}{g_{rr}}\right)$$

$$\delta_r = -\frac{1}{2}x_{rr}\left(\frac{f_{rr'}}{f_{rr}} + \frac{g_{rr'}^{(e)}}{g_{rr}}\right) \qquad (2.51)$$

Although these equations give an estimate for the effects of the potential energy bilinear coupling term expansion and for the effects of the use of anharmonic oscillator matrix elements instead of harmonic oscillator matrix elements for nondiagonal cases, they are less useful in practice because some other neglected Hamiltonian terms have effects of the same order of magnitude [12].

In the HCAO model, different basis functions are used for diagonal and off-diagonal matrix elements. However, following the procedure given in Ref. [21] for bent XH_2 molecules, it is shown that the HCAO model is obtained using only harmonic oscillator matrix elements with perturbation theory. The zeroth-order Hamiltonian is taken to be

$$H_0 = \tfrac{1}{2}g_{rr}(p_{r_1}^2 + p_{r_2}^2) + \tfrac{1}{2}f_{rr}(r_1^2 + r_2^2) \qquad (2.52)$$

where

$$f_{rr} = \left(\frac{\partial^2 V}{\partial r_1^2}\right)_e = \left(\frac{\partial^2 V}{\partial r_2^2}\right)_e$$

Two perturbation terms are added: H_1 converts the potential energy function to a sum of two Morse functions and H_2 includes the bilinear kinetic and potential energy coupling terms between bond oscillators. These two operators are given as

$$H_1 = -\tfrac{1}{2}f_{rr}(r_1^2 + r_2^2) + D_e(y_1^2 + y_2^2) \qquad (2.53)$$

and

$$H_2 = g_{rr'}^{(e)}p_{r_1}p_{r_2} + f_{rr'}r_1 r_2 \qquad (2.54)$$

Harmonic oscillator basis functions consistent with H_0 are used with the Hamiltonian $H = H_0 + H_1 + H_2$, where H_0 gives rise only to the diagonal matrix elements and H_1 possesses only diagonal matrix elements within each overtone manifold $v = v_{r_1} + v_{r_2} = \text{const}$. The operator H_1 is expanded as Taylor series including terms up to the fourth power of r_i. The fourth-power terms are treated with first-order perturbation theory and contribute to the anharmonic term in x_{rr} on diagonal. The cubic terms in the expanded H_1 are responsible for the couplings between different manifolds with selection rules $\Delta v_{r_1} = \pm 1, \pm 3$ and $\Delta v_{r_2} = 0$ or $\Delta v_{r_1} = 0$ and $\Delta v_{r_2} = \pm 1, \pm 3$.

The corresponding off-diagonal matrix elements are removed with second-order perturbation theory. This also contributes to the diagonal term in x_{rr}. The operator H_2 gives rise to the coupling terms in λ_r as before. The end result is such that the HCAO model is recovered. In the traditional normal mode model, the diagonal anharmonic terms are also obtained using second-order perturbation theory and harmonic oscillator matrix element [21]. Both models, the HCAO model and the normal mode model with Darling–Dennison resonances, can be constructed using harmonic oscillator matrix elements throughout with second-order perturbation theory.

III. SIMPLE STRETCHING AND BENDING VIBRATIONAL MODELS

A. Bent Triatomic Molecules

The most serious disadvantage of the original local mode model is the neglect of nonstretching vibrations. In order to include other types of vibrational degrees of freedom, two somewhat different types of effects may be dealt with: harmonic and anharmonic effects. These are now discussed in well-bent XY_2 triatomic molecules.

The normal coordinates are defined in such a way that there are no harmonic kinetic or potential energy terms which couple different normal modes. Thus, in the case of normal mode models with Darling–Dennison resonances included, direct harmonic coupling effects are absent and the different modes are coupled only via anharmonic potential energy terms. On the other hand, the local mode models are formulated in terms of internal coordinates. In this approach, there may exist harmonic couplings between different modes. For bent XY_2 molecules, the model Hamiltonian, which includes the bending degree of freedom characterized by the bending curvilinear internal displacement coordinate θ and its conjugate momentum $p_\theta = -i\hbar(\partial/\partial\theta)$, is given as

$$H = T + V = H_{\text{stretch}} + \tfrac{1}{2}g_{\theta\theta}^{(e)}p_\theta^2 + \tfrac{1}{2}f_{\theta\theta}\theta^2$$
$$+ g_{r\theta}^{(e)}(p_{r_1} + p_{r_2})p_\theta + f_{r\theta}a_r^{-1}(y_1 + y_2)\theta \qquad (3.1)$$

where H_{stretch} is the stretching vibrational Hamiltonian in Eq. (2.1). The kinetic energy coefficients $g_{\theta\theta}^{(e)} = 2[m_X + m_Y(1 - \cos\alpha_e)]/(m_X m_Y R_e^2)$ and $g_{r\theta}^{(e)} = -\sin\alpha_e/(m_X R_e)$ are evaluated at the equilibrium geometry [17]. Here R_e is the equilibrium XY bond length. The harmonic force constants

$$f_{\theta\theta} = \left(\frac{\partial^2 V}{\partial \theta^2}\right)_e \quad \text{and} \quad f_{r\theta} = \left(\frac{\partial^2 V}{\partial r_1 \, \partial \theta}\right) = \left(\frac{\partial^2 V}{\partial r_2 \, \partial \theta}\right)_e$$

are potential energy parameters. All other quantities are defined using. Eq. (2.1).

Second-order Van Vleck perturbation theory can be employed to take the effects of the harmonic coupling terms into account in the HCAO model [3,13]. There are second-order contributions to both diagonal and off-diagonal matrix elements. The quantum number dependencies of all these contributions, as well as of the corresponding matrix elements in the original HCAO model, are identical. Just the physical interpretation of the original model is slightly changed. By subtracting the zero-point energy from the diagonal terms and employing harmonic oscillator formulas, the effective Hamiltonian matrix elements become

$$\langle v_{r_1} v_{r_2}, v_\theta | \frac{H}{hc_0} | v_{r_1} v_{r_2}, v_\theta \rangle$$

$$= \omega_r'(v_{r_1} + v_{r_2}) + x_{rr}[v_{r_1}(v_{r_1} + 1) + v_{r_2}(v_{r_2} + 1)] + \omega_\theta' v_\theta$$

$$\langle (v_{r_1} + 1)(v_{r_2} - 1), v_\theta | \frac{H}{hc_0} | v_{r_1} v_{r_2}, v_\theta \rangle = \lambda_r'[(v_{r_1} + 1)v_{r_2}]^{1/2} \qquad (3.2)$$

$$\langle (v_{r_1} - 1)(v_{r_2} + 1), v_\theta | \frac{H}{hc_0} | v_{r_1} v_{r_2}, v_\theta \rangle = \lambda_r'[v_{r_1}(v_{r_2} + 1)]^{1/2}$$

where $|v_{r_1} v_{r_2}, v_\theta\rangle$ denotes a product basis $|v_{r_1}\rangle |v_{r_2}\rangle |v_\theta\rangle$, v_θ is the bending quantum number, and

$$\omega_r' = \omega_r - \frac{1}{8} \omega_r \left(\frac{g_{rr'}^{(e)}}{g_{rr}} - \frac{f_{rr'}}{f_{rr}}\right)^2 + \frac{1}{4}\left(\frac{d_{r\theta}(1)^2}{\omega_r - \omega_\theta} - \frac{d_{r\theta}(2)^2}{\omega_r + \omega_\theta}\right)$$

$$\omega_\theta' = \omega_\theta - \frac{1}{2}\left(\frac{d_{r\theta}(1)^2}{\omega_r - \omega_\theta} + \frac{d_{r\theta}(2)^2}{\omega_r + \omega_\theta}\right) \qquad (3.3)$$

$$\lambda_r' = \lambda_r + \frac{1}{4}\left(\frac{d_{r\theta}(1)^2}{\omega_r - \omega_\theta} - \frac{d_{r\theta}(2)^2}{\omega_r + \omega_\theta}\right)$$

with

$$\omega_\theta = \frac{1}{2\pi c_0} (f_{\theta\theta} g_{\theta\theta}^{(e)})^{1/2}$$

$$d_{r\theta}(1) = \frac{\hbar^2}{hc_0} g_{r\theta}^{(e)} \alpha_r^{1/2} \alpha_\theta^{1/2} + \frac{1}{hc_0} f_{r\theta} \alpha_r^{-1/2} \alpha_\theta^{-1/2}$$

$$= (\omega_r \omega_\theta)^{1/2} \left[\frac{g_{r\theta}^{(e)}}{(g_{rr} g_{\theta\theta}^{(e)})^{1/2}} + \frac{f_{r\theta}}{(f_{rr} f_{\theta\theta})^{1/2}} \right] \qquad (3.4)$$

$$d_{r\theta}(2) = -\frac{\hbar^2}{hc_0} g_{r\theta}^{(e)} \alpha_r^{1/2} \alpha_\theta^{1/2} + \frac{1}{hc_0} f_{r\theta} \alpha_r^{-1/2} \alpha_\theta^{-1/2}$$

$$= (\omega_r \omega_\theta)^{1/2} \left[-\frac{g_{r\theta}^{(e)}}{(g_{rr} g_{\theta\theta}^{(e)})^{1/2}} + \frac{f_{r\theta}}{(f_{rr} f_{\theta\theta})^{1/2}} \right]$$

and

$$\alpha_r = \frac{4\pi^2 c_0 \omega_r}{hg_{rr}} = \frac{(f_{rr}/g_{rr})^{1/2}}{\hbar} \qquad \alpha_\theta = \frac{4\pi^2 c_0 \omega_\theta}{hg_{\theta\theta}^{(e)}} = \frac{(f_{\theta\theta}/g_{\theta\theta}^{(e)})^{1/2}}{\hbar} \qquad (3.5)$$

Another possible way to take the effects of other vibrations into account is to use adiabatic approximations where locallike stretching vibrations are separated from other normallike modes in such a way that the electronic motion is separated from the nuclear motion in the Born–Oppenheimer approximation. A simple adiabatic model with a single stretching degree of freedom coupled to a bend has been published [79]. The bend is treated as the slow motion (corresponding to the motion of a nucleus in electronic problems). A simple expression is obtained for the vibrational anharmonic constant x_{sb} where the index s refers to the stretch and the index b to the bend. Moreover, a relationship is obtained between x_{sb} constants of isotopically related molecules. In symmetrical bent triatomic molecules with a heavy central mass, the Darling–Dennison empirical formula [20]

$$\frac{x_{s2}^*}{x_{s2}} = \frac{\omega_s^* \omega_2^*}{\omega_s \omega_2} \qquad (3.6)$$

is shown to be obeyed well. Here, $s = 1, 3$ is the index for the symmetric or for the antisymmetric stretching vibration and 2 is the index for the bend, and x_{s2}^* is x_{s2} for an isotopically related molecule. Equation (3.6) is also valid in the case of the stretching anharmonic parameter x_{13} in water-type molecules because, according to Eq. (2.12), $x_{13} = 2x_{rr} = -\omega_r^2/4D_e$ and because the Morse harmonic wavenumber is close to the normal mode harmonic wavenumbers ($\omega_r^2 \approx \omega_1\omega_3$) for molecules close to the local mode limit [3]. Table IX gives calculated x^* constants for D_2O, D_2S, and D_2Se using Eq. (3.6). A comparison with experimental values [38,39,69,80] given in Table IX shows a good agreement between this simple model and experiment.

Adiabatic approximations can be applied to the HCAO model to investigate the effects of bending vibrations. This has been done by assuming that the **g** matrix elements depend on the locallike stretching coordinates but are evaluated at the equilibrium values of the normallike coordinates [79]. In this revised model for bent XY_2 molecules, the local model splittings possess a dependence on the bending quantum number, unlike in the original HCAO model where the vibrational energy level splittings do not depend on bending excitations.

Approximate methods based on perturbation theory cannot be used when interactions between locallike and normallike modes are strong. The most common of these types of interactions is Fermi resonance, which occurs, for example, in water between the upper vibrational states of the symmetric stretching fundamental v_1 and the first bending overtone $2v_2$ transitions. It is weak due to the relatively large energy level separation of the interacting vibrational states, $v_1 = 3657$ cm^{-1} and $2v_2 = 3152$ cm^{-1}, but it becomes significant at high energies because the stretching and bending levels get closer due to large stretching anharmonicity.

In the customary vibration–rotation theory, cubic potential energy terms are responsible for the Fermi resonance interactions. In water, for example, this kind of term takes the form $H_{\text{Fermi}}/hc_0 = \frac{1}{2}\phi_{122}\,q_1q_2^2$, where

$$\phi_{122} = \frac{1}{hc_0}\left(\frac{\partial^3 V}{\partial q_1\,\partial q_2^2}\right)_e$$

is the potential energy coefficient (a cubic force constant in the dimensionless normal coordinate representation) which describes the strength of the resonance. When harmonic oscillator formulas (see Table II) are used, the matrix elements of this cubic term are

$$\langle v_1 - 1, v_2 + 2, v_3 | \frac{H_{\text{Fermi}}}{hc_0} | v_1, v_2, v_3 \rangle = \tfrac{1}{4}\phi_{122}[\tfrac{1}{2}v_1(v_2 + 1)(v_2 + 2)]^{1/2}$$

$$\langle v_1 + 1, v_2 - 2, v_3 | \frac{H_{\text{Fermi}}}{hc_0} | v_1, v_2, v_3 \rangle = \tfrac{1}{4}\phi_{122}[\tfrac{1}{2}(v_1 + 1)v_2(v_2 - 1)]^{1/2}$$

$$(3.7)$$

TABLE IX

Anharmonicity Parameters (cm^{-1}) of D_2O, D_2S, and D_2Se Calculated with Darling's and Dennison's Empirical Formula[a]

Parameter	H_2O, Observed	D_2O		$H_2{}^{32}S$, Observed	$D_2{}^{32}S$		$H_2{}^{80}Se$ Observed	$D_2{}^{80}Se$ Calculated
		Observed	Calculated		Observed	Calculated		
x_{12}	−15.9	−7.6	−8.6	−19.7	−9.1	−10.3	−17.7	−9.1
x_{13}	−165.8	−87.2	−89.9	−94.7	−49.5	−49.9	−84.9	−44.0
x_{23}	−20.3	−10.6	−11.1	−21.1	−11.1	−11.1	−20.2	−10.4

[a] Observed x constants are taken from Refs. [69] (H_2O), [38] ($H_2{}^{32}S$), and [39] ($H_2{}^{80}Se$). The x parameters of $D_2{}^{32}S$ are determined from the potential energy surface in Ref. [80]. Fundamental wavenumbers taken from the results of Ref. [80] are used in applying the Darling–Dennison empirical formula.

where $|v_1, v_2, v_3\rangle$ denotes the product basis $|v_1\rangle|v_2\rangle|v_3\rangle$. Clearly, in water, using normal mode labels for the interacting states v_1 would be coupled to $2v_2$ via Fermi resonance matrix elements.

The internal coordinate formulation of the Fermi resonance problem is best started from an approximate vibrational Hamiltonian [81,82]

$$H = \tfrac{1}{2}\mathbf{p}_r^T \mathbf{g}(\mathbf{r})\mathbf{p}_r + V(\mathbf{r}) \tag{3.8}$$

which is derived in Section IV. In this equation $\mathbf{r}$ includes stretches and bends. Here, $p_{r_j} = -i\hbar(\partial/\partial r_j)$ is the momentum operator conjugate to the internal coordinate r_j. The $\mathbf{g}(\mathbf{r})$ matrix, which is a function of curvilinear internal coordinates and atomic masses, can be formed for any polyatomic molecule using the method described in Section IV [17]. The usual potential energy function $V(\mathbf{r})$ is expressed in terms of curvilinear internal coordinates. For bent XY_2 molecules, the Hamiltonian may be written as

$$
\begin{aligned}
H = T + V = {} & \tfrac{1}{2}p_{r_1} g_{rr}\, p_{r_1} + \tfrac{1}{2}p_{r_2} g_{rr}\, p_{r_2} \\
& + \tfrac{1}{2}p_{r_1} g_{rr'}\, p_{r_2} + \tfrac{1}{2}p_{r_2} g_{rr'}\, p_{r_1} \\
& + \tfrac{1}{2}p_\theta g_{\theta\theta}\, p_\theta + \tfrac{1}{2}p_\theta g_{r_1\theta}\, p_{r_1} + \tfrac{1}{2}p_{r_1} g_{r_1\theta}\, p_\theta \\
& + \tfrac{1}{2}p_\theta g_{r_2\theta}\, p_{r_2} + \tfrac{1}{2}p_{r_2} g_{r_2\theta}\, p_\theta + V(r_1, r_2, \theta)
\end{aligned}
\tag{3.9}
$$

The $\mathbf{g}$ matrix elements are given in Table X. Both the $\mathbf{g}$ matrix elements and the potential energy function $V(r_1, r_2, \theta)$ are expanded as a Taylor series around the equilibrium configuration [83–87]. In this way, the local mode Hamiltonian in Eq. (2.1) and the modified local mode Hamiltonian including bending vibration at the harmonic coupling level as given in Eq. (3.1) are obtained. The Fermi resonance operators are also obtained by this procedure by noticing that they must be of the same form as previously given in the dimensionless normal coordinate representation; that is, they must be one-power operators in stretching variables p_{r1}, p_{r2}, r_1, or r_2 and two-power operators in bending variables p_θ and θ. There are two kinetic energy Fermi resonance operators in addition to the Fermi resonance potential energy term. The Fermi resonance terms are [83]

$$
\begin{aligned}
H_{\text{Fermi}} = {} & \frac{1}{2}\, a_r^{-1}\left(\frac{\partial g_{\theta\theta}}{\partial r}\right)_e (y_1 + y_2)p_\theta^2 \\
& + \frac{1}{2}\left(\frac{\partial g_{r\theta}}{\partial \theta}\right)_e (p_{r_1} + p_{r_2})(p_\theta \theta + \theta p_\theta) \\
& + \frac{1}{2}\, a_r^{-1} f_{r\theta\theta}(y_1 + y_2)\theta^2
\end{aligned}
\tag{3.10}
$$

TABLE X
g Matrix Elements for Bent XY$_2$ Molecules[a]

$$g_{rr} = \mu_X + \mu_Y = \frac{1}{m_X} + \frac{1}{m_Y}$$

$$g_{\theta\theta} = \mu_Y\left(\frac{1}{R_1^2} + \frac{1}{R_2^2}\right) + \mu_X\left(\frac{1}{R_1^2} + \frac{1}{R_2^2} - \frac{2\cos\alpha}{R_1 R_2}\right)$$

$$g_{rr'} = \mu_X \cos\alpha$$

$$g_{r_1\theta} = -\frac{\mu_X \sin\alpha}{R_2}$$

$$g_{r_2\theta} = -\frac{\mu_X \sin\alpha}{R_1}$$

[a] $m_X = 1/\mu_X$ and $m_Y = 1/\mu_Y$ are masses of the X and Y atoms, respectively; R_i ($i = 1, 2$) is the instantaneous bond length of the ith bond and α is the instantaneous valence angle.

where

$$\left(\frac{\partial g_{\theta\theta}}{\partial r}\right)_e = \left(\frac{\partial g_{\theta\theta}}{\partial r_1}\right)_e = \left(\frac{\partial g_{\theta\theta}}{\partial r_2}\right)_e = \left(-\frac{2}{R_e^3}\right)[\mu_Y + \mu_X(1 - \cos\alpha_e)]$$

$$\left(\frac{\partial g_{r\theta}}{\partial \theta}\right)_e = \left(\frac{\partial g_{r_1\theta}}{\partial \theta}\right)_e = \left(\frac{\partial g_{r_2\theta}}{\partial \theta}\right)_e = -\frac{\mu_X \cos\alpha_e}{R_e} \qquad (3.11)$$

$$f_{r\theta\theta} = \left(\frac{\partial^3 V}{\partial r_1\, \partial\theta^2}\right)_e = \left(\frac{\partial^3 V}{\partial r_2\, \partial\theta^2}\right)_e$$

with $\mu_X = 1/m_X$ and $\mu_Y = 1/m_Y$. The Morse variable y_i ($i = 1, 2$) has been used instead of the stretching displacement coordinate r_i in order to ensure realistic asymptotic limits at large-amplitude displacements. The Fermi resonance matrix elements in the harmonic approximation give rise to the resonance selection rules $\Delta v_{r_1} = \pm 1$, $\Delta v_{r_2} = 0$, $\Delta v_\theta = \mp 2$ or $\Delta v_{r_1} = 0$, $\Delta v_{r_2} = \pm 1$, $\Delta v_\theta = \mp 2$. From the results given in Table II, the following nonzero matrix elements are obtained:

$$\langle v_{r_1} - 1)v_{r_2}, (v_\theta + 2)| \frac{H_{\text{Fermi}}}{hc_0} |v_{r_1}v_{r_2}, v_\theta\rangle$$

$$= \tfrac{1}{4}\phi_{r\theta\theta}[\tfrac{1}{2}v_{r_1}(v_\theta + 1)(v_\theta + 2)]^{1/2}$$

$$\langle (v_{r_1} + 1)v_{r_2}, (v_\theta - 2)| \frac{H_{\text{Fermi}}}{hc_0} |v_{r_1}v_{r_2}, v_\theta\rangle$$
$$= \tfrac{1}{4}\phi_{r\theta\theta}[\tfrac{1}{2}(v_{r_1} + 1)v_\theta(v_\theta - 1)]^{1/2}$$

$$\langle v_{r_1}(v_{r_2} - 1), (v_\theta + 2)| \frac{H_{\text{Fermi}}}{hc_0} |v_{r_1}v_{r_2}, v_\theta\rangle$$
$$= \tfrac{1}{4}\phi_{r\theta\theta}[\tfrac{1}{2}v_{r_2}(v_\theta + 1)(v_\theta + 2)]^{1/2}$$

$$\langle v_{r_1}(v_{r_2} + 1), (v_\theta - 2)| \frac{H_{\text{Fermi}}}{hc_0} |v_{r_1}v_{r_2}, v_\theta\rangle$$
$$= \tfrac{1}{4}\phi_{r\theta\theta}[\tfrac{1}{2}(v_{r_1} + 1)v_\theta(v_\theta - 1)]^{1/2} \quad (3.12)$$

where

$$\phi_{r\theta\theta} = \left(-\frac{\hbar^2}{hc_0}\right)\left[\left(\frac{\partial g_{\theta\theta}}{\partial r}\right)_e \alpha_r^{-1/2}\alpha_\theta - 2\left(\frac{\partial g_{r\theta}}{\partial \theta}\right)_e \alpha_r^{1/2}\right]$$
$$+ (hc_0)^{-1}f_{r\theta\theta}\,\alpha_r^{-1/2}\alpha_\theta^{-1} \quad (3.13)$$

with α_r and α_θ defined in Eqs. (3.5).

The Fermi resonance matrix elements in Eqs. (3.12) together with the diagonal matrix element

$$G(v_{r_1}v_{r_2}, v_\theta) = \langle v_{r_1}v_{r_2}, v_\theta| \frac{H}{hc_0} |v_{r_1}v_{r_2}, v_\theta\rangle$$
$$= \omega_r'(v_{r_1} + v_{r_2}) + x_{rr}[v_{r_1}(v_{r_1} + 1) + v_{r_2}(v_{r_2} + 1)]$$
$$+ \omega_\theta' v_\theta + x_{\theta\theta} v_\theta(v_\theta + 1)$$
$$+ x_{r\theta}[v_\theta(v_{r_1} + v_{r_2} + 1) + \tfrac{1}{2}(v_{r_1} + v_{r_2})] \quad (3.14)$$

and the local mode coupling matrix elements

$$\langle (v_{r_1} + 1)(v_{r_2} - 1), v_\theta| \frac{H}{hc_0} |v_{r_1}v_{r_2}, v_\theta\rangle = \lambda_r'[(v_{r_1} + 1)v_{r_2}]^{1/2}$$
$$\quad (3.15)$$
$$\langle (v_{r_1} - 1)(v_{r_2} + 1), v_\theta| \frac{H}{hc_0} |v_{r_1}v_{r_2}. v_\theta\rangle = \lambda_r'[v_{r_1}(v_{r_2} + 1)]^{1/2}$$

constitute a model which includes both Fermi resonance and local mode effects. The zero-point energy has been subtracted from the diagonal elements. The term in $x_{\theta\theta}$ describes the anharmonicity of the bending vibration and the term in $x_{r\theta}$ describes the anharmonic nonresonance interaction between stretching and bending vibrations. Theoretical expressions for these parameters are similar to that given for $\phi_{r\theta\theta}$ in Eq. (3.13). The model which should be used to derive these expressions also includes higher order terms than those given in Eqs. (2.1) and (3.1) (see Refs. [12] for more details). By employing symmetrized basis functions for the stretches as given in Eqs. (2.7), symmetry factorization ($+$ and $-$ matrices, i.e., A_1 and B_1 matrices) is achieved for each $v = v_{r_1} + v_{r_2} + \frac{1}{2}v_\theta$. For example, for $v = 2$

$$H^+(v = 2; A_1) =$$

$$
\begin{pmatrix}
G(20, 0) & 2\lambda'_r & \frac{\sqrt{2}}{4}\phi_{r\theta\theta} & 0 \\[2ex]
2\lambda'_r & G(11, 0) & \frac{\sqrt{2}}{4}\phi_{r\theta\theta} & 0 \\[2ex]
\frac{\sqrt{2}}{4}\phi_{r\theta\theta} & \frac{\sqrt{2}}{4}\phi_{r\theta\theta} & G(10, 2) + \lambda'_r & \frac{\sqrt{3}}{2}\phi_{r\theta\theta} \\[2ex]
0 & 0 & \frac{\sqrt{3}}{2}\phi_{r\theta\theta} & G(00, 4)
\end{pmatrix}
$$

$$
=
\begin{pmatrix}
2\omega'_r + 6x_{rr} + x_{r\theta} & 2\lambda'_r & \frac{\sqrt{2}}{4}\phi_{r\theta\theta} & 0 \\[2ex]
2\lambda'_r & 2\omega'_r + 4x_{rr} + x_{r\theta} & \frac{\sqrt{2}}{4}\phi_{r\theta\theta} & 0 \\[2ex]
\frac{\sqrt{2}}{4}\phi_{r\theta\theta} & \frac{\sqrt{2}}{4}\phi_{r\theta\theta} & \omega'_r + 2x_{rr} + \lambda'_r + 2\omega_\theta + 6x_{\theta\theta} + \frac{9}{2}x_{r\theta} & \frac{\sqrt{3}}{2}\phi_{r\theta\theta} \\[2ex]
0 & 0 & \frac{\sqrt{3}}{2}\phi_{r\theta\theta} & 4\omega_\theta + 20x_{\theta\theta} + 4x_{r\theta}
\end{pmatrix}
$$

$$H^-(v = 2; B_1) =$$

$$\begin{pmatrix} G(20, 0) & \dfrac{\sqrt{2}}{4}\,\phi_{r\theta\theta} \\[2mm] \dfrac{\sqrt{2}}{4}\,\theta_{r\theta\theta} & G(10, 2) - \lambda_r' \end{pmatrix}$$

$$= \begin{pmatrix} 2\omega_r' + 6x_{rr} + x_{r\theta} & \dfrac{\sqrt{2}}{4}\,\phi_{r\theta\theta} \\[2mm] \dfrac{\sqrt{2}}{4}\,\phi_{r\theta\theta} & \omega_r' + 2x_{rr} - \lambda_r' + 2\omega_\theta + 6x_{\theta\theta} + \tfrac{9}{2}x_{r\theta} \end{pmatrix} \quad (3.16)$$

are obtained.

The corresponding model using the conventional approach is obtained from the Fermi resonance coupling matrix elements given in Eqs. (3.7) (these elements are diagonal in v_3 without any dependence on v_3), from the diagonal matrix element

$$G(v_1, v_2, v_3) = \langle v_1, v_2, v_3 | \frac{H}{hc_0} | v_1, v_2, v_3 \rangle$$

$$= \omega_1 v_1 + \omega_2 v_2 + \omega_3 v_3 + x_{11} v_1 (v_1 + 1) + x_{22}^* v_2(v_2 + 1)$$
$$+ x_{33} v_3 (v_3 + 1) + x_{12}^* [v_1 v_2 + \tfrac{1}{2}(v_1 + v_2)]$$
$$+ x_{13}[v_1 v_3 + \tfrac{1}{2}(v_1 + v_3)] + x_{23}[v_2 v_3 + \tfrac{1}{2}(v_2 + v_3)]$$

$$(3.17)$$

and from the following vibrationally off-diagonal Darling–Dennison resonance matrix elements [compare with Eqs. (2.10) for the case where the bending vibration was excluded]:

$$\langle v_1 - 2, v_2, v_3 + 2 | \frac{H}{hc_0} | v_1, v_2, v_3 \rangle$$

$$= \tfrac{1}{4} K_{11;33}[v_1(v_1 - 1)(v_3 + 1)(v_3 + 2)]^{1/2}$$

$$\langle v_1 + 2, v_2, v_3 - 2 | \frac{H}{hc_0} | v_1, v_2, v_3 \rangle$$

$$= \tfrac{1}{4} K_{11;33}[(v_1 + 1)(v_1 + 2)v_3(v_3 - 1)]^{1/2} \quad (3.18)$$

Zero-point energy has been subtracted from the diagonal element. An asterisk in the anharmonicity parameters x^*_{22} and x^*_{12} indicates that the contribution from Fermi resonance terms to the traditional theoretical expressions of these parameters is excluded. For the overtone chromophore $v = v_1 + v_3 + \frac{1}{2}v_2 = 2$, the A_1 and B_1 Hamiltonian matrices are

$$H(v = 2; A_1)$$

$$= \begin{pmatrix} G(2v_1) & \frac{1}{2}K_{11;33} & \frac{\sqrt{2}}{4}\phi_{122} & 0 \\[2mm] \frac{1}{2}K_{11;33} & G(2v_3) & 0 & 0 \\[2mm] \frac{\sqrt{2}}{4}\phi_{122} & 0 & G(v_1 + 2v_2) & \frac{\sqrt{6}}{4}\phi_{122} \\[2mm] 0 & 0 & \frac{\sqrt{6}}{4}\phi_{122} & G(4v_2) \end{pmatrix}$$

$$= \begin{pmatrix} 2\omega_1 + 6x_{11} + x^*_{12} + x_{13} & \frac{1}{2}K_{11;33} \\[3mm] \frac{1}{2}K_{11;33} & 2\omega_3 + 6x_{33} + x_{13} + x_{23} \\[3mm] \frac{\sqrt{2}}{4}\phi_{122} & 0 \\[3mm] 0 & 0 \end{pmatrix}$$

$$\begin{matrix} \frac{\sqrt{2}}{4}\phi_{122} & 0 \\[3mm] 0 & 0 \\[3mm] \omega_1 + 2\omega_2 + 2x_{11} + 6x^*_{22} + \frac{7}{2}x^*_{12} + \frac{1}{2}x_{13} + x_{23} & \frac{\sqrt{6}}{4}\phi_{122} \\[3mm] \frac{\sqrt{6}}{4}\phi_{122} & 4\omega_2 + 20x^*_{22} + 2x^*_{12} + 2x_{23} \end{matrix}$$

$$H(v = 2; B_1) = \begin{pmatrix} G(v_1 + v_3) & \frac{1}{4}\phi_{122} \\ \frac{1}{4}\phi_{122} & G(2v_2 + v_3) \end{pmatrix}$$

$$= \begin{pmatrix} \omega_1 + \omega_3 + 2x_{11} + 2x_{33} + \frac{1}{2}x_{12}^* + 2x_{13} + \frac{1}{2}x_{23} \\ \frac{1}{4}\phi_{122} \end{pmatrix}$$

$$\begin{pmatrix} \frac{1}{4}\phi_{122} \\ 2\omega_2 + \omega_3 + 6x_{22}^* + 2x_{33} + x_{12}^* + \frac{1}{2}x_{13} + \frac{7}{2}x_{23} \end{pmatrix}$$

$$(3.19)$$

where the standard notation is used in G, that is for example $G(2v_1) = G(2, 0, 0)$.

The two different models, which include Fermi resonance interaction terms, produce identical eigenvalues if the x–K relations given in Eqs. (2.12), together with relations $\omega_2 = \omega_\theta$, $x_{12}^* = x_{23} = x_{r\theta}$, $x_{22}^* = x_{\theta\theta}$, and $\phi_{122} = \sqrt{2}\phi_{r\theta\theta}$, are applied between the vibrational parameters. This can be shown by using the shift operator technique discussed before [25,26,34]. Due to the equivalence of these two approaches, it seems that there is little to choose between them. However, the model formulated in terms of curvilinear internal coordinates has the advantage that it is more easily related to the concept of Born–Oppenheimer potential energy surface. The model parameters are directly expressible in terms of internal coordinate force constants. Thus, at least in principle, the internal coordinate model can be used simultaneously with the same potential energy parameters for different isotopic species. It should be borne in mind that in some cases the various approximations may introduce an effective element to the potential energy surface obtained. In the normal mode model, it is customary to make a nonlinear transformation between force constants in the normal coordinate and internal coordinate representation [88]. This is a complicated way to interpret the spectroscopic parameters in terms of Born–Oppenheimer potential energy force constants. Of course, the derived x–K relations (and their possible extensions for cases where x–K relations are not well obeyed) could be used to tie the traditional spectroscopic parameters directly to Born–Oppenheimer force constants.

Table XI contains a set of spectroscopic parameters for $^{32}H_2S$ used to calculate vibrational energy levels, and these levels, along with experimental values for comparison, are given in Table XII [89–91]. The model is in good agreement with experiment. The spectroscopic parameters provide information about the hydrogen sulfide potential energy surface. For example, with Eq. (3.13), $f_{r\theta\theta} = -0.373$ cm^{-1} is obtained. This agrees well with values given in Table XXIV in Section IV. In Refs. [12], the model has been extended to include second-order perturbation theory contributions to the theoretical expressions of the spectroscopic parameters. Using surfaces given in Ref. [80] with the full theoretical formula of $\phi_{r\theta\theta}$ from Refs. [12], the values 119.4 and 58.04 cm^{-1} for H_2O and H_2S, respectively, are obtained. These numbers should be compared with the values 114.6 and 55.2 cm^{-1} calculated with Eq. (3.13). Clearly, the second-order perturbation theory corrections make only small contributions. This is pleasing in the sense that the simple model is useful in practical applications. In spite of the success of this treatment, it should be observed that triatomic molecules are small enough to be treated exactly variationally. This aspect is discussed in Section IV.

B. Pyramidal XH$_3$ Molecules

Fermi resonances between the upper states of stretching and bending overtone transitions can also be significant in XH_3-type pyramidal molecules. Fundamental wavenumbers with their customary normal mode notations and interpretations for NH_3, PH_3, AsH_3, and SbH_3 are given in Table XIII [9,92–96]. Apart from ammonia, Fermi resonances should be considered between $v_1(A_1)$ and $2v_2^0(A_1)$, between $v_1(A_1)$ and $2v_4^0(A_1)$, between $v_3^{\pm 1}(E)$ and $2v_4^{\mp 2}(E)$, and between $v_3^{\pm 1}(E)$ and $v_2 + v_4^{\pm 1}(E)$ in the stretching fundamental region, where the normal mode labels refer to interacting states. The superscripts give the vibrational angular momentum state. These resonances are weak in the case of the stretching fundamentals, but they become strong at high energies due to large stretching anharmonicities (i.e., the interacting states get close as energy increases).

The local mode model for stretching vibrations, combined with the Fermi resonance model to include the bending vibration in bent XH_2-type triatomics, can be extended to pyramidal XH_3 molecules. One possibility to proceed is to treat the bending vibrations as coupled valence angle oscillators in the same way as the stretching vibrations in the local mode model are treated as coupled anharmonic oscillators. The Fermi resonance Hamil-

TABLE XI
Spectroscopic Parameters for H_2S

Parameter	Value[a] (cm^{-1})
ω_r'	2727.55 (98)
x_{rr}	-48.14 (24)
λ_r'	-6.40 (60)
ω_θ'	1217.00 (170)
$x_{\theta\theta}$	-5.81 (28)
$x_{r\theta}$	-21.37 (62)
$\phi_{r\theta\theta}$	35.8 (76)

[a] Values in parentheses are one standard errors in the least significant digit. All data as given in Table XII were given unit weights.

TABLE XII
Observed and Calculated Vibrational Term Values (cm^{-1}) for H_2S^a

$2v$	$v_{r_1}v_{r_2} \pm v_\theta$	$v_1 v_2 v_3$	Observed	Observed − Calculated
1	00 + 1	010	1182.57	−1.44
2	00 + 2	020	2353.96	−1.82
	10 + 0	100	2614.14	−0.41
	10 − 0	001	2628.46	1.47
3	00 + 3	030	3513.79	−1.54
	10 + 1	110	3779.17	0.48
	10 − 1	011	3789.27	−0.37
4	00 + 4	040	4661.68	−0.99
	10 + 2	120	4932.70	2.51
	10 − 2	021	4939.10	−0.77
	20 + 0	200	5144.99	0.80
	20 − 0	101	5147.22	1.51
	11 + 0	002	5243.10	−0.15
5	00 + 5	050	5797.24	−0.59
	10 + 3	130	6074.58	5.22
	10 − 3	031	6077.60	−0.14
	20 + 1	210	6288.15	0.88
	20 − 1	111	6289.17	0.66
6	20 − 2	121	7420.09	1.51
	30 + 0	300	7576.38	2.43
	30 − 0	201	7576.55	2.53
	21 + 0	102	7752.26	−2.63
	21 − 0	003	7779.32	0.23
7	30 + 1	310	8697.14	0.12
	30 − 1	211	8697.16	0.08
8	40 + 0	400	9911.02	3.52
	40 − 0	301	9911.02	3.52
	31 + 0	202	10188.30	−2.84
	31 − 0	103	10194.45	−1.11
9	40 + 1	410	11008.68	−3.50
	40 − 1	311	11008.68	−3.50
10	50 + 0	500	12149.46	0.78
	50 − 0	401	12149.46	0.78
	41 + 0	302	12524.63	−2.04
	41 − 0	203	12525.20	−1.75

^{a} Observed data are from Refs. [89–91].

tonian is obtained as before by expanding both the kinetic and the potential energy parts as Taylor series and by collecting the terms containing two-power operators in bending variables and one-power operators in stretching variables. The three internal bond displacement coordinates are denoted by r_1, r_2, and r_3 and the corresponding three internal valence angle displacement coordinates are denoted θ_1, θ_2, and θ_3. The conventional

TABLE XIII
Fundamentals (cm^{-1}) for Pyramidal XH$_3$ Molecules[a]

	NH$_3$	PH$_3$	AsH$_3$	SbH$_3$
$v_1(A_1)$, symmetric stretch	3336.2	2321.1	2115.2	1890.5
$v_2(A_1)$, symmetric bend	931.6, 968.1	992.1	906.8	782.2
$v_3(E)$, antisymmetric stretch	3443.4	2326.9	2126.4	1894.5
$v_4(E)$, antisymmetric bend	1626.4	1118.3	999.2	827.9

[a] Data are taken from Refs. [9] and [92–96]. In ammonia, two numbers are given for v_2 due to inversion doubling. For the other ammonia fundamentals, the inversion doublings are very much smaller and an average fundamental wavenumber is given. For the other molecules, no inversion doublings have been observed.

notation used is such that θ_i describes the displacement of the valence angle between the bonds j and k, where $j \neq i$ and $k \neq i$. The Hamiltonian is expressed in terms of these variables and their conjugate momenta $p_{r_j} = -i\hbar(\partial/\partial r_j)$ and $p_{\theta_j} = -i\hbar(\partial/\partial \theta_j)$. The pure stretching Hamiltonian analogous to Eq. (2.1) is

$$H_{\text{stretch}} = \sum_{i=1}^{3} (\tfrac{1}{2}g_{rr} p_{r_i}^2 + D_e y_i^2)$$
$$+ \sum_{i<j} (g_{rr'}^{(e)} p_{r_i} p_{r_j} + a_r^{-2} f_{rr'} y_i y_j) \qquad (3.20)$$

where the definitions of the coefficients and the Morse variable y_i are analogous to those given for bent triatomic molecules. The bending Hamiltonian is

$$H_{\text{bend}} = \sum_{i=1}^{3} (\tfrac{1}{2}g_{\theta\theta}^{(e)} p_{\theta_i}^2 + \tfrac{1}{2}f_{\theta\theta} \theta_i^2)$$
$$+ \sum_{i<j} (g_{\theta\theta'}^{(e)} p_{\theta_i} p_{\theta_j} + f_{\theta\theta'} \theta_i \theta_j) + H_{\text{anh}} \qquad (3.21)$$

where H_{anh} contains anharmonic pure bending kinetic and potential energy terms. The meaning of the other terms and their coefficients is obvious. The bilinear stretch–bend interaction Hamiltonian is

$$H_{\text{stretch–bend}} = \sum_{i=1}^{3} (g_{r\theta}^{(e)} p_{r_i} p_{\theta_i} + a_r^{-1} f_{r\theta} y_i \theta_i)$$
$$+ \sum_{i \neq j} (g_{r\theta'}^{(e)} p_{r_i} p_{\theta_j} + a_r^{-1} f_{r\theta'} y_i \theta_j) \qquad (3.22)$$

The Fermi resonance operator is

$$
\begin{aligned}
H_{\text{Fermi}} = \sum_{i \neq j} \Bigg[& a_r^{-1} \left(\frac{\partial g_{\theta\theta'}}{\partial r} \right)_e y_i p_{\theta_i} p_{\theta_j} + \left(\frac{\partial g_{r\theta}}{\partial \theta} \right)_e \theta_i p_{r_j} p_{\theta_j} \\
& + \frac{1}{2} \left(\frac{\partial g_{r\theta'}}{\partial \theta'} \right)_e p_{r_i} (\theta_j p_{\theta_j} + p_{\theta_j} \theta_j) + a_r^{-1} f_{r\theta\theta'} y_i \theta_i \theta_j \\
& + \frac{1}{2} a_r^{-1} \left(\frac{\partial g_{\theta\theta}}{\partial r'} \right)_e y_i p_{\theta_j}^2 + a_r^{-1} \frac{1}{2} f_{r\theta'\theta'} y_i \theta_j^2 \Bigg] \\
& + \sum_{i \neq j \neq k \neq i, \; j < k} \left[a_r^{-1} \left(\frac{\partial g_{\theta\theta'}}{\partial r''} \right)_e y_i p_{\theta_j} p_{\theta_k} + a_r^{-1} f_{r\theta'\theta''} y_i \theta_j \theta_k \right] \\
& + \sum_i \left[\frac{1}{2} \left(\frac{\partial g_{r\theta}}{\partial \theta} \right)_e p_{r_i} (\theta_i p_{\theta_i} + p_{\theta_i} \theta_i) + \frac{1}{2} a_r^{-1} f_{r\theta\theta} y_i \theta_i^2 \right]
\end{aligned}
\tag{3.23}
$$

Explicit expressions for the **g** matrix elements and their derivatives are given in Ref. [56]. As before, apart from the Morse oscillator energy expression in the diagonal, harmonic oscillator matrix elements are used both for the stretches and for the bends.

By denoting the local stretching vibrational quantum numbers v_{r_1}, v_{r_2}, and v_{r_3} and locallike bending vibrational quantum numbers v_{θ_1}, v_{θ_2}, and v_{θ_3}, the diagonal matrix element in this model is

$$
\begin{aligned}
G(v_{r_1} & v_{r_2} v_{r_3}, \; v_{\theta_1} v_{\theta_2} v_{\theta_3}) \\
& = \langle v_{r_1} v_{r_2} v_{r_3}, \; v_{\theta_1} v_{\theta_2} v_{\theta_3} | \frac{H}{hc_0} | v_{r_1} v_{r_2} v_{r_3}, \; v_{\theta_1} v_{\theta_2} v_{\theta_3} \rangle \\
& = \omega_r'(v_{r_1} + v_{r_2} + v_{r_3}) + x_{rr}(v_{r_1}^2 + v_{r_1} + v_{r_2}^2 + v_{r_2} + v_{r_3}^2 + v_{r_3}) \\
& \quad + \omega_\theta'(v_{\theta_1} + v_{\theta_2} + v_{\theta_3}) + x_{\theta\theta}(v_{\theta_1}^2 + v_{\theta_1} + v_{\theta_2}^2 + v_{\theta_2} + v_{\theta_3}^2 + v_{\theta_3}) \\
& \quad + x_{r\theta'}(v_{r_2} v_{\theta_1} + v_{r_3} v_{\theta_1} + v_{r_1} v_{\theta_2} + v_{r_3} v_{\theta_2} + v_{r_1} v_{\theta_3} \\
& \qquad + v_{r_2} v_{\theta_3} + v_{r_1} + v_{r_2} + v_{r_3} + v_{\theta_1} + v_{\theta_2} + v_{\theta_3}) \\
& \quad + x_{r\theta}[v_{r_1} v_{\theta_1} + v_{r_2} v_{\theta_2} + v_{r_3} v_{\theta_3} \\
& \qquad + \tfrac{1}{2}(v_{r_1} + v_{r_2} + v_{r_3} + v_{\theta_1} + v_{\theta_2} + v_{\theta_3})]
\end{aligned}
\tag{3.24}
$$

As in XH_2 molecules, theoretical expressions for the coefficients are obtained using both first- and second-order perturbation theory. This leads to the expressions

$$
\omega_r' = \omega_r - \frac{1}{4}\,\omega_r\left(\frac{g_{rr'}^{(e)}}{g_{rr}} - \frac{f_{rr'}}{f_{rr}}\right)^2
$$

$$
+ \frac{1}{4}\left(\frac{d_{r\theta}(1)^2 + 2d_{r\theta'}(1)^2}{\omega_r - \omega_\theta} - \frac{d_{r\theta}(2)^2 + 2d_{r\theta'}(2)^2}{\omega_r + \omega_\theta}\right)
$$

$$
\omega_\theta' = \omega_\theta - \frac{1}{4}\,\omega_\theta\left(\frac{g_{\theta\theta'}^{(e)}}{g_{\theta\theta}^{(e)}} - \frac{f_{\theta\theta'}}{f_{\theta\theta}}\right)^2
$$

$$
- \frac{1}{4}\left(\frac{d_{r\theta}(1)^2 + 2d_{r\theta'}(1)^2}{\omega_r - \omega_\theta} + \frac{d_{r\theta}(2)^2 + 2d_{r\theta'}(2)^2}{\omega_r + \omega_\theta}\right)
$$

(3.25)

where

$$
d_{r\theta'}(1) = \frac{\hbar^2}{hc_0}\,g_{r\theta'}^{(e)}\,\alpha_r^{1/2}\alpha_\theta^{1/2} + \frac{1}{hc_0}\,f_{r\theta'}\,\alpha_r^{-1/2}\alpha_\theta^{-1/2}
$$

$$
= (\omega_r\,\omega_\theta)^{1/2}\left[\frac{g_{r\theta'}^{(e)}}{(g_{rr}\,g_{\theta\theta}^{(e)})^{1/2}} + \frac{f_{r\theta'}}{(f_{rr}\,f_{\theta\theta})^{1/2}}\right]
$$

$$
d_{r\theta'}(2) = - \frac{\hbar^2}{hc_0}\,g_{r\theta'}^{(e)}\,\alpha_r^{1/2}\alpha_\theta^{1/2} + \frac{1}{hc_0}\,f_{r\theta'}\,\alpha_r^{-1/2}\alpha_\theta^{-1/2}
$$

$$
= (\omega_r\,\omega_\theta)^{1/2}\left[-\frac{g_{r\theta'}^{(e)}}{(g_{rr}\,g_{\theta\theta}^{(e)})^{1/2}} + \frac{f_{r\theta'}}{(f_{rr}\,f_{\theta\theta})^{1/2}}\right]
$$

(3.26)

Formulas for the harmonic wavenumbers ω_r and ω_θ, for the Morse anharmonicity parameter x_{rr}, and for $d_{r\theta}(1)$ and $d_{r\theta}(2)$ are identical to those given for XH_2 molecules. For the other anharmonicity parameters, no results are given in this context. They can be derived with second-order perturbation theory. In $x_{\theta\theta}$, cubic and quartic bending terms contribute to the second and to the first orders, respectively. The Fermi resonance terms make a second-order contribution. In the case of the anharmonicity coefficients $x_{r\theta}$ and $x_{r\theta'}$, it is necessary to include quartic stretch–bend coupling Hamiltonian terms which contribute to the first order. The Fermi resonance terms contribute to the second order.

The local-mode-type off-diagonal matrix elements are

$$
\langle(v_{r_1}-1)(v_{r_2}+1)v_{r_3},\,v_{\theta_1}v_{\theta_2}v_{\theta_3}|\,\frac{H}{hc_0}\,|v_{r_1}v_{r_2}v_{r_3},\,v_{\theta_1}v_{\theta_2}v_{\theta_3}\rangle
$$

$$
= \lambda_r'[v_{r_1}(v_{r_2}+1)]^{1/2} \quad (3.27)
$$

and

$$\langle v_{r_1} v_{r_2} v_{r_3}, (v_{\theta_1} - 1)(v_{\theta_2} + 1)v_{\theta_3} | \frac{H}{hc_0} | v_{r_1} v_{r_2} v_{r_3}, v_{\theta_1} v_{\theta_2} v_{\theta_3}\rangle$$

$$= \lambda_\theta'[v_{\theta_1}(v_{\theta_2} + 1)]^{1/2} \quad (3.28)$$

with

$$\lambda_r' = \frac{1}{2}\,\omega_r\left(\frac{g_{rr'}^{(e)}}{g_{rr}} + \frac{f_{rr'}}{f_{rr}}\right)$$

$$+ \frac{1}{4}\left(\frac{2d_{r\theta}(1)d_{r\theta'}(1) + d_{r\theta'}(1)^2}{\omega_r - \omega_\theta} - \frac{2d_{r\theta}(2)d_{r\theta'}(2) + d_{r\theta'}(2)^2}{\omega_r + \omega_\theta}\right)$$

$$\lambda_\theta' = \frac{1}{2}\,\omega_r\left(\frac{g_{\theta\theta'}^{(e)}}{g_{\theta\theta}^{(e)}} + \frac{f_{\theta\theta'}}{f_{rr}}\right) \quad (3.29)$$

$$+ \frac{1}{4}\left(\frac{2d_{r\theta}(1)d_{r\theta'}(1) + d_{r\theta'}(1)^2}{\omega_r - \omega_\theta} - \frac{2d_{r\theta}(2)d_{r\theta'}(2) + d_{r\theta'}(2)^2}{\omega_r + \omega_\theta}\right)$$

The Fermi resonance operator gives rise to several off-diagonal matrix elements which are of the type (these are obtainable from the matrix elements given in Table II)

$$\langle (v_{r_1} - 1)v_{r_2} v_{r_3}, (v_{\theta_1} + 2)v_{\theta_2} v_{\theta_3} | \frac{H_{\text{Fermi}}}{hc_0} | v_{r_1} v_{r_2} v_{r_3}, v_{\theta_1} v_{\theta_2} v_{\theta_3}\rangle$$

$$= d_{r\theta\theta}[v_{r_1}(v_{\theta_1} + 1)(v_{\theta_1} + 2)]^{1/2}$$

$$\langle (v_{r_1} - 1)v_{r_2} v_{r_3}, v_{\theta_1}(v_{\theta_2} + 2)v_{\theta_3} | \frac{H_{\text{Fermi}}}{hc_0} | v_{r_1} v_{r_2} v_{r_3}, v_{\theta_1} v_{\theta_2} v_{\theta_3}\rangle$$

$$= d_{r\theta'\theta'}[v_{r_1}(v_{\theta_2} + 1)(v_{\theta_2} + 2)]^{1/2}$$

$$\langle (v_{r_1} - 1)v_{r_2} v_{r_3}, (v_{\theta_1} + 1)(v_{\theta_2} + 1)v_{\theta_3} | \frac{H_{\text{Fermi}}}{hc_0} | v_{r_1} v_{r_2} v_{r_3}, v_{\theta_1} v_{\theta_2} v_{\theta_3}\rangle$$

$$= d_{r\theta\theta'}[v_{r_1}(v_{\theta_1} + 1)(v_{\theta_2} + 1)]^{1/2}$$

$$\langle (v_{r_1} - 1)v_{r_2} v_{r_3}, v_{\theta_1}(v_{\theta_2} + 1)(v_{\theta_3} + 1) | \frac{H_{\text{Fermi}}}{hc_0} | v_{r_1} v_{r_2} v_{r_3}, v_{\theta_1} v_{\theta_2} v_{\theta_3}\rangle$$

$$= d_{r\theta'\theta''}[v_{r_1}(v_{\theta_2} + 1)(v_{\theta_3} + 1)]^{1/2} \quad (3.30)$$

where

$$d_{r\theta\theta} = \frac{\hbar^2}{2\sqrt{2}hc_0}\left(\frac{\partial g_{r\theta}}{\partial\theta}\right)_e \alpha_r^{1/2} + \frac{1}{4\sqrt{2}hc_0}f_{r\theta\theta}\,\alpha_r^{-1/2}\alpha_\theta^{-1}$$

$$d_{r\theta'\theta'} = -\frac{\hbar^2}{4\sqrt{2}hc_0}\left(\frac{\partial g_{\theta\theta}}{\partial r'}\right)_e \alpha_r^{-1/2}\alpha_\theta + \frac{\hbar^2}{2\sqrt{2}hc_0}\left(\frac{\partial g_{r\theta'}}{\partial\theta'}\right)_e \alpha_r^{1/2}$$

$$+ \frac{1}{4\sqrt{2}hc_0}f_{r\theta'\theta'}\,\alpha_r^{-1/2}\alpha_\theta^{-1}$$

$$d_{r\theta\theta'} = -\frac{\hbar^2}{2\sqrt{2}hc_0}\left(\frac{\partial g_{\theta\theta'}}{\partial r}\right)_e \alpha_r^{-1/2}\alpha_\theta + \frac{\hbar^2}{2\sqrt{2}hc_0}\left(\frac{\partial g_{r\theta}}{\partial\theta'}\right)_e \alpha_r^{1/2}$$

$$+ \frac{1}{2\sqrt{2}hc_0}f_{r\theta\theta'}\,\alpha_r^{-1/2}\alpha_\theta^{-1}$$

$$d_{r\theta'\theta''} = -\frac{\hbar^2}{2\sqrt{2}hc_0}\left(\frac{\partial g_{\theta\theta'}}{\partial r''}\right)_e \alpha_r^{-1/2}\alpha_\theta$$

$$+ \frac{1}{2\sqrt{2}hc_0}f_{r\theta'\theta''}\,\alpha_r^{-1/2}\alpha_\theta^{-1} \tag{3.31}$$

with α_r and α_θ defined in Eqs. (3.5).

The Hamiltonian factorizes to smaller matrices such that only states with the same $v = v_{r_1} + v_{r_2} + v_{r_3} + \frac{1}{2}(v_{\theta_1} + v_{\theta_2} + v_{\theta_3})$ are coupled. Here, v remains a good quantum number. The matrices can be factorized further by symmetrization of the product basis

$$|v_{r_1}v_{r_2}v_{r_3}, v_{\theta_1}v_{\theta_2}v_{\theta_3}\rangle = |v_{r_1}v_{r_2}v_{r_3}\rangle|v_{\theta_1}v_{\theta_2}v_{\theta_3}\rangle$$

$$= |v_{r_1}v_{r_2}v_{r_3}\rangle_s|v_{\theta_1}v_{\theta_2}v_{\theta_3}\rangle_b$$

$$= |v_{r_1}\rangle|v_{r_2}\rangle|v_{r_3}\rangle|v_{\theta_1}\rangle|v_{\theta_2}\rangle|v_{\theta_3}\rangle$$

where the subindex s refers to the stretch and b to the bend. This is done by symmetrizing the stretching and the bending parts separately and combining both parts in an appropriate way. The symmetrization of the bending part is done in the same way as for the stretching part (see Section II.B). One technical difficulty occurs when both the stretching and the bending functions span the doubly degenerate representation E. In this case, the symmetrized A_1, A_2, and E basis functions ($E \times E = A_1 + A_2 + E$) can be obtained by the vector coupling coefficient method as described in Section II.B. This is explained in more detail in Ref. [56]. As an example, the Hamiltonian matrices are given for $v = 1$. In this case, there exist an A_1 block and two E blocks of which the E blocks yield degenerate eigenvalues. For the A_1 block, the symmetrized basis functions are

$$|100, 000; A_1\rangle = |100A_1\rangle_s|000A_1\rangle_b$$

$$= \frac{1}{\sqrt{3}}(|100\rangle_s + |010\rangle_s + |001\rangle_s)|000\rangle_b$$

$$|000, 200; A_1\rangle = |000A_1\rangle_s|200A_1\rangle_b$$

$$= \frac{1}{\sqrt{3}}|000\rangle_s(|200\rangle_b + |020\rangle_b + |002\rangle_b) \qquad (3.32)$$

$$|000, 110; A_1\rangle = |000A_1\rangle_s|110A_1\rangle_b$$

$$= \frac{1}{\sqrt{3}}|000\rangle_s(|110\rangle_b + |101\rangle_b + |011\rangle_b)$$

and the Hamiltonian matrix in wavenumber units is

$$H(v = 1; A_1)$$

$$= \begin{pmatrix} G(100, 000) + 2\lambda_r' & \sqrt{2}(d_{r\theta\theta} + 2d_{r\theta'\theta'}) & d_{r\theta'\theta''} + 2d_{r\theta\theta'} \\ \sqrt{2}(d_{r\theta\theta} + 2d_{r\theta'\theta'}) & G(000, 200) & 2\sqrt{2}\lambda_\theta' \\ d_{r\theta'\theta''} + 2d_{r\theta\theta'} & 2\sqrt{2}\lambda_\theta' & G(000, 110) + 2\lambda_\theta' \end{pmatrix}$$

$$= \begin{pmatrix} \omega_r' + 2x_{rr} + x_{r\theta} + \frac{1}{2}x_{r\theta'} + 2\lambda_r' & \sqrt{2}(d_{r\theta\theta} + 2d_{r\theta'\theta'}) \\ \sqrt{2}(d_{r\theta\theta} + 2d_{r\theta'\theta'}) & 2\omega_\theta' + 6x_{\theta\theta} + 2x_{r\theta} + x_{r\theta'} \\ d_{r\theta'\theta''} + 2d_{r\theta\theta'} & 2\sqrt{2}\lambda_\theta' \end{pmatrix}$$

$$\begin{pmatrix} & & d_{r\theta'\theta''} + 2d_{r\theta\theta'} \\ & & 2\sqrt{2}\lambda_\theta' \\ & & 2\omega_\theta' + 4x_{\theta\theta} + 2x_{r\theta} + x_{r\theta'} + 2\lambda_\theta' \end{pmatrix} \qquad (3.33)$$

The E_a functions are

$$|100, 000; E_a\rangle = |100E_a\rangle_s|000A_1\rangle_b$$

$$= \frac{1}{\sqrt{6}}(2|100\rangle_s - |010\rangle_s - |001\rangle_s)|000\rangle_b$$

$$|000, 200; E_a\rangle = |000A_1\rangle_s|200E_a\rangle_b$$

$$= \frac{1}{\sqrt{6}}|000\rangle_s(2|200\rangle_b - |020\rangle_b - |002\rangle_b)$$

$$|000, 110; E_a\rangle = |000A_1\rangle_s|110E_a\rangle_b$$

$$= \frac{1}{\sqrt{6}}|000\rangle_s(2|011\rangle_b - |101\rangle_b - |110\rangle_b) \quad (3.34)$$

and the corresponding E_b functions take the forms

$$|100, 000; E_b\rangle = |100E_b\rangle_s|000A_1\rangle_b = \frac{1}{\sqrt{2}}(|010\rangle_s - |001\rangle_s)|000\rangle_b$$

$$|000, 200; E_b\rangle = |000A_1\rangle_s|200E_b\rangle_b = \frac{1}{\sqrt{2}}|000\rangle_s(|020\rangle_b - |002\rangle_b) \quad (3.35)$$

$$|000, 110; E_b\rangle = |000A_1\rangle_s|110E_b\rangle_b = \frac{1}{\sqrt{2}}|000\rangle_s(|101\rangle_b - |110\rangle_b)$$

The two identical E Hamiltonian blocks become

$$H(v = 1; E)$$

$$= \begin{pmatrix} G(100, 000) - \lambda_r' & \sqrt{2}(d_{r\theta\theta} - d_{r\theta'\theta'}) & d_{r\theta'\theta''} - d_{r\theta\theta'} \\ \sqrt{2}(d_{r\theta\theta} - d_{r\theta'\theta'}) & G(000, 200) & -\sqrt{2}\lambda_\theta' \\ d_{r\theta'\theta''} - d_{r\theta\theta'} & -\sqrt{2}\lambda_\theta' & G(000, 110) - \lambda_\theta' \end{pmatrix}$$

$$= \begin{pmatrix} \omega_r' + 2x_{rr} + x_{r\theta} + \frac{1}{2}x_{r\theta'} - \lambda_r' & \sqrt{2}(d_{r\theta\theta} - d_{r\theta'\theta'}) \\ \sqrt{2}(d_{r\theta\theta} - d_{r\theta'\theta'}) & 2\omega_\theta' + 6x_{\theta\theta} + 2x_{r\theta} + x_{r\theta'} \\ d_{r\theta'\theta''} - d_{r\theta\theta'} & -\sqrt{2}\lambda_\theta' \end{pmatrix}$$

$$\begin{pmatrix} d_{r\theta'\theta''} - d_{r\theta\theta'} \\ -\sqrt{2}\lambda_\theta' \\ 2\omega_\theta' + 4x_{\theta\theta} + 2x_{r\theta} + x_{r\theta'} - \lambda_\theta' \end{pmatrix} \quad (3.36)$$

The corresponding model using the standard vibration–rotation spectroscopy is also easily derived. When the ground-state energy is subtracted, the vibrationally diagonal matrix element possesses the usual form

(Table VI)

$$G(v_1, v_2, v_3^{l_3}, v_4^{l_4}) = \omega_1 v_1 + \omega_2 v_2 + \omega_3 v_3 + \omega_4 v_4$$

$$+ x_{11}v_1(v_1 + 1) + x_{22}^* v_2(v_2 + 1)$$

$$+ x_{33} v_3(v_3 + 2) + x_{44}^* v_4(v_4 + 2)$$

$$+ x_{12}^*[v_1 v_2 + \tfrac{1}{2}(v_1 + v_2)]$$

$$+ x_{13}(v_1 v_3 + v_1 + \tfrac{1}{2}v_3)$$

$$+ x_{14}^*(v_1 v_4 + v_1 + \tfrac{1}{2}v_4)$$

$$+ x_{23}^*(v_2 v_3 + v_2 + \tfrac{1}{2}v_3)$$

$$+ x_{24}^*(v_2 v_4 + v_2 + \tfrac{1}{2}v_4)$$

$$+ x_{34}^*(v_3 v_4 + v_3 + v_4)$$

$$+ g_{33} l_3^2 + g_{44}^* l_4^2 + g_{34}^* l_3 l_4 \qquad (3.37)$$

where v_1, v_2, v_3, and v_4 are the standard normal mode vibrational quantum numbers for the symmetric stretch, symmetric bend, antisymmetric stretch, and the antisymmetric bend, respectively. Here, l_3 ($= -v_3$, $-v_3 + 2, \ldots, v_3$) and l_4 ($= -v_4$, $-v_4 + 2$, $\ldots$, v_4) are vibrational angular momentum quantum numbers of the doubly degenerate antisymmetric stretch and doubly degenerate antisymmetric bend, respectively. The coefficients ω, x, and g are standard vibrational parameters [9,22,23]. The anharmonicity coefficients x and g are functions of quadratic, cubic, and quartic force constants in the normal coordinate representation [see similar equation for triatomics in Eqs. (2.30) and (2.31)] [22]. Some of the coefficients have been marked with an asterisk. The resonance contributions due to Fermi terms should be left out when standard theoretical formulas expressed in terms of quadratic, cubic, and quartic force constants in the normal coordinate representation are used to interpret these coefficients. As in XH_2-type molecules, there are Darling–Dennison resonance interactions between stretching overtone and combination states. Corresponding resonance terms between bending overtones and combinations should also be included. These type of quartic resonances are less important between stretching and bending overtones of phosphine, arsine, and stibine. The operators responsible for the important resonances are $\tfrac{1}{2}K_{ss;\,tt}[(q_s^+)^2(q_{t-}^- q_{t+}^- + q_{t+}^- q_{t-}^-) + (q_s^-)^2(q_{t-}^+ q_{t+}^+ + q_{t+}^+ q_{t-}^+)]$ and

$$\tfrac{1}{2}K_{st;\,tt}\{q_s^+\,[(q_{t+}^-)^2 q_{t+}^+ + q_{t+}^-\,q_{t+}^+\,q_{t+}^- + q_{t+}^+(q_{t+}^-)^2]$$

$$+ q_s^+\,[(q_{t-}^-)^2 q_{t+}^+ + q_{t-}^-\,q_{t-}^+\,q_{t-}^- + q_{t-}^+(q_{t-}^-)^2]$$

$$+ q_s^-\,[(q_{t-}^+)^2 q_{t-}^- + q_{t-}^+\,q_{t-}^-\,q_{t-}^+ + q_{t-}^-(q_{t-}^+)^2]$$

$$+ q_s^-\,[(q_{t+}^+)^2 q_{t+}^- + q_{t+}^+\,q_{t+}^-\,q_{t+}^+ + q_{t+}^-(q_{t+}^+)^2]\}.$$

The individual operators are given in Table II, and they are defined in terms of dimensionless normal coordinates q_s and $q_t = (q_{ta}, q_{tb})$. Here, $s = 1$ and $t = 3$ for the stretching part and $s = 2$ and $t = 4$ for the bending part. The appropriate matrix elements can be derived using the formulas in Table II, and the final results are in Table VI in 1(d). Altogether, four different Darling–Dennison matrix elements appear with coefficients $K_{11;\,33}$, $K_{13;\,33}$, $K_{22;\,44}$, and $K_{24;\,44}$. These are functions of quadratic, cubic, and quartic force constants in the normal coordinate representation [see a similar expression for triatomics in Eq. (2.32)] [21].

There are four different kinds of Fermi resonance matrix elements. Cubic potential energy terms

$$\frac{H_{\text{Fermi}}^{(1)}}{hc_0} = \tfrac{1}{2}\phi_{122}\,q_1 q_2^2 \tag{3.38}$$

$$\frac{H_{\text{Fermi}}^{(2)}}{hc_0} = \tfrac{1}{2}\phi_{144}\,q_1(q_{4a}^2 + q_{4b}^2) = \tfrac{1}{2}\phi_{144}\,q_1 q_{4+} q_{4-} \tag{3.39}$$

$$\frac{H_{\text{Fermi}}^{(3)}}{hc_0} = \tfrac{1}{2}\phi_{344}(q_{3a}\,q_{4a}^2 - q_{3a}\,q_{4b}^2 - 2q_{3b}\,q_{4a}\,q_{4b})$$

$$= \tfrac{1}{4}\phi_{344}(q_{3+}\,q_{4+}^2 + q_{3-}\,q_{4-}^2) \tag{3.40}$$

and

$$\frac{H_{\text{Fermi}}^{(4)}}{hc_0} = \phi_{234}\,q_2(q_{3a}\,q_{4a} + q_{3b}\,q_{4b}) = \tfrac{1}{2}\phi_{234}\,q_2(q_{3+}\,q_{4-} + q_{3-}\,q_{4+}) \tag{3.41}$$

are responsible for Fermi resonances, for example, between a pair of states (upper states) such as $v_1(A_1)$ and $2v_2^0(A_1)$, $v_1(A_1)$ and $2v_4^0(A_1)$, $v_3^{\pm 1}(E)$ and $2v_4^{\mp 2}(E)$, and $v_3^{\pm 1}(E)$ and $v_2 + v_4^{\pm 1}(E)$, respectively. The corresponding matrix elements are ($H_{\text{Fermi}} = \sum_i H_{\text{Fermi}}^{(i)}$)

$$\langle v_1 - 1, v_2 + 2 | \frac{H_{\text{Fermi}}}{hc_0} | v_1, v_2 \rangle = \frac{1}{4\sqrt{2}} \phi_{122}[v_1(v_2 + 1)(v_2 + 2)]^{1/2} \quad (3.42)$$

$$\langle v_1 - 1, v_4 + 2, l_4 | \frac{H_{\text{Fermi}}}{hc_0} | v_1, v_4, l_4 \rangle$$

$$= \frac{1}{4\sqrt{2}} \phi_{144}[v_1(v_4 + l_4 + 2)(v_4 - l_4 + 2)]^{1/2} \quad (3.43)$$

$$\langle v_3 - 1, l_3 \mp 1, v_4 + 2, l_4 \mp 2 | \frac{H_{\text{Fermi}}}{hc_0} | v_3, l_3, v_4, l_4 \rangle$$

$$= \frac{1}{8\sqrt{2}} \phi_{344}[(v_3 \pm l_3)(v_4 \mp l_4 + 4)(v_4 \mp l_4 + 2)]^{1/2} \quad (3.44)$$

and

$$\langle v_2 + 1, v_3 - 1, l_3 \mp 1, v_4 + 1, l_4 \pm 1 | \frac{H_{\text{Fermi}}}{hc_0} | v_2, v_3, l_3, v_4, l_4 \rangle$$

$$= \frac{1}{4\sqrt{2}} \phi_{234}[(v_2 + 1)(v_3 \pm l_3)(v_4 \pm l_4 + 2)]^{1/2} \quad (3.45)$$

respectively. These results are obtained from the matrix elements given in Table II. The notation for the basis set is, for example, $|v_1, v_4, l_4\rangle = |v_1\rangle|v_4, l_4\rangle$, where $|v_1\rangle$ is a one- and $|v_4, l_4\rangle$ a two-dimensional harmonic oscillator eigenfunction in the normal coordinate representation. The ϕ coefficients are third-order derivatives of the potential energy function with respect to dimensionless normal coordinates evaluated at the equilibrium configuration

$$\phi_{122} = \frac{1}{hc_0} \left(\frac{\partial^3 V}{\partial q_1 \, \partial q_2^2} \right)_e$$

$$\phi_{244} = \phi_{24a4a} = \frac{1}{hc_0} \left(\frac{\partial^3 V}{\partial q_2 \, \partial q_{4a}^2} \right)_e = \phi_{24b4b}$$

$$\phi_{344} = \phi_{3a4a4a} = \frac{1}{hc_0} \left(\frac{\partial^3 V}{\partial q_{3a} \, \partial q_{4a}^2} \right)_e$$

$$= -\phi_{3a4b4b} = -\tfrac{1}{2}\phi_{3a4a4b}$$

$$\phi_{234} = \phi_{23a4a} = \frac{1}{hc_0} \left(\frac{\partial^3 V}{\partial q_2 \, \partial q_{3a} \, \partial q_{4a}} \right)_e = \phi_{23b4b}$$

Using the equations given above, the following matrices for the interacting $v = v_1 + v_3 + \frac{1}{2}(v_2 + v_4) = 1$ states can be derived:

$H(v = 1, A_1)$

$$
= \begin{pmatrix}
G(v_1) & \frac{1}{4}\phi_{122} & \frac{1}{2\sqrt{2}}\phi_{144} \\[2mm]
\frac{1}{4}\phi_{122} & G(2v_2) & \frac{1}{\sqrt{2}}K_{22;44} \\[2mm]
\frac{1}{2\sqrt{2}}\phi_{144} & \frac{1}{\sqrt{2}}K_{22;44} & G(2v_4^0)
\end{pmatrix}
$$

$$
= \begin{pmatrix}
\omega_1 + 2x_{11} + \frac{1}{2}x_{12}^* + x_{13} + x_{14}^* & \frac{1}{4}\phi_{122} & \frac{1}{2\sqrt{2}}\phi_{144} \\[2mm]
\frac{1}{4}\phi_{122} & 2\omega_2 + 6x_{22}^* + x_{12}^* + 2x_{23}^* + 2x_{24}^* & \frac{1}{\sqrt{2}}K_{22;44} \\[2mm]
\frac{1}{2\sqrt{2}}\phi_{144} & \frac{1}{\sqrt{2}}K_{22;44} & 2\omega_4 + 8x_{44}^* + x_{14}^* + x_{24}^* + 2x_{34}^*
\end{pmatrix}
$$

$H(v = 1, E)$

$$
= \begin{pmatrix}
G(v_3^{\pm 1}) & \frac{1}{2\sqrt{2}}\phi_{344} & \frac{1}{2\sqrt{2}}\phi_{234} \\[2mm]
\frac{1}{2\sqrt{2}}\phi_{344} & G(2v_4^{\mp 2}) & \frac{3}{2}K_{24;44} \\[2mm]
\frac{1}{2\sqrt{2}}\phi_{234} & \frac{3}{2}K_{24;44} & G(v_2 + v_4^{\pm 1})
\end{pmatrix}
$$

$$
= \begin{pmatrix}
\omega_3 + 3x_{33} + \frac{1}{2}x_{13} + \frac{1}{2}x^*_{23} + x^*_{34} + g_{33} & \frac{1}{2\sqrt{2}}\phi_{344} \\[2em]
\frac{1}{2\sqrt{2}}\phi_{344} & 2\omega_4 + 8x^*_{44} + x^*_{14} + x^*_{24} + 2x^*_{34} + 4g^*_{44} \\[2em]
\frac{1}{2\sqrt{2}}\phi_{234} & \frac{3}{2}K_{24;\,44}
\end{pmatrix}
$$

$$
\left.
\begin{matrix}
\frac{1}{2\sqrt{2}}\phi_{234} \\[2em]
\frac{3}{2}K_{24;\,44} \\[2em]
\omega_2 + \omega_4 + 2x^*_{22} + 3x^*_{44} + \frac{1}{2}x^*_{12} + \frac{1}{2}x^*_{14} \\
+ x^*_{23} + \frac{5}{2}x^*_{24} + x^*_{34} + g^*_{44}
\end{matrix}
\right) \quad (3.46)
$$

The two models for the Fermi resonance coupled stretch–bend problem produce identical eigenvalues once generalized x–K relations are satisfied between the vibrational parameters. These relations can be derived using creation and annihilation operators both for stretching and bending degrees of freedom. This derivation is similar to the one given earlier for XH_2-type molecules (see also Ref. [76]). The final results obtained are [97]

$$
\omega_1 = \omega'_r + 2\lambda'_r \qquad \omega_3 = \omega'_r - \lambda'_r
$$

$$
x_{11} = \tfrac{2}{3}x_{33} = \tfrac{1}{4}x_{13} = -2g_{33} = \tfrac{1}{3}x_{rr}
$$

$$
K_{11;\,33} = \frac{3}{\sqrt{2}}\,K_{13;\,33} = \tfrac{4}{3}x_{rr}
$$

$$
\omega_2 = \omega'_\theta + 2\lambda'_\theta \qquad \omega_4 = \omega'_\theta - \lambda'_\theta
$$

$$
x^*_{22} = \tfrac{2}{3}x^*_{44} = \tfrac{1}{4}x^*_{24} = -2g^*_{44} = \tfrac{1}{3}x_{\theta\theta}
$$

$$
K_{22;\,44} = \frac{3}{\sqrt{2}}\,K_{24;\,44} = \tfrac{4}{3}x_{\theta\theta}
$$

$$
x^*_{12} = x^*_{14} = x^*_{23} = x^*_{34} = x_{r\theta} = x_{r\theta'}
$$

$$g_{34}^* = 0$$

$$\phi_{122} = \frac{4\sqrt{6}}{3} \left(d_{r\theta\theta} + 2d_{r\theta'\theta'} + 2d_{r\theta\theta'} + d_{r\theta'\theta''} \right)$$

$$\phi_{144} = \frac{2\sqrt{6}}{3} \left(2d_{r\theta\theta} + 4d_{r\theta'\theta'} - 2d_{r\theta\theta'} - d_{r\theta'\theta''} \right)$$

$$\phi_{344} = \frac{4\sqrt{3}}{3} \left(d_{r\theta\theta} - d_{r\theta'\theta'} - d_{r\theta\theta'} + d_{r\theta'\theta''} \right)$$

$$\phi_{234} = \frac{2\sqrt{6}}{3} \left(2d_{r\theta\theta} - 2d_{r\theta'\theta'} + d_{r\theta\theta'} - d_{r\theta'\theta''} \right) \tag{3.47}$$

As expected, the purely stretching vibrational x–K relations are the same as those given in Table VII for XH_3-type pyramidal molecules. Due to the symmetry of the problem, the purely bending vibrational x–K relations are of similar form as the purely stretching vibrational relations.

It is of interest to know how well the extended relations are satisfied in practice. Although the amount of experimental data are limited for arsine and stibine, all diagonal x parameters are available from an ab initio calculation [98]. The stretching vibrational x parameters satisfy well the theoretical relations but the same cannot be said about the bending and stretch–bend x parameters. The situation is similar in acetylene, where a HCAO model has been applied to doubly degenerate bending vibrations [99]. The result should not be taken as a failure of the local mode/Fermi resonance model, which can be made more sophisticated by adding higher order terms. This makes it equivalent to a normal coordinate model where the x–K relations are not applied. There is another possible modification which improves the model. The bending part of the molecular Hamiltonian could be expressed in terms of symmetrized internal coordinates, which leads to a model where the x–K relations between the bending parameters do not have to be obeyed. The bending vibrational parameters obtained are identical to the standard normal mode parameters. As a by-product a simple theoretical interpretation of the normal mode bending vibrational parameters is obtained. This improved model is discussed more in the context of the XH_4 tetrahedra below.

In spite of the reservations given above, the simple local mode/Fermi resonance model has been applied with success to the vibrational spectra of arsine and stibine [56,57]. In these cases, the model was made simpler by disregarding bending anharmonicities completely and including only part of the stretch–bend diagonal anharmonicity. Table XIV gives the observed and calculated vibrational term values both for AsH_3 and $^{121}SbH_3$,

TABLE XIV

Observed and Calculated Vibrational Term Values for AsH_3 and $^{121}SbH_3$ [a]

$v_{r_1}v_{r_2}v_{r_3}$, $v_{\theta_1}v_{\theta_2}v_{\theta_3}$; Γ	$v_2 v_4^{l_4}$	AsH_3 (cm^{-1})		$^{121}SbH_3$ (cm^{-1})	
		Observed	—	Observed	—
		v_{obs}	Calculated	v_{obs}	Calculated
000, 100; A_1	10^0	906.752	1.97	782.24	1.95
000, 100; E	$01^{\pm1}$	999.225	0.31	827.85	1.05
000, 110; A_1	20^0	1806.149	−1.59	1559.0	−0.67
000, 200; E	$11^{\pm1}$	1904.115	0.53	—	—
000, 200; A_1	02^0	1990.998	−6.25	1652.7	−0.77
000, 110; E	$02^{\pm2}$	2003.483	5.92	—	—
100, 000; A_1	00^0	2115.164	0.14	1890.502	0.82
100, 000; E	00^0	2126.423	0.34	1894.497	−0.31
100, 100; A_1	10^0	3013	0.64	2661	−0.61
100, 100; E	$01^{\pm1}$	3102	−0.16	2705	0.50
200, 000; A_1	00^0	4168.5	2.49	3719.933	0.42
200, 000; E	00^0	4168.5	1.34	3719.860	0.08
200, 100; A_1	10^0	5057	5.48	4545	4.87
200, 100; E	10^0	5057	5.11	4513	−5.54
110, 100; A_1	10^0	5128	2.67	—	—
110, 100; E	10^0	5128	−4.42	—	—
200, 100; A_1	$01^{\pm1}$	5158	0.42	—	—
200, 100; E	$01^{\pm1}$	5158	1.16	—	—
300, 000; A_1	00^0	6136.316	−0.63	5480.285	0.02
300, 000; E	00^0	6136.310	−0.69	5480.235	−0.04
210, 000; A_1	00^0	6276	−1.37	5607	−0.45
210, 000; E	00^0	6295	−1.00	—	—
400, 000; A_1	00^0	8029.2	9.87	7173.799	−0.93
400, 000; E	00^0	8029.2	9.90	7173.783	−0.95
600, 000; A_1	00^0	—	—	10358	−10.93
600, 000; E	00^0	—	—	10358	−10.93
510, 000; A_1	00^0	—	—	10691.5	−2.80
510, 000; E	00^0	—	—	10691.5	−2.80
700, 000; A_1	00^0	—	—	11843.5	−17.14
700, 000; E	00^0	—	—	11843.5	−17.14

[a] The calculated values are from Refs. [56] and [57]. The experimental data are from these same sources, where references to the original experimental work are found, and from Ref. [100].

together with local mode quantum labels $[v_{r_1}v_{r_2}v_{r_3}$, $v_{\theta_1}v_{\theta_2}v_{\theta_3}$; $\Gamma]$ for stretching and bending energy levels [56,57,100]. The conventional normal model labels $(v_2 v_4^{l_4})$ for bends are also given for reference, but the corresponding labels for the stretching vibrations are absent because they do not give a physically good description at higher excitations. It is worth pointing out that the model was successful in providing quantum labels for some previously unassigned overtone bands. Table XV gives potential energy coefficients for both molecules [56,57]. These were obtained with the non-linear least squares method. A comparison with the ab initio results [98] given in Table XV shows a good agreement with experiment with the exception of the higher order force constant $f_{rr\theta'\theta'}$ (this force constant has not been discussed in this chapter; see Ref. [56] for details [56,57]. It is fair to say that many of the empirical model force constants were constrained to ab initio values. The failure of the model in reproducing the ab initio values of $f_{rr\theta'\theta'}$ should not be regarded as a serious problem because most likely this coefficient is significantly affected by the neglect of higher order terms such as bending anharmonicity. It could be that $f_{rr\theta'\theta'}$ is an effective parameter without a deep physical meaning. It is also unclear how reliable an SCF calculation is in reproducing small quartic interaction force constants.

Ammonia is different from arsine and stibine in the sense that the symmetric bend $[v_2(A_1)]$, which is associated with the inversion motion, is so low in wavenumber that Fermi resonance interaction between stretching and symmetrical bending modes is weak. In a recent paper, a local mode/Fermi resonance model which excluded explicitly the symmetric bend was applied with success to ammonia [101]. The effects of this mode, as well as those of the bilinear stretch–bend coupling terms, were taken into account

TABLE XV
Potential Energy Parameters for Arsine and Stibine[a]

Parameter	AsH$_3$		^{121}SbH$_3$	
	Fermi	Ab initio	Fermi	Ab initio
f_{rr}, aJ Å^{-2}	2.876	2.829	2.293	2.243
$f_{rr'}$, aJ Å^{-2}	−0.0100	−0.0097	−0.0042	−0.0037
f_{rrr}, aJ Å^{-3}	−13.088	−14.192	−9.656	−9.887
f_{rrrr}, aJ Å^{-4}	46.32	54.40	31.62	35.20
$f_{\theta\theta}$, aJ	0.6470	0.642	0.5798	0.5703
$f_{\theta\theta'}$, aJ	−0.0334	−0.027	−0.0162	−0.0117
$f_{r\theta}$, aJ Å^{-1}	—	0.0147	—	0.0137
$f_{r\theta'}$, aJ Å^{-1}	—	0.0617	—	0.0407
$f_{r\theta'\theta'}$, aJ Å^{-1}	−0.2341[b]	−0.2341	−0.2001[b]	−0.2001
$f_{r\theta\theta}$, aJ Å^{-1}	0.0662[b]	0.0662	0.0504[b]	0.0504
$f_{r\theta'\theta''}$, aJ Å^{-1}	0.0686[b]	0.0686	0.0562[b]	0.0562
$f_{r\theta\theta}$, aJ Å^{-1}	0.0403[b]	0.0403	0.0308[b]	0.0308
$f_{rr\theta'\theta'}$, aJ Å^{-2}	−3.634	0.291	−2.936	0.267

[a] The surfaces have been taken from Refs. [56] (Fermi, AsH$_3$), [57] (Fermi, SbH$_3$), and [98] (ab initio). The ab initio surfaces contain more cubic force constants than given here.

[b] Transferred from the appropriate ab initio surface.

by second-order perturbation theory. The symmetric bend was excluded by symmetrizing the bending part of the vibrational Hamiltonian and constraining the symmetry coordinate associated with the symmetric bend to the equilibrium value. The doubly degenerate antisymmetric bending mode included in the analysis was modeled in the symmetrized Cartesian representation. When compared with AsH_3 and SbH_3, higher order terms such as anharmonic bending terms were included. The approach adopted reproduced well all observed vibrational energy levels (excluding those associated with the symmetric bend) of three isotopic species: NH_3, ND_3, and NT_3. The potential surface obtained with the least-squares optimization method agrees well with electronic structure calculations. The data set used in the optimization calculation included levels up to $v = 6$. The same surface is used for the three isotopic species. This should be contrasted with a traditional normal coordinate concept based model where similar vibrational energy level fits were obtained just for the most abundant isotopic species [70,102].

C.   Tetrahedral XH₄ Molecules

An extension of the local mode/Fermi resonance models to tetrahedral XH_4 molecules in the internal valence coordinate representation was discussed in detail in Ref. [103], where references to earlier work are found. When comparing XH_2- and XH_3-type molecules with tetrahedral molecules, there is an additional problem. The six valence angle displacement coordinates are not independent. There exists a redundancy which mathematically can be expressed in the form [104–106]

$$\begin{vmatrix} 1 & \cos \alpha_{12} & \cos \alpha_{13} & \cos \alpha_{14} \\ \cos \alpha_{12} & 1 & \cos \alpha_{23} & \cos \alpha_{24} \\ \cos \alpha_{13} & \cos \alpha_{23} & 1 & \cos \alpha_{34} \\ \cos \alpha_{14} & \cos \alpha_{24} & \cos \alpha_{34} & 1 \end{vmatrix} = 0 \qquad (3.48)$$

where α_{ij} is the valence angle between bonds i and j. The redundancy condition given above is obtained by considering four unit vectors e_k ($k = 1, 2, 3$, and 4) along the four different bonds. Because only three vectors are required to span a three-dimensional space, there exists a linear relation between the four unit vectors, that is, $\sum_{k=1}^{4} c_k e_k = 0$. A set of linear equations are obtained by forming dot products between this relation and the four unit vectors. These equations possess a nontrivial solution for the c_k coefficients when Eq. (3.48) is valid. By expanding the cosine functions around the equilibrium configuration, the redundancy relation takes the form

$$\frac{1}{\sqrt{6}} (\theta_{12} + \theta_{13} + \theta_{14} + \theta_{23} + \theta_{24} + \theta_{34})$$

$$+ \frac{\sqrt{3}}{8} (\theta_{12}^2 + \theta_{13}^2 + \theta_{14}^2 + \theta_{23}^2 + \theta_{24}^2 + \theta_{34}^2) + \frac{1}{2\sqrt{3}} (\theta_{12}\theta_{13}$$

$$+ \theta_{12}\theta_{14} + \theta_{12}\theta_{23} + \theta_{12}\theta_{24} + \theta_{13}\theta_{14} + \theta_{13}\theta_{23} + \theta_{13}\theta_{34}$$

$$+ \theta_{14}\theta_{24} + \theta_{14}\theta_{34} + \theta_{23}\theta_{24} + \theta_{23}\theta_{34} + \theta_{24}\theta_{34}) + \cdots = 0 \qquad (3.49)$$

where $\theta_{ij} = \alpha_{ij} - \alpha_e$ is the valence angle displacement coordinate and α_e is the tetrahedral angle. It is clear from this equation that for large-amplitude valence angle displacements it is not practical to express one of these coordinates as a function of the five others and then express the vibrational Hamiltonian in terms of the five remaining coordinates. Another type of solution to this problem must be looked for.

The fundamentals of the XH_4 tetrahedra are labeled in the customary vibrational spectroscopy as $\nu_1(A_1)$, $\nu_2(E)$, $\nu_3(F_2)$, and $\nu_4(F_2)$, which are non-degenerate symmetric stretch, doubly degenerate bend, triply degenerate antisymmetric stretch, and triply degenerate bend, respectively. Symmetrized curvilinear internal coordinates which span the same symmetry species are defined as [71]

$$S_1 = \tfrac{1}{2}(r_1 + r_2 + r_3 + r_4)$$

$$S_{2a} = \frac{1}{\sqrt{12}} (2\theta_{12} - \theta_{13} - \theta_{14} - \theta_{23} - \theta_{24} + 2\theta_{34})$$

$$S_{2b} = \tfrac{1}{2}(\theta_{13} - \theta_{14} - \theta_{23} + \theta_{24})$$

$$S_{3x} = \tfrac{1}{2}(r_1 - r_2 + r_3 - r_4)$$

$$S_{3y} = \tfrac{1}{2}(r_1 - r_2 - r_3 + r_4)$$

$$S_{3z} = \tfrac{1}{2}(r_1 + r_2 - r_3 - r_4) \qquad (3.50)$$

$$S_{4x} = \frac{1}{\sqrt{2}} (\theta_{24} - \theta_{13})$$

$$S_{4y} = \frac{1}{\sqrt{2}} (\theta_{23} - \theta_{14})$$

$$S_{4z} = \frac{1}{\sqrt{2}} (\theta_{34} - \theta_{12})$$

Usually, the potential energy surface is expressed in terms of these coordinates, which consist of five coordinates for the angle bends [71,75,106]. What remains to be done is to express the expanded valence coordinate kinetic energy Hamiltonian operator in terms of these symmetrized coordinates and their conjugate momenta. Note that although Eqs. (3.50) appear to be linear, they are not, in the sense that their inverse is not linear, but in the general case it can be written as [88]

$$R_i = V_i^{(j)} S_j + \tfrac{1}{2} V_i^{(jk)} S_j S_k + \tfrac{1}{6} V_i^{(jkl)} S_j S_k S_l + \cdots \tag{3.51}$$

where R_i is r_j or θ_{lk} and Einstein's summation convention has been adopted. It follows from the classical definition of the momentum that the transformation from momenta conjugate to symmetrized coordinates are related in a simple way to the momenta conjugate to internal coordinates

$$P_{R_i} = P_{S_j} U_j^{(i)} \tag{3.52}$$

where the $\mathbf{U}$ matrix elements have already been given in Eqs. (3.50). Note that again this equation is not linear in the sense that its inverse is nonlinear. Reference [88] contains an explanation of how to derive expressions for the $\mathbf{V}$ tensor elements in Eq. (3.51) by using the redundancy expansion, Eq. (3.49). In going up to the second order, the nonlinear transformation from symmetrized coordinates to internal valence coordinates is

$$r_1 = \tfrac{1}{2}(S_1 + S_{3x} + S_{3y} + S_{3z})$$

$$r_2 = \tfrac{1}{2}(S_1 - S_{3x} - S_{3y} + S_{3z})$$

$$r_3 = \tfrac{1}{2}(S_1 + S_{3x} - S_{3y} - S_{3z})$$

$$r_4 = \tfrac{1}{2}(S_1 - S_{3x} + S_{3y} - S_{3z})$$

$$\theta_{12} = \frac{2}{\sqrt{12}} S_{2a} - \frac{1}{\sqrt{2}} S_{4z} + \frac{1}{24\sqrt{2}} (S_{2a}^2 + S_{2b}^2)$$

$$- \frac{1}{8\sqrt{2}} (S_{4x}^2 + S_{4y}^2 + S_{4z}^2)$$

$$\theta_{13} = - \frac{1}{\sqrt{12}} S_{2a} + \frac{1}{2} S_{2b} - \frac{1}{\sqrt{2}} S_{4x} + \frac{1}{24\sqrt{2}} (S_{2a}^2 + S_{2b}^2)$$

$$- \frac{1}{8\sqrt{2}} (S_{4x}^2 + S_{4y}^2 + S_{4z}^2)$$

$$\theta_{14} = -\frac{1}{\sqrt{12}} S_{2a} - \frac{1}{2} S_{2b} - \frac{1}{\sqrt{2}} S_{4y} + \frac{1}{24\sqrt{2}} (S_{2a}^2 + S_{2b}^2)$$

$$-\frac{1}{8\sqrt{2}} (S_{4x}^2 + S_{4y}^2 + S_{4z}^2)$$

$$\theta_{23} = -\frac{1}{\sqrt{12}} S_{2a} - \frac{1}{2} S_{2b} + \frac{1}{\sqrt{2}} S_{4y} + \frac{1}{24\sqrt{2}} (S_{2a}^2 + S_{2b}^2)$$

$$-\frac{1}{8\sqrt{2}} (S_{4x}^2 + S_{4y}^2 + S_{4z}^2)$$

$$\theta_{24} = -\frac{1}{\sqrt{12}} S_{2a} + \frac{1}{2} S_{2b} + \frac{1}{\sqrt{2}} S_{4x} + \frac{1}{24\sqrt{2}} (S_{2a}^2 + S_{2b}^2)$$

$$-\frac{1}{8\sqrt{2}} (S_{4x}^2 + S_{4y}^2 + S_{4z}^2)$$

$$\theta_{34} = \frac{2}{\sqrt{12}} S_{2a} + \frac{1}{\sqrt{2}} S_{4z} + \frac{1}{24\sqrt{2}} (S_{2a}^2 + S_{2b}^2)$$

$$-\frac{1}{8\sqrt{2}} (S_{4x}^2 + S_{4y}^2 + S_{4z}^2) \tag{3.53}$$

and similarly for the conjugate momenta

$$p_{r_1} = \tfrac{1}{2}(P_1 + P_{3x} + P_{3y} + P_{3z})$$

$$p_{r_2} = \tfrac{1}{2}(P_1 - P_{3x} - P_{3y} + P_{3z})$$

$$p_{r_3} = \tfrac{1}{2}(P_1 + P_{3x} - P_{3y} - P_{3z})$$

$$P_{r_4} = \tfrac{1}{2}(P_1 - P_{3x} + P_{3y} - P_{3z})$$

$$p_{\theta_{12}} = \frac{2}{\sqrt{12}} P_{2a} - \frac{1}{\sqrt{2}} P_{4z}$$

$$p_{\theta_{13}} = -\frac{1}{\sqrt{12}} P_{2a} + \tfrac{1}{2} P_{2b} - \frac{1}{\sqrt{2}} P_{4x}$$

$$P_{\theta_{14}} = -\frac{1}{\sqrt{12}} P_{2a} - \tfrac{1}{2} P_{2b} - \frac{1}{\sqrt{2}} P_{4y}$$

$$p_{\theta_{23}} = -\frac{1}{\sqrt{12}} P_{2a} - \tfrac{1}{2} P_{2b} + \frac{1}{\sqrt{2}} P_{4y}$$

$$p_{\theta 24} = -\frac{1}{\sqrt{12}}\, P_{2a} + \tfrac{1}{2} P_{2b} + \frac{1}{\sqrt{2}}\, P_{4x}$$

$$p_{\theta 34} = \frac{2}{\sqrt{12}}\, P_{2a} + \frac{1}{\sqrt{2}}\, P_{4z} \tag{3.54}$$

Note that in transforming a kinetic energy which is expanded up the fourth order, the quadratic terms in Eqs. (3.53) are only needed for the quartic bending terms.

The stretching vibrational Hamiltonian for the XH_4 tetrahedra is

$$H_{\text{strech}} = \sum_{i=1}^{4} (\tfrac{1}{2} g_{rr}\, p_{ri}^2 + D_e\, y_i^2)$$

$$+ \sum_{i<j} (g_{rr'}^{(e)}\, p_{ri} p_{rj} + a_r^{-2} f_{rr'}\, y_i y_j) \tag{3.55}$$

where, as in XH_2 and XH_3, the Morse variable y_i is used instead of the usual r_i coordinate, in order to ensure realistic asymptotic behavior at large stretching displacements. As discussed above, the bending Hamiltonian is expressed in terms of symmetrized curvilinear internal coordinates and their conjugate momenta and takes the form

$$H_{\text{bend}} = \tfrac{1}{2} G_{22}^{(e)}(P_{2a}^2 + P_{2b}^2) + \tfrac{1}{2} F_{22}(S_{2a}^2 + S_{2b}^2)$$

$$+ \tfrac{1}{2} G_{44}^{(e)}(P_{4x}^2 + P_{4y}^2 + P_{4z}^2) + \tfrac{1}{2} F_{44}(S_{4x}^2 + S_{4y}^2 + S_{4z}^2) + H_{\text{anh}} \tag{3.56}$$

where $G_{22}^{(e)} = g_{\theta\theta}^{(e)} - 2g_{\theta\theta'}^{(e)} + g_{\theta\theta''}^{(e)}$ and $G_{44}^{(e)} = g_{\theta\theta}^{(e)} - g_{\theta\theta''}^{(e)}$. The equilibrium $\mathbf{g}$ matrix elements in tetrahedral molecules possess simple forms (Table XVI). The operator H_{anh} contains cubic and quartic bending terms

$$F_{22} = F_{2a2a} = F_{2b2b} = \left(\frac{\partial^2 V}{\partial S_{2a}^2}\right)_e = \left(\frac{\partial^2 V}{\partial S_{2b}^2}\right)_e$$

and

$$F_{44} = F_{4x4x} = F_{4y4y} = F_{4z4z} = \left(\frac{\partial^2 V}{\partial S_{4x}^2}\right)_e = \left(\frac{\partial^2 V}{\partial S_{4y}^2}\right)_e = \left(\frac{\partial^2 V}{\partial S_{4z}^2}\right)_e$$

are harmonic force constants (potential energy parameters). The stretch–

TABLE XVI
Kinetic Energy Coefficients for XH_4 Tetrahedra[a]

$$g_{rr} = g_{rr}^{(e)} = \mu_X + \mu_H$$

$$g_{rr'}^{(e)} = -\tfrac{1}{3}\mu_X$$

$$G_{22}^{(e)} = g_{\theta\theta}^{(e)} - 2g_{\theta\theta'}^{(e)} + g_{\theta\theta''}^{(e)} = \frac{3\mu_H}{R_e^2}$$

$$G_{44}^{(e)} = g_{\theta\theta}^{(e)} - g_{\theta\theta''}^{(e)} = \frac{2(\mu_H + \tfrac{8}{3}\mu_X)}{R_e^2}$$

$$G_{34}^{(e)} = \sqrt{2}(-g_{r\theta'}^{(e)} + g_{r\theta}^{(e)}) = \frac{\tfrac{8}{3}\mu_X}{R_e}$$

$$g_1 = \frac{1}{2}\left(\frac{\partial g_{\theta\theta}}{\partial r'}\right)_e + \left(\frac{\partial g_{\theta\theta''}}{\partial r}\right)_e - \left(\frac{\partial g_{\theta\theta'}}{\partial r}\right)_e - \frac{1}{2}\left(\frac{\partial g_{\theta\theta'}}{\partial r''}\right)_e = -\frac{\tfrac{3}{2}\mu_H}{R_e^3}$$

$$g_2 = \frac{1}{2}\left[\left(\frac{\partial g_{r\theta}}{\partial \theta}\right)_e - \left(\frac{\partial g_{r\theta}}{\partial \theta'}\right)_e + \left(\frac{\partial g_{r\theta'}}{\partial \theta}\right)_e\right] = 0$$

$$g_3 = \frac{1}{2}\left(\frac{\partial g_{\theta\theta}}{\partial r'}\right)_e + \left(\frac{\partial g_{\theta\theta''}}{\partial r}\right)_e = -\frac{\mu_H + \tfrac{8}{3}\mu_X}{R_e^3}$$

$$g_4 = \frac{1}{2}\left[\left(\frac{\partial g_{r\theta}}{\partial \theta}\right)_e + \left(\frac{\partial g_{r\theta'}}{\partial \theta'}\right)_e\right] = \frac{\tfrac{2}{3}\mu_X}{R_e}$$

$$g_5 = -2\left(\frac{\partial g_{\theta\theta'}}{\partial r}\right)_e + \left(\frac{\partial g_{\theta\theta'}}{\partial r''}\right)_e = \frac{\mu_H + \tfrac{8}{3}\mu_X}{R_e^3}$$

$$g_6 = -\left(\frac{\partial g_{r\theta}}{\partial \theta'}\right)_e = -\frac{\tfrac{4}{3}\mu_X}{R_e}$$

$$g_7 = \frac{1}{2}\left[\left(\frac{\partial g_{\theta\theta}}{\partial r'}\right)_e - \left(\frac{\partial g_{\theta\theta'}}{\partial r''}\right)_e\right] = -\frac{\tfrac{1}{2}(3\mu_H + 4\mu_X)}{R_e^3}$$

$$g_8 = \frac{1}{2}\left[\left(\frac{\partial g_{r\theta}}{\partial \theta}\right)_e - \left(\frac{\partial g_{r\theta}}{\partial \theta'}\right)_e - \left(\frac{\partial g_{r\theta'}}{\partial \theta'}\right)_e\right] = -\frac{\tfrac{1}{3}\mu_X}{R_e}$$

$$g_9 = \frac{1}{2}\left[\left(\frac{\partial g_{r\theta}}{\partial \theta}\right)_e + \left(\frac{\partial g_{r\theta}}{\partial \theta'}\right)_e - \left(\frac{\partial g_{r\theta'}}{\partial \theta'}\right)_e\right] = \frac{\mu_X}{R_e}$$

[a] $\mu_X = 1/m_X$ and $\mu_H = 1/m_H$; R_e is the equilibrium bond length. The **g** matrix elements and their derivatives are from Ref. [103]

bend bilinear harmonic coupling Hamiltonian is

$$
\begin{aligned}
H_{stretch-bend} = {} & G_{34}^{(e)}(P_{3x}P_{4x} + P_{3y}P_{4y} + P_{3z}P_{4z}) \\
& + F_{34}(S_{3x}S_{4x} + S_{3y}S_{4y} + S_{3z}S_{4z})
\end{aligned}
\tag{3.57}
$$

where, for the sake of simplicity, S_{3i} ($i = x, y, z$) is defined in terms of internal coordinates r_j ($j = 1, 2, 3, 4$) and not in terms of the Morse variable

$y_j = 1 - e^{-a_r r_j}$. Here, $G_{34}^{(e)} = \sqrt{2}(-g_{r\theta'}^{(e)} + g_{r\theta}^{(e)})$. Its explicit expression is given in Table XVI. And F_{34} is a harmonic interaction force constants which is defined similar to F_{22} and F_{44} above.

The fundamentals of $^{12}CH_4$ are located as follows: $v_1(A_1)$ at 2918 cm^{-1}, $v_2(E)$ at 1533 cm^{-1}, $v_3(F_2)$ at 3019 cm^{-1}, and $v_4(F_2)$ at 1311 cm^{-1} [71]. This implies that in the stretching fundamental region Fermi resonance interactions between the $v_1(A_1)$ and $2v_2(A_1)$, between $v_1(A_1)$ and $2v_4(A_1)$, between $v_3(F_2)$ and $2v_4(F_2)$, and between $v_3(F_2)$ and $v_2 + v_4(F_2)$ should be included in the model. The normal mode symbols refer to interacting states. Thus, four different types of Fermi resonance operators are needed. The Hamiltonian operator is

$$
\begin{aligned}
H = {}& g_1 S_1(P_{2a}^2 + P_{2b}^2) + g_2 P_1 \\
& \times [(P_{2a}S_{2a} + S_{2a}P_{2a}) + (P_{2b}S_{2b} + S_{2b}P_{2b})] \\
& + \tfrac{1}{2}F_{122}S_1(S_{2a}^2 + S_{2b}^2) \\
& + g_3 S_1(P_{4x}^2 + P_{4y}^2 + P_{4z}^2) \\
& + g_4 P_1[(P_{4x}S_{4x} + S_{4x}P_{4x}) \\
& + (P_{4y}S_{4y} + S_{4y}P_{4y}) + (P_{4z}S_{4z} + S_{4z}P_{4z})] \\
& + \tfrac{1}{2}F_{144}S_1(S_{4x}^2 + S_{4y}^2 + S_{4z}^2) \\
& + g_5(S_{3x}P_{4y}P_{4z} + S_{3y}P_{4x}P_{4z} + S_{3z}P_{4x}P_{4y}) \\
& + g_6[P_{3x}(S_{4y}P_{4z} + P_{4y}S_{4z}) \\
& + P_{3y}(S_{4x}P_{4z} + P_{4x}S_{4z}) + P_{3z}(S_{4x}P_{4y} + P_{4x}S_{4y})] \\
& + F_{344}(S_{3x}S_{4y}S_{4z} + S_{3y}S_{4x}S_{4z} + S_{3z}S_{4x}S_{4y}) \\
& + \sqrt{\tfrac{2}{3}}g_7[S_{3x}(P_{2a} - \sqrt{3}P_{2b})P_{4x} \\
& + S_{3y}(P_{2a} + \sqrt{3}P_{2b})P_{4y} - 2S_{3z}P_{2a}P_{4z}] \\
& - \sqrt{\tfrac{2}{3}}g_8[P_{3x}(S_{2a} - \sqrt{3}S_{2b})P_{4x} \\
& + P_{3y}(S_{2a} + \sqrt{3}S_{2b})P_{4y} - 2P_{3z}S_{2a}P_{4z}] \\
& - \sqrt{\tfrac{2}{3}}g_9[P_{3x}(P_{2a} - \sqrt{3}P_{2b})S_{4x} \\
& + P_{3y}(P_{2a} + \sqrt{3}P_{2b})S_{4y} - 2P_{3z}P_{2a}S_{4z}] \\
& - \tfrac{1}{2}F_{234}[S_{3x}(S_{2a} - \sqrt{3}S_{2b})S_{4x} \\
& + S_{3y}(S_{2a} + \sqrt{3}S_{2b})S_{4y} - 2S_{3z}S_{2a}S_{4z}] \quad\quad (3.58)
\end{aligned}
$$

where $g_1, g_2, \ldots, g_9$ are linear combinations of the appropriate g matrix derivatives. Their expressions are given in Table XVI. The coefficients F_{122}, F_{144}, F_{344}, and F_{234} are the usual Fermi resonance force constants [71].

The Hamiltonian matrices are formed in a product basis, $|v_{r_1}v_{r_2}v_{r_3}v_{r_4},$
$v_{2a}v_{2b}, v_{4x}v_{4y}v_{4z}\rangle = |v_{r_1}v_{r_2}v_{r_3}v_{r_4}\rangle |v_{2a}v_{2b}\rangle |v_{4x}v_{4y}v_{4z}\rangle$, where $|v_{r_1}v_{r_2}v_{r_3}v_{r_4}\rangle$
$= |v_{r_1}\rangle |v_{r_2}\rangle |v_{r_3}\rangle |v_{r_4}\rangle$ is a local mode product basis function for the four
stretching degrees of freedom; $|v_{2a}v_{2b}\rangle = |v_{2a}\rangle |v_{2b}\rangle$ and $|v_{4x}v_{4y}v_{4z}\rangle$
$= |v_{4x}\rangle |v_{4y}\rangle |v_{4z}\rangle$ are two- and three-dimensional harmonic oscillator basis
functions in the Cartesian representation for the bending degrees of
freedom. Apart from the Morse oscillator diagonal terms, all matrix elements are evaluated within the harmonic approximation. For the vibrationally diagonal term,

$$G(v_{r_1}v_{r_2}v_{r_3}v_{r_4}, v_{2a}v_{2b}, v_{4x}v_{4y}v_{4z})$$

$$= \langle v_{r_1}v_{r_2}v_{r_3}v_{r_4}, v_{2a}v_{2b}, v_{4x}v_{4y}v_{4z}|$$

$$\times \frac{H}{hc_0} |v_{r_1}v_{r_2}v_{r_3}v_{r_4}, v_{2a}v_{2b}, v_{4x}v_{4y}v_{4z}\rangle$$

$$= \omega'_r(v_{r_1} + v_{r_2} + v_{r_3} + v_{r_4})$$

$$+ x_{rr}(v_{r_1}^2 + v_{r_1} + v_{r_2}^2 + v_{r_2} + v_{r_3}^2 + v_{r_3} + v_{r_4}^2 + v_{r_4})$$

$$+ \omega_2 v_2 + x_{22} v_2(v_2 + 2) + \omega'_4 v_4$$

$$+ x_{44} v_4(v_4 + 3) + x_{24}(v_2 v_4 + \tfrac{3}{2} v_2 + v_4)$$

$$+ x_{r2}[v_2(v_{r_1} + v_{r_2} + v_{r_3} + v_{r_4} + 2)$$

$$+ (v_{r_1} + v_{r_2} + v_{r_3} + v_{r_4})]$$

$$+ x_{r4}[v_4(v_{r_1} + v_{r_2} + v_{r_3} + v_{r_4} + 2)$$

$$+ \tfrac{3}{2}(v_{r_1} + v_{r_2} + v_{r_3} + v_{r_4})] \tag{3.59}$$

is obtained, where $v_2 = v_{2a} + v_{2b}$ and $v_4 = v_{4x} + v_{4y} + v_{4z}$. Note that these
should not be confused with corresponding normal mode quantum
numbers. Theoretical expressions for ω'_r and ω'_4 are

$$\omega'_r = \omega_r - \frac{3}{8} \omega_r \left(\frac{g_{rr'}^{(e)}}{g_{rr}} - \frac{f_{rr'}}{f_{rr}}\right)^2$$

$$+ \frac{3}{16} \left(\frac{d_{34}(1)^2}{\omega_r - \omega_4} - \frac{d_{34}(2)^2}{\omega_r + \omega_4}\right) \tag{3.60}$$

$$\omega'_4 = \omega_4 - \frac{1}{4} \left(\frac{d_{34}(1)^2}{\omega_r - \omega_4} + \frac{d_{34}(2)^2}{\omega_r + \omega_4}\right)$$

where

$$d_{34}(1) = \frac{\hbar^2}{hc_0} G_{34}^{(e)} \alpha_r^{1/2} \alpha_4^{1/2} + \frac{1}{hc_0} F_{34} \alpha_r^{-1/2} \alpha_4^{-1/2}$$

$$= (\omega_r \omega_4)^{1/2} \left[\frac{G_{34}^{(e)}}{(g_{rr} G_{44}^{(e)})^{1/2}} + \frac{F_{34}}{(f_{rr} F_{44})^{1/2}} \right]$$

$$d_{34}(2) = - \frac{\hbar^2}{hc_0} G_{34}^{(e)} \alpha_r^{1/2} \alpha_4^{1/2} + \frac{1}{hc_0} F_{34} \alpha_r^{-1/2} \alpha_4^{-1/2}$$

$$= (\omega_r \omega_4)^{1/2} \left[- \frac{G_{34}^{(e)}}{(g_{rr} G_{44}^{(e)})^{1/2}} + \frac{F_{34}}{(f_{rr} F_{44})^{1/2}} \right] \qquad (3.61)$$

Bending vibrational harmonic wavenumbers are defined as

$$\omega_i = \frac{1}{2\pi c_0} (F_{ii} G_{ii}^{(e)})^{1/2}$$

and

$$\alpha_i = \frac{4\pi^2 c_0 \omega_i}{h G_{ii}^{(e)}} = \frac{(F_{ii}/G_{ii}^{(e)})^{1/2}}{\hbar}$$

where $i = 2$, 4. The stretching vibrational harmonic wavenumber and anharmonicity parameter are defined as before. For the other anharmonicity parameters, no theoretical expressions are given. They could be obtained by applying first-order perturbation theory to quartic terms and second-order perturbation theory to cubic bending terms and to Fermi resonance terms.

The local mode coupling matrix elements are of the type

$$\langle (v_{r_1} + 1)(v_{r_2} - 1) v_{r_3} v_{r_4} | \frac{H}{hc_0} | v_{r_1} v_{r_2} v_{r_3} v_{r_4} \rangle = \lambda_r' [(v_{r_1} + 1) v_{r_2}]^{1/2} \qquad (3.62)$$

where

$$\lambda_r' = \lambda_r + \frac{1}{16} \left(- \frac{d_{34}(1)^2}{\omega_r - \omega_4} + \frac{d_{34}(2)^2}{\omega_r + \omega_4} \right) \qquad (3.63)$$

and

$$\lambda_r = \frac{1}{2} \omega_r \left(\frac{g_{rr'}^{(e)}}{g_{rr}} + \frac{f_{rr'}}{f_{rr}} \right) \qquad (3.64)$$

Four distinctly different kinds of Fermi resonance matrix elements exist.
These are

$$\langle (v_{r_1} + 1)v_{r_2} v_{r_3} v_{r_4}, (v_{2a} - 2)v_{2b}, v_{4x} v_{4y} v_{4z} |$$

$$\times \frac{H}{hc_0} | v_{r_1} v_{r_2} v_{r_3} v_{r_4}, v_{2a} v_{2b}, v_{4x} v_{4y} v_{4z} \rangle$$

$$= d_{r22}[(v_{r_1} + 1)v_{2a}(v_{2a} - 1)]^{1/2}$$

$$\langle (v_{r_1} + 1)v_{r_2} v_{r_3} v_{r_4}, v_{2a} v_{2b}, (v_{4x} - 2)v_{4y} v_{4z} |$$

$$\times \frac{H}{hc_0} | v_{r_1} v_{r_2} v_{r_3} v_{r_4}, v_{2a} v_{2b}, v_{4x} v_{4y} v_{4z} \rangle$$

$$= d_{r44}[(v_{r_1} + 1)v_{4x}(v_{4x} - 1)]^{1/2}$$

$$\langle (v_{r_1} + 1)v_{r_2} v_{r_3} v_{r_4}, v_{2a} v_{2b}, (v_{4x} - 1)(v_{4y} - 1)v_{4z} | \tag{3.65}$$

$$\times \frac{H}{hc_0} | v_{r_1} v_{r_2} v_{r_3} v_{r_4}, v_{2a} v_{2b}, v_{4x} v_{4y} v_{4z} \rangle$$

$$= d'_{r44}[(v_{r_1} + 1)v_{4x} v_{4y}]^{1/2}$$

$$\langle (v_{r_1} + 1)v_{r_2} v_{r_3} v_{r_4}, (v_{2a} - 1)v_{2b}, (v_{4x} - 1)v_{4y} v_{4z} |$$

$$\times \frac{H}{hc_0} | v_{r_1} v_{r_2} v_{r_3} v_{r_4}, v_{2a} v_{2b}, v_{4x} v_{4y} v_{4z} \rangle$$

$$= d_{r24}[(v_{r_1} + 1)v_{2a} v_{4x}]^{1/2}$$

There exist many matrix elements of these types with the same or with
slightly different numerical coefficients on the right-hand side. The d coefficients are given by

$$d_{r22} = -\frac{\hbar^2}{4\sqrt{2}hc_0} g_1 \alpha_r^{-1/2}\alpha_2 + \frac{\hbar^2}{2\sqrt{2}hc_0} g_2 \alpha_r^{1/2}$$

$$+ \frac{1}{8\sqrt{2}hc_0} F_{122} \alpha_r^{-1/2}\alpha_2^{-1}$$

$$d_{r44} = -\frac{\hbar^2}{4\sqrt{2}hc_0} g_3 \alpha_r^{-1/2}\alpha_4 + \frac{\hbar^2}{2\sqrt{2}hc_0} g_4 \alpha_r^{1/2}$$

$$+ \frac{1}{8\sqrt{2}hc_0} F_{144} \alpha_r^{-1/2}\alpha_4^{-1}$$

$$d'_{r44} = -\frac{\hbar^2}{4\sqrt{2}hc_0} g_5 \alpha_r^{-1/2}\alpha_4 + \frac{\hbar^2}{2\sqrt{2}hc_0} g_6 \alpha_r^{1/2}$$

$$+ \frac{1}{4\sqrt{2hc_0}} F_{344} \alpha_r^{-1/2} \alpha_4^{-1}$$

$$d_{r24} = - \frac{\hbar^2}{4\sqrt{3hc_0}} g_7 \alpha_r^{-1/2} \alpha_2^{1/2} \alpha_4^{1/2}$$

$$- \frac{\hbar^2}{4\sqrt{3hc_0}} g_8 \alpha_r^{1/2} \alpha_2^{-1/2} \alpha_4^{1/2}$$

$$- \frac{\hbar^2}{4\sqrt{3hc_0}} g_9 \alpha_r^{1/2} \alpha_2^{1/2} \alpha_4^{-1/2}$$

$$- \frac{1}{8\sqrt{2hc_0}} F_{234} \alpha_r^{-1/2} \alpha_2^{-1/2} \alpha_4^{-1/2} \qquad (3.66)$$

The two- and three-dimensional harmonic oscillator basis functions in the Cartesian representation can be symmetrized by extending the promotion operator technique used before for the pure stretching vibrational basis functions. A two-dimensional promotion operator $[\sigma_a(E_a), \sigma_b(E_b)]$ and a three-dimensional promotion operator $[\sigma_x(F_{2x}), \sigma_y(F_{2y}), \sigma_z(F_{2z})]$ are defined as

$$\sigma_a(E_a)|v_{2a}v_{2b}\rangle = \sqrt{(v_{2a}+1)}|(v_{2a}+1)v_{2b}\rangle$$

$$\sigma_b(E_b)|v_{2a}v_{2b}\rangle = \sqrt{(v_{2b}+1)}|v_{2a}(v_{2b}+1)\rangle \qquad (3.67)$$

and

$$\sigma_x(F_{2x})|v_{4x}v_{4y}v_{4z}\rangle = \sqrt{(v_{4x}+1)}|(v_{4x}+1)v_{4y}v_{4z}\rangle$$

$$\sigma_y(F_{2y})|v_{4x}v_{4y}v_{4z}\rangle = \sqrt{(v_{4y}+1)}|v_{4x}(v_{4y}+1)v_{4z}\rangle \qquad (3.68)$$

$$\sigma_z(F_{2z})|v_{4x}v_{4y}v_{4z}\rangle = \sqrt{(v_{4z}+1)}|v_{4x}v_{4y}(v_{4z}+1)\rangle$$

Using the vector coupling coefficients from Ref. [51], all required basis functions can be derived. If the same symmetry species appears more than once for one particular type of basis function set, the Gram–Schmidt orthogonalization technique is useful in forming the final orthonormal basis functions. This can be done by hand (at least up to moderately high quantum numbers). One example of the procedure is given by operating with the $[\sigma_a(E_a), \sigma_b(E_b)]$ operator pair to a pair of basis functions ($|10\rangle$, $|10\rangle$) = ($|10E_a\rangle$, $|01E_b\rangle$). The result is

$$-\frac{1}{\sqrt{2}}\,\sigma_a(E_a)|10E_a\rangle + \frac{1}{\sqrt{2}}\,\sigma_b(E_b)|01E_b\rangle$$

$$= -|20\rangle + |02\rangle = \sqrt{2}|20E_a\rangle \tag{3.69}$$

The $1/\sqrt{2}$ factors on the left-hand side are vector coupling coefficients (see Table V). Similarly

$$\frac{1}{\sqrt{2}}\,\sigma_a(E_a)|01E_b\rangle + \frac{1}{\sqrt{2}}\,\sigma_b(E_b)|10E_a\rangle = \sqrt{2}|11\rangle = \sqrt{2}|11E_b\rangle \tag{3.70}$$

Different types of basis functions are combined using vector coupling coefficients [56]. For example, $|0000, 10, 100; F_{2x}\rangle$ is obtained as

$$|0000, 10, 100; F_{2x}\rangle$$
$$= |0000A_1\rangle(-\tfrac{1}{2}|10E_a\rangle|100F_{2x}\rangle + \tfrac{1}{2}\sqrt{3}|01E_b\rangle|100F_{2x}\rangle)$$
$$= |0000\rangle(-\tfrac{1}{2}|10\rangle|100\rangle + \tfrac{1}{2}\sqrt{3}|01\rangle|100\rangle)$$
$$= -\tfrac{1}{2}|0000, 10, 100\rangle + \tfrac{1}{2}\sqrt{3}|0000, 01, 100\rangle \tag{3.71}$$

where the coupling coefficients are from Table V. The $v = (v_{r_1} + v_{r_2} + v_{r_3} + v_{r_4}) + \frac{1}{2}(v_2 + v_4) = 1$ basis functions are given below:

$$|1000, 00, 000; A_1\rangle = \tfrac{1}{2}(|1_1\rangle + |1_2\rangle + |1_3\rangle + |1_4\rangle)|00\rangle|000\rangle$$

$$|0000, 20, 000; A_1\rangle = \frac{1}{\sqrt{2}}|0000\rangle(|20\rangle + |02\rangle)|000\rangle$$

$$|0000, 00, 200; A_1\rangle = \frac{1}{\sqrt{3}}|0000\rangle|00\rangle(|200\rangle + |020\rangle + |002\rangle)$$

$$|0000, 20, 000; E_a\rangle = \frac{1}{\sqrt{2}}|0000\rangle(-|20\rangle + |02\rangle)|000\rangle$$

$$|0000, 00, 200; E_a\rangle = \frac{1}{\sqrt{6}}|0000\rangle|00\rangle(|200\rangle + |020\rangle - 2|002\rangle)$$

$$|0000, 20, 000; E_b\rangle = |0000\rangle|11\rangle|000\rangle$$

$$|0000, 00, 200; E_b\rangle = \frac{1}{\sqrt{2}}|0000\rangle|00\rangle(-|200\rangle + |020\rangle)$$

$$|0000, 10, 100; F_{1x}\rangle = |0000\rangle(-\tfrac{1}{2}\sqrt{3}|10\rangle - \tfrac{1}{2}|01\rangle)|100\rangle$$

$$|0000, 10, 100; F_{1y}\rangle = |0000\rangle(\tfrac{1}{2}\sqrt{3}|10\rangle - \tfrac{1}{2}|01\rangle)|010\rangle$$

$$|0000, 10, 100; F_{1z}\rangle = |0000\rangle|01\rangle|001\rangle$$

$$|1000, 00, 000; F_{2x}\rangle = \tfrac{1}{2}(|1_1\rangle - |1_2\rangle + |1_3\rangle - |1_4\rangle)|00\rangle|000\rangle$$

$$|0000, 10, 100; F_{2x}\rangle = |0000\rangle(-\tfrac{1}{2}|10\rangle + \tfrac{1}{2}\sqrt{3}|01\rangle)|100\rangle$$

$$|0000, 00, 110; F_{2x}\rangle = -|0000\rangle|00\rangle|011\rangle$$

$$|1000, 00, 000; F_{2y}\rangle = \tfrac{1}{2}(|1_1\rangle - |1_2\rangle - |1_3\rangle + |1_4\rangle)|00\rangle|000\rangle$$

$$|0000, 10, 100; F_{2y}\rangle = |0000\rangle(-\tfrac{1}{2}|10\rangle - \tfrac{1}{2}\sqrt{3}|01\rangle)|010\rangle$$

$$|0000, 00, 110; F_{2y}\rangle = -|0000\rangle|00\rangle|101\rangle$$

$$|1000, 00, 000; F_{2z}\rangle = \tfrac{1}{2}(|1_1\rangle + |1_2\rangle - |1_3\rangle - |1_4\rangle)|00\rangle|000\rangle$$

$$|0000, 10, 100; F_{2z}\rangle = |0000\rangle|10\rangle|001\rangle$$

$$|0000, 00, 110; F_{2z}\rangle = -|0000\rangle|00\rangle|110\rangle \tag{3.72}$$

There is off-diagonal Fermi resonance coupling only in the A_1 and F_2 Hamiltonian matrix blocks for $v = 1$. The A_1 matrix is given as

$$\begin{pmatrix} G(1000, 00, 000) + 3\lambda_r' & 4d_{r22} & 2\sqrt{6}d_{r44} \\ 4d_{r22} & G(0000, 20, 000) & 0 \\ 2\sqrt{6}d_{r44} & 0 & G(0000, 00, 200) \end{pmatrix}$$

$$= \begin{pmatrix} \begin{array}{c} \omega_r' + 2x_{rr} + x_{r2} \\ + \tfrac{3}{2}x_{r4} + 3\lambda_r' \end{array} & 4d_{r22} & 2\sqrt{6}d_{r44} \\ 4d_{r22} & \begin{array}{c} 2\omega_2 + 8x_{22} \\ + 4x_{r2} + 3x_{24} \end{array} & 0 \\ 2\sqrt{6}d_{r44} & 0 & \begin{array}{c} 2\omega_4' + 10x_{44} \\ + 4x_{r4} + 2x_{24} \end{array} \end{pmatrix} \tag{3.73}$$

The three F_2 matrices are identical and they are

$$\begin{pmatrix} G(1000, 00, 000) - \lambda_r' & -4d_{r24} & -2d_{r44}' \\ -4d_{r24} & G(0000, 10, 100) & 0 \\ -2d_{r44}' & 0 & G(0000, 00, 110) \end{pmatrix}$$

$$
= \begin{pmatrix}
\begin{array}{l} \omega_r' + 2x_{rr} + x_{r2} \\ \quad + \frac{3}{2}x_{r4} - \lambda_r' \end{array} & -4d_{r24} & -2d_{r44}' \\[2ex]
-4d_{r24} & \begin{array}{l} \omega_2 + \omega_4' + 3x_{22} + 4x_{44} \\ \quad + 2x_{r2} + 2x_{r4} + \frac{7}{2}x_{24} \end{array} & 0 \\[2ex]
-2d_{r44}' & 0 & \begin{array}{l} 2\omega_4' + 10x_{44} \\ \quad + 4x_{r4} + 2x_{24} \end{array}
\end{pmatrix} \tag{3.74}
$$

TABLE XVII

Internal Coordinate Spectroscopic Parameters for $^{12}CH_4$ [a]

Parameter	Value
ω_r	3147.18 cm^{-1}
ω_2	1565.18 cm^{-1}
ω_4	1365.59 cm^{-1}
x_{rr}	-58.44 cm^{-1}
x_{22}	-1.98 cm^{-1}
x_{44}	-2.28 cm^{-1}
x_{r2}	-10.90 cm^{-1}
x_{r4}	-15.68 cm^{-1}
x_{24}	-3.78 cm^{-1}
λ_r	-38.69 cm^{-1}
d_{r22}	5.90 cm^{-1}
d_{r44}	14.94 cm^{-1}
d_{r44}'	-28.20 cm^{-1}
d_{r24}	6.67 cm^{-1}
G_{22}	1.33 cm^{-1}
G_{44}	3.23 cm^{-1}
T_{44}	0.138 cm^{-1}
T_{24}	0.71 cm^{-1}
T_{23}	0 cm^{-1}
G_{34}	4.03 cm^{-1}
S_{34}	0.159 cm^{-1}
T_{34}	0.385 cm^{-1}
$K_{22, 44A_1}$	0 cm^{-1}
$K_{22, 44E}$	0 cm^{-1}
$K_{24, 44}$	0 cm^{-1}
λ_{r2}	4.89 cm^{-1}
λ_{r4}	1.69 cm^{-1}
F_{34}	-0.221 aJ Å^{-1}

[a] $\omega_r' = 3133.57$ cm^{-1}, $\omega_4' = 1348.53$ cm^{-1}, and $\lambda_r' = -34.44$ cm^{-1}.

TABLE XVIII
Potential Energy Parameters for CH_4 [a]

Force constant	Fermi	Anharmonic force field	Ab initio
f_{rr}, aJ Å^{-2}	5.426	5.392	5.401
f_{rrr}, aJ Å^{-3}	−29.2	−31.47	−31.01
f_{rrrr}, aJ Å^{-4}	122	173	162
F_{22}, aJ	0.572	0.5845	0.5770
F_{44}, aJ	0.533	0.5480	0.5323
$f_{rr'}$, aJ Å^{-2}	0.007	0.014	0.024
F_{34}, aJ Å^{-1}	−0.221	−0.221	−0.211
F_{122}, aJ Å^{-1}	−0.299	−0.299	−0.254
F_{144}, aJ Å^{-1}	−0.110	−0.110	−0.226
F_{344}, aJ Å^{-1}	−0.100	−0.101	−0.096
F_{234}, aJ Å^{-1}	0.159	0.160	0.180

[a] The surfaces have been taken from Refs. [103] (Fermi), [71] (anharmonic force field), and [75] (ab initio).

Table XVII contains model parameters which have been used to calculate the overtone spectrum of $^{12}CH_4$ [103]. Some potential energy force constants derived from the spectroscopic parameters are in Table XVIII. Table XIX contains a comparison between observed and calculated vibrational term values up to $v = 2$ [103]. The results show a good agreement between theory and experiment. Bending vibrational phenomenological quartic coupling terms (Darling–Dennison resonance-type and Hecht-type terms), which can be obtained by symmetry arguments, have been added between the bending states in order to reduce deviations [103]. The appropriate matrix elements are given in Table VI (they are of the same form as the corresponding normal mode matrix elements). All these terms have their origins in the same kind of quartic terms in the original internal coordinate Hamiltonian as the terms in x_{22}, x_{44}, and x_{24}. In addition, it is found that Hecht-type quartic coupling terms between stretching and bending vibrations play an important role in the overtone energy level region. Some of these had to be added to the model [103]. Finally, it has also been necessary to include terms in λ_{r2} and λ_{r4} which describe bending vibrational dependence of the interbound coupling parameter λ'_r (see Refs. [12], where an analogous term has been included in XH_2). They also come from quartic terms in the first order. The importance of such additional coupling indicates that quartic off-diagonal bending operators must be included in the internal coordinate Hamiltonian if a good agreement with experiment is desired. In any case, it is pleasing that a simple model works well with such a large number of high-precision experimental observations.

TABLE XIX
Observed and Calculated Vibrational Energy Levels for $^{12}CH_4$ [a]

	Local	Normal		
v	$v_{r_1}v_{r_2}v_{r_3}v_{r_4}$, $v_{2a}v_{2b}$, $v_{4x}v_{4y}v_{4y}$; Γ	$v_1v_2v_3v_4$	v_{obs} (cm^{-1})	v_{calc} (cm^{-1})
$\frac{1}{2}$	0000, 00, 100; F_2	0001	1310.8	1310.7
	0000, 10, 000; E	0100	1533.3	1533.1
1	0000, 00, 200; A_1	0002	2587.0	2586.8
	0000, 00, 110; F_2	0002	2614.3	2614.1
	0000, 00, 200; E	0002	2624.6	2625.0
	0000, 10, 100; F_2	0101	2830.3	2830.7
	0000, 10, 100; F_1	0101	2846.1	2845.7
	1000, 00, 000; A_1	1000	2916.5	2914.8
	1000, 00, 000; F_2	1000	3019.5	3021.1
	0000, 20, 000; A_1	0200	3063.6	3063.3
	0000, 11, 000; E	0200	3065.1	3064.9
$\frac{3}{2}$	0000, 00, 210; F_2	0003	3870.5	3870.8
	0000, 00, 111; A_1	0003	3909.2	3909.1
	0000, 00, 210; F_1	0003	3920.5	3920.3
	0000, 00, 300; F_2	0003	3930.9	3931.0
	0000, 10, 200; E	0102	4105.2	4104.8
	0000, 10, 110; F_1	0102	4128.6	4128.8
	0000, 10, 200; A_1	0102	4133.0	4132.2
	0000, 10, 110; F_2	0102	4142.9	4143.0
	0000, 10, 200; E	0102	4151.2	4151.6
	0000, 10, 200; A_2	0102	4161.9	4161.9
	1000, 00, 100; F_2	1001	4223.5	4223.3
	1000, 00, 100; F_2	0011	4319.2	4319.2
	1000, 00, 100; E	0011	4322.1	4322.2
	1000, 00, 100; F_1	0011	4322.6	4322.7
	1000, 00, 100; A_1	0011	4322.7	4322.7
	0000, 20, 100; F_2	0201	4348.8	4349.0
	0000, 11, 100; F_1	0201	4363.3	4364.0
	0000, 20, 100; F_2	0201	4379.1	4378.5
	1000, 10, 000; E	1100	4446.4	4447.2
	1000, 10, 000; F_1	0110	4537.6	4538.9
	1000, 10, 000; F_2	0110	4543.8	4542.4
	0000, 30, 000; E	0300	4592.0	4592.5
	0000, 30, 000; A_2	0300	4595.3	4595.3
	0000, 30, 000; A_1	0300	4595.6	4595.3
2	2000, 00, 000; A_1	2000	5790	5790.7
	1100, 00, 000; A_1	0020	5968.1	5967.1
	1100, 00, 000; E	0020	6043.8	6040.3
	1100, 00, 000; F_2	0020	6004.6	6007.9

[a] Calculated values and experimental data are from Ref. [103], where references to original experimental papers are found.

The derivation of an equivalent normal mode model for XH_4 molecules is easier than for XH_3 molecules because the bending part of the Hamiltonian is of the same form as the normal mode Hamiltonian. For example, the bending vibrational parameters ω_2, ω_4, x_{22}, x_{44}, x_{24}, G_{22}, G_{44}, T_{44}, T_{24}, $K_{24;\,44}$, $K_{22;\,44A_1}$, and $K_{22;\,44E}$ can be interpreted to be equal to the corresponding normal mode vibrational parameters. It is necessary to apply the creation and annihilation operator technique only to the stretching operator [25,26,76]. The following relations are obtained:

$$\omega_1 = \omega_r' + 3\lambda_r' \qquad \omega_3 = \omega_r' - \lambda_r'$$

$$x_{11} = \tfrac{1}{4}x_{13} = \tfrac{5}{9}x_{33} = \tfrac{1}{4}x_{rr}$$

$$G_{33} = 3T_{33} = -\tfrac{3}{20}x_{rr}$$

$$K_{11;\,33} = \tfrac{1}{4}K_{13;\,33} = x_{rr} \tag{3.75}$$

$$x_{12} = x_{23} = x_{r2} \qquad x_{14} = x_{34} = x_{r4}$$

$$\phi_{122} = 8\sqrt{2}d_{r22} \qquad \phi_{144} = 8\sqrt{2}d_{r44}$$

$$\phi_{344} = 4\sqrt{2}d_{r44}' \qquad \phi_{234} = -8\sqrt{2}d_{r24}$$

For the Fermi resonance coefficients ϕ_{stu}, analogous nomenclature and symmetry relations apply as for internal coordinate F_{ijk} force constants [71]. The calculated normal coordinate parameters using the relations given above are given in Table XX together with values from Refs. [71] and [75].

D.   Extensions to Other Molecular Species

A large number of papers have been published on the analysis of overtone spectra of linear molecules (e.g., HCN, HCP, HCCH, HCCD, DCCD, HCCX, and DCCX with X = F, Cl, Br, and I) [107]. The results have been analyzed in terms of normal mode vibrational models, including Darling–Dennison resonance terms both between stretching vibrational states and between stretching and bending vibrations. In some cases Fermi resonances between stretching and bending vibrations have been found to be important. No general internal-coordinate-based models which include both stretching and bending vibrations appear to exist.

In non linear molecules, Fermi resonances have been incorporated into local mode models in an empirical way, where the bending vibrations (or other low-wavenumber fundamentals) have been treated as normal modes. Mixed basis sets have been used with local mode basis functions for the stretches and normal mode basis sets for the bends. Equivalently, the same eigenvalues are obtained using a full normal mode model with x–K relations and their appropriate extensions. These models are similar to the one

TABLE XX
Calculated Vibrational Normal Mode Parameters (cm^{-1}) for $^{12}CH_4$ [a]

Parameter	Fermi	Anharmonic force field	Ab initio
ω_1	3030.2	3025.5	3036.2
ω_2	1565.2	1582.7	1570.4
ω_3	3168.0	3156.8	3157.1
ω_4	1365.9	1367.4	1314.0
x_{11}	−14.61	−11.00	−12.48
x_{12}	−10.90	−16.33	−18.49
x_{13}	−58.44	−50.32	−52.49
x_{14}	−15.68	6.12	−0.71
x_{22}	−1.98	−2.98	−0.30
x_{23}	−10.90	−10.69	−12.85
x_{24}	−3.79	−12.09	−5.21
x_{33}	−26.30	−29.18	−28.34
x_{34}	−15.68	−3.98	−8.24
x_{44}	−2.31	−15.63	−5.85
G_{22}	1.33	0.92	−0.43
G_{33}	8.77	12.13	11.54
G_{34}	4.03	−1.25	−0.74
G_{44}	3.23	9.27	5.02
T_{23}	0	0.41	0.41
T_{24}	0.71	−0.18	−0.91
T_{33}	2.92	3.40	3.25
T_{34}	0.38	0.20	0.18
T_{44}	0.14	2.00	0.54
S_{34}	−0.16	−0.11	0.02
$K_{11;33}$	−58.44		
$K_{13;33}$	−233.74		
$K_{22;44A_1}$	0		
$K_{22;44E}$	0		
$K_{24;44}$	0		
ϕ_{122}	67	42	
ϕ_{144}	169	229	
ϕ_{344}	−160	−153	
ϕ_{234}	−75	−147	

[a] Spectroscopic parameters are from Refs. [103] (Fermi), [71] (anharmonic force field), and 75 (ab initio).

used for methane described in this chapter. For a more thorough review of these models see Ref. [108] (see also Ref. [109] for another perspective and for additional references). These kinds of approaches have been applied with success to cyclobutene [110], $CHDCl_2$ [60], CHD_2Cl [60], C_2H_4 and its various isotopic species [111], CH_3Cl [112], CH_3Br [113], CH_3I [114], CH_3CD_3 [115], and mono- and fully hydrogenated ethanes, propanes, and butanes [116]. Although these models have been successful in assigning experimental observations, some of the molecular parameters, such as the Fermi resonance coefficients, have not been given physical interpretations as functions of the Born–Oppenheimer potential energy surface of the molecule in question. It will remain a task of the future to produce theoretical expressions which would directly link these parameters to Born–Oppenheimer surfaces. In this context, the work on some alkanes (n-butane, n-pentane, n-heptane, n-$C_{36}H_{74}$), cyclohexene-3,3-6,6-d_4, monohydrogenated cyclopentene, and monohydrogenated nitromethane should also be mentioned [117]. In these studies the **g** matric elements are expanded as Taylor series in order to model Fermi resonances in single CH groups or in methylenic CH_2 groups. These approaches are similar to those used for H_2X-type molecules and for CH_2Cl_2 [12,80,83,118,119].

IV. GENERAL VIBRATIONAL HAMILTONIANS AND THE VARIATIONAL METHOD

A. Molecular Hamiltonians

The simple models presented so far used to understand highly excited vibrational states can be obtained from more general theories where an exact vibration–rotation kinetic energy is expressed in terms of a suitable coordinate system. Two possible sets of coordinate systems—curvilinear internal valence and rectilinear normal coordinates—have already been discussed. There are also some other choices which are used with vibrational Hamiltonians. In addition to the above-mentioned coordinates, a recent review by Bacic and Light [120] discusses and gives references to close coupled coordinates (atom–diatom, Jacobi, or scattering coordinates), hyperspherical coordinates, and Radau coordinates. However, as the present review deals with molecular vibrations of semirigid molecules, the internal valence and normal coordinates provide physically the most suitable coordinate systems.

In this section, the vibrational problem is treated variationally. The rotational motion as dealt with in Section V is included by using perturbation theory. The main emphasis in this section is on vibrational motion. The classical Cartesian Hamiltonian for molecular vibrations in polyatomic molecules is expressed as

$$H = T + V = \sum_{\substack{i = 1, \ldots, N \\ \xi = x, y, z}} \frac{1}{2} m_i \dot{x}_{i\xi}^2 + V(\mathbf{x}) = \sum_{\substack{i = 1, \ldots, N \\ \xi = x, y, z}} \frac{1}{2m_i} p_{x_{i\xi}}^2 + V(\mathbf{x}) \quad (4.1)$$

where $x_{i\xi}$ is the Cartesian coordinate of the ith atom, $p_{x_{i\xi}} = \partial L/\partial \dot{x}_{i\xi} = \partial(T - V)/\partial \dot{x}_{i\xi} = \partial T/\partial \dot{x}_{i\xi}$ is the canonical momentum of $x_{i\xi}$, and N is the number of atoms. Using the chain rule for partial differentials, the momentum may be expressed as

$$p_{x_{i\xi}} = \frac{\partial L}{\partial \dot{x}_{i\xi}} = \sum_k \frac{\partial L}{\partial \dot{t}_k} \frac{\partial \dot{t}_k}{\partial \dot{x}_{i\xi}} = \sum_k \frac{\partial L}{\partial \dot{t}_k} \frac{\partial t_k}{\partial x_{i\xi}} = \sum_k p_k \frac{\partial t_k}{\partial x_{i\xi}} \tag{4.2}$$

where the summation is over coordinates which consist of three translational, three rotational, and a set of internal coordinates of the molecule in question. This equation is also valid in cases where redundancies exist between internal coordinates. This follows from the observation that total differentials can be expressed either in terms of independent coordinates or in terms of dependent ones [121]. By inserting Eq. (4.2) into Eq. (4.1), the Hamilton function is obtained as

$$H = \frac{1}{2} \sum_{k, k'} p_k g^{(kk')} p_{k'} + V(\mathbf{t}) \tag{4.3}$$

The contravariant matrix tensor $g^{(kk')}$ is defined as

$$g^{(kk')} = \sum_{i\xi} \frac{1}{m_i} \frac{\partial t_k}{\partial x_{i\xi}} \frac{\partial t_{k'}}{\partial x_{i\xi}} \tag{4.4}$$

The next task is to show how to obtain the quantum-mechanical Hamiltonian from the classical counterpart. The rule is to retain the order of the terms in the kinetic energy part as given in Eq. (4.3) and to make the substitutions [122–124]

$$p_k \Rightarrow - i\hbar g^{-1/4} \frac{\partial}{\partial t_k} g^{1/4} \quad \text{(left)}$$

$$p_{k'} \Rightarrow - i\hbar g^{1/4} \frac{\partial}{\partial t_{k'}} g^{-1/4} \quad \text{(right)} \tag{4.5}$$

The quantum-mechanical Hamiltonian becomes

$$H\left(\mathbf{t}, \frac{\partial}{\partial \mathbf{t}}\right) = - \frac{\hbar^2}{2} \sum_{kk'} g^{-1/4} \frac{\partial}{\partial t_k} g^{1/4} g^{(kk')}(\mathbf{t}) g^{1/4} \frac{\partial}{\partial t_{k'}} g^{-1/4} + V(\mathbf{t}) \tag{4.6}$$

where g denotes the determinant

$$
g = \begin{vmatrix}
g_{11} & g_{12} & \cdots & g_{1n} \\
g_{21} & g_{22} & \cdots & g_{2n} \\
\vdots & \vdots & \ddots & \vdots \\
g_{n1} & g_{n2} & \cdots & g_{nn}
\end{vmatrix}
\tag{4.7}
$$

and its square root is called the functional determinant or Jacobian of the coordinates $\mathbf{x}$ with respect to coordinates $\mathbf{t}$. Here, n is the number of coordinates $\mathbf{t}$ and the covariant metrical tensor element $g_{kk'}$ is defined as

$$
g_{kk'} = \sum_{i\xi} m_i \frac{\partial x_{i\xi}}{\partial t_k} \frac{\partial x_{i\xi}}{\partial t_{k'}}
\tag{4.8}
$$

The matrices obtained from $g_{kk'}$ and $g^{(kk')}$ elements are their inverses.

How can the substitutions given in Eqs. (4.5) be justified? The answer to this is found by starting from the quantum-mechanical counterpart of the Cartesian Hamiltonian on the right-hand side of Eq. (4.1). Using the Cartesian momenta $p_{x_{j\xi}} = -i\hbar \partial/\partial x_{j\xi}$ and the Hermitian property of the momentum operator, the matrix element of the kinetic energy becomes

$$
\langle a | T | b \rangle = \frac{\hbar^2}{2} \int dV \sum_{i\xi} \frac{1}{m_i} \frac{\partial \psi_{xa}}{\partial x_{i\xi}} \frac{\partial \psi_{xb}}{\partial x_{i\xi}}
\tag{4.9}
$$

where the Cartesian volume element is given by the usual expression $dV = dx_{1x}\, dx_{1y}\, dx_{1z}\, dx_{2x}\, dx_{2y}\, dx_{2z} \cdots$. Using the chain rule

$$
\frac{\partial \psi_{xi}}{\partial x_{j\xi}} = \sum_k \frac{\partial \psi_{xi}}{\partial t_k} \frac{\partial t_k}{\partial x_{j\xi}} \qquad (i = a, b)
$$

Eq. (4.4), and the volume element $dV = g(\mathbf{t})^{1/2} dt_1\, dt_2\, dt_3 \cdots dt_n$, the kinetic energy matrix element becomes

$$
\langle a | T | b \rangle = \tfrac{1}{2}\hbar^2 \int dt_1\, dt_2\, dt_3 \cdots dt_n \sum_{kk'} \frac{\partial \psi_{xa}}{\partial t_k} g^{1/4} g^{(kk')} g^{1/4} \frac{\partial \psi_{xb}}{\partial t_{k'}}
$$

$$
= -\tfrac{1}{2}\hbar^2 \int dt_1\, dt_2\, dt_3 \cdots dt_n
$$

$$
\times \sum_{kk'} \psi_{xa} \frac{\partial}{\partial t_k} g^{1/4} g^{(kk')} g^{1/4} \frac{\partial}{\partial t_{k'}} \psi_{xb}
\tag{4.10}
$$

The last equality is obtained by partial integration. According to Podolsky, the normalization condition is [123]

$$\int dt_1 \, dt_2 \, \cdots \, dt_n \, g^{1/2} \psi_x^*(\mathbf{t}) \psi_x(\mathbf{t}) = \int dt_1 \, dt_2 \, \cdots \, dt_n \, \psi_t^*(\mathbf{t}) \psi_t(\mathbf{t}) = 1 \quad (4.11)$$

Thus,

$$\psi_x(\mathbf{t}) = g^{-1/4} \psi_t(\mathbf{t}) \quad (4.12)$$

and Eq. (4.10) becomes

$$\langle a \, | \, T \, | \, b \rangle = -\tfrac{1}{2} \hbar^2 \int dt_1 \, dt_2 \, dt_3 \, \cdots \, dt_n \, \psi_{ta}$$

$$\times \left(\sum_{kk'} g^{-1/4} \frac{\partial}{\partial t_k} g^{1/4} g^{(kk')} g^{1/4} \frac{\partial}{\partial t_{k'}} g^{-1/4} \right) \psi_{tb} \quad (4.13)$$

which justifies Eqs. (4.5) and (4.6).

By rearranging the operators, Eq. (4.6) can also be written in an equivalent form [81,82,122]

$$H = -\frac{\hbar^2}{2} \sum_{kk'} \left(\frac{\partial}{\partial t_k} + \frac{1}{4} \frac{\partial \ln g}{\partial t_k} \right) g^{(kk')} \left(\frac{\partial}{\partial t_{k'}} - \frac{1}{4} \frac{\partial \ln g}{\partial t_{k'}} \right) + V(\mathbf{t})$$

$$= -\frac{\hbar^2}{2} \sum_{kk'} \frac{\partial}{\partial t_k} g^{(kk')} \frac{\partial}{\partial t_{k'}} + V'(\mathbf{t}) + V(\mathbf{t})$$

$$= \sum_{kk'} p_k g^{(kk')} p_{k'} + V'(\mathbf{t}) + V(\mathbf{t}) \quad (4.14)$$

where the pseudopotential term is

$$V'(\mathbf{t}) = \frac{\hbar^2}{8} \sum_{kk'} \left(\frac{\partial g^{(kk')}}{\partial t_k} \frac{\partial \ln g}{\partial t_{k'}} + g^{(kk')} \frac{\partial^2 \ln g}{\partial t_k \, \partial t_{k'}} + \frac{1}{4} g^{(kk')} \frac{\partial \ln g}{\partial t_k} \frac{\partial \ln g}{\partial t_{k'}} \right) \quad (4.15)$$

and the momenta p_k are defined as $p_k = -i\hbar(\partial/\partial t_k)$.

The translational motion can be separated from vibrations and rotations. This means that if the translational coordinates are defined as

$$t_\xi = \sum_{i=1}^{N} m_i x_{i\xi} \Big/ \sum_{i=1}^{N} m_i \quad (4.16)$$

($\xi = x$, y, z) both in the kinetic and potential energy part of the Hamiltonian, cross terms between the translational and the rovibrational parts disappear. The translational Hamiltonian can be disregarded for the present purposes. On the other hand, rotation and vibration cannot be separated. In the pure vibrational problem (i.e., the total angular momentum $\hat{\mathbf{J}} = 0$), the momenta conjugate to rotational coordinates can be set equal to zero and the summations in Eq. (4.14), for example, are only over vibrational coordinates. In this case, the Hamiltonian (the pure vibrational Hamiltonian) does not depend on how molecular axes are tight to the molecule [81]. In the matrix notation, the quantum-mechanical vibrational Hamiltonian can be written as

$$H = \tfrac{1}{2}\mathbf{p}_r^T \mathbf{g(r)} \mathbf{p}_r + V'(\mathbf{r}) + V(\mathbf{r}) \tag{4.17}$$

where $p_{r_j} = -i\hbar(\partial/\partial r_j)$ is the momentum conjugate to the internal valence coordinate r_j and the eigenfunctions of H are normalized so that $\int \psi^*(\mathbf{r})\psi(\mathbf{r})\, dr_1\, dr_2 \cdots = 1$. The matrix $\mathbf{g(r)}$ consists of the contravariant tensor elements given in Eq. (4.4) and should not be confused with g defined in Eq. (4.7). By defining Wilson's $\bar{s}_{j(i)}$ vector ($\bar{s}$ vector of the internal coordinate r_j on the atom i) as

$$\bar{s}_{j(i)} = \overline{\nabla}_i r_j = \left(\frac{\partial r_j}{\partial x_{ix}}, \frac{\partial r_j}{\partial x_{iy}}, \frac{\partial r_j}{\partial x_{iz}} \right)$$

the $\mathbf{g}$ matrix element $g^{(jk)}$ as given in Eq. (4.4) is expressed as

$$g^{(jk)} = \sum_{i=1}^{N} \frac{1}{m_i} (\overline{\nabla}_i r_j \cdot \overline{\nabla}_i r_k) = \sum_{i=1}^{N} \frac{1}{m_i} (\bar{s}_{j(i)} \cdot \bar{s}_{k(i)}) \tag{4.18}$$

This provides a practical way to calculate the elements. In Ref. [17] many of the most common matrix elements are tabulated. Note that in earlier sections the superscripts of the $\mathbf{g}$ matrix elements have been given as subscripts. These should not be confused with the covariant tensor elements defined in Eq. (4.8).

The pseudopotential term $V'(\mathbf{r})$ does not contain derivative operators [i.e., being more precise after taking the derivatives in Eq. (4.15)]. It could be regarded as part of the potential energy surface, but its mass dependence would make the potential energy function effective. Its effect on overtone transitions in water has been found to be small [83]. It is a reasonable approximation to omit it except in the most accurate calculations. If $V'(\mathbf{r})$ is set equal to zero, then formally Eq. (4.17) looks the same as the classical analogue. However, it must be borne in mind that the order of the terms in the quantum-mechanical Hamiltonian is relevant, unlike in the classical counterpart.

An alternative way to derive quantum-mechanical vibration–rotation Hamiltonians for polyatomic molecules has often been employed [125–127]. It starts from the Cartesian representation of the kinetic energy operator and uses the chain rule for the derivatives to obtain Hamiltonians in terms of internal coordinates and Euler angles. The partial derivatives with respect to Euler angles are expressed in terms of the molecule fixed Cartesian components $\hat{J}_x$, $\hat{J}_y$, and $\hat{J}_z$ of the total angular momentum $\hat{\mathbf{J}}$. In this method, Eckart conditions [128] are not used to tie the molecule fixed axes to the molecule. Interactions between vibration and rotation are not minimized. This is not important if the eigenvalues are obtained variationally with a large basis set. The method described leads to complicated intermediate expressions, although the final Hamiltonians are simpler. It has been necessary to use computer algebra programs to perform necessary mathematics. Hamiltonians have been derived for triatomic molecules, acetylene, formaldehyde, and other species [126]. This complicated algebra is avoided by using infinitesimal rotational coordinates ε_x, ε_y, and ε_z instead of Euler angles [129]. For these coordinates $\partial/\partial\varepsilon_x = (i/\hbar)\hat{J}_x$. The simple $\bar{s}$ vector method used by Wilson [17] to derive vibrational Hamiltonians [see Eq. (4.18)] can be used both for the infinitesimal rotational coordinates and for the vibrational coordinates to derive exact vibration–rotation Hamiltonians for polyatomic molecules. This approach is simple and fast and calculations can be performed even by hand. References [129] also give transformations to be used when Hamiltonians are derived from Hamiltonians of another coordinate system. It might be advisable to minimize Coriolis interactions if the aim is to model rotational motion with perturbation theory in the internal coordinate representation [81,86,130].

There exists a rather different kind of internal coordinate Hamiltonian model called MORBID for triatomic molecules proposed by Jensen [131]. In this approach, interactions between vibrations and rotations are minimized via Eckart conditions [128]. The kinetic energy used is taken from a model [132] where the bend is treated as a large-amplitude motion. The MORBID Hamiltonian is expanded as a power series in the Morse variable y_i, and the eigenvalues are computed variationally. Numerically integrated basis functions (obtained with the Numerov–Cooley method [133]) are used for the bend, and Morse oscillator eigenfunctions are used for the stretches. This method is particularly successful in treating floppy triatomic molecules such as CH_2, where the traditional variational methods require a large number of basis functions to converge. This model can be applied more easily to high J levels than the conventional variational methods.

Exact vibration–rotation Hamiltonians in terms of normal coordinates q_s and their conjugate momenta p_s for nonlinear molecules have been shown to be of the form [134]

$$\frac{H}{hc_0} = \frac{1}{2} \sum_{\alpha\beta} \hbar^2 \mu_{\alpha\beta}(\hat{J}_\alpha - \pi_\alpha)(\hat{J}_\beta - \pi_\beta) + \frac{1}{2} \sum_s \omega_s p_s^2 + U + V(\mathbf{q}) \quad (4.19)$$

where $\hat{J}_\alpha$ ($\alpha = x, y, z$) is the molecule fixed Cartesian component of the total angular momentum $\hat{\mathbf{J}}$ and π_α is the vibrational angular momentum defined as

$$\pi_\alpha = \sum_{ss'} \zeta_{ss'}^{(\alpha)} q_s p_{s'} \left(\frac{\omega_{s'}}{\omega_s}\right)^{1/2} \quad (4.20)$$

where $\zeta_{ss'}^{(\alpha)}$ is the Coriolis coupling coefficient between the normal modes s and s' through rotation about the α axis. The tensor element $\mu_{\alpha\beta}$ is the reciprocal of $I'_{\alpha\beta}$, which in turn is given by

$$I'_{\alpha\beta} = I_{\alpha\beta} - \sum_{stu} \zeta_{su}^{(\alpha)} \zeta_{tu}^{(\beta)} q_s q_t (\gamma_s \gamma_t)^{-1/2} \quad (4.21)$$

where $\gamma_i = 2\pi c_0 \omega_i / \hbar$. Here, $I_{\alpha\beta}$ is the element of the instantaneous moment of inertia. The usual potential energy surface $V(\mathbf{q})$ is expressed in terms of dimensionless normal coordinates, and $U = -\frac{1}{8}\hbar^2 \sum_\alpha \mu_{\alpha\alpha}$ is a mass-dependent kinetic energy term which does not involve momentum operators. Its effect is usually small. This form of the vibration–rotation Hamiltonian is suitable for problems which can be modeled by the small-amplitude vibrational approximation. Note that, if desired, the general vibration–rotation Hamiltonian in the curvilinear internal coordinate representation can also be expressed formally in a similar type of form [81].

In the case of zero total angular momentum, the vibrational normal coordinate Hamiltonian takes the simple form

$$\frac{H}{hc_0} = \frac{1}{2} \sum_s \omega_s p_s^2 + V(\mathbf{q}) \quad (4.22)$$

if U and the term involving vibrational angular momentum are neglected. The simple form of the kinetic energy operator in the normal coordinate representation can be immediately anticipated from both Eqs. (4.19) and (4.22). This is an advantage when compared with the internal-coordinate-based models. However, the physical interpretation of $V(\mathbf{q})$ is more complicated. Ultimately one is interested in $V(\mathbf{r})$, that is, in the potential energy surface expressed in terms of internal coordinates, which according to the Born–Oppenheimer approximation is isotope independent.

B.   The Eigenvalue Problem

There are various ways to obtain eigenvalues of vibration–rotation Hamiltonians. A common choice is to perform variational calculations where the eigenfunctions of the original Hamiltonian are expressed as linear combinations of suitable basis functions such as one-, two-, or three-dimensional harmonic oscillator eigenfunctions [17,18,105,135,136], Morse oscillator eigenfunctions [3,14,15,80], or tridiagonal Morse oscillator eigenfunctions [137,138] for the stretching degrees of freedom, Legendre or Jacobi polynomials for the bending degrees of freedom [125,139,140], and rigid symmetric rotor eigenfunctions for the rotation [141,142]. The choice of the basis sets depends on the particular problem in question. Generally speaking, a good convergence with the smallest possible number of basis functions is desired. Therefore, for example, the Morse oscillator eigenfunctions are usually better suited for stretching vibrations than harmonic oscillator eigenfunctions are. On the other hand, the Morse oscillator basis set for states near the bond dissociation cannot be used because the neglected continuum may make a significant contribution. Harmonic oscillator and tridiagonal Morse basis sets are free from this drawback. See, for example, Ref. [125] for additional details on customary variational methods in molecular rovibrational problems.

There is another kind of variational method where the eigenfunctions are expanded in a basis of localized functions of coordinates. The method is called the distributed Gaussian basis method where Gaussian functions localized around coordinate points are employed [120,143]. In the discrete variable representation (DVR) method, the localized functions are approximate eigenfunctions of coordinate operators [120,143,144]. These localized functions can be obtained by diagonalization of coordinate matrices in primitive basis set representations. If the primitive basis set $\{\phi_i(x)\}$ is chosen to consist of orthogonal polynomials $\{f_i(x)\}$ times the square root of a positive weight function $w(z)$, that is, $\phi_i(z) = h_i^{-1/2}w(z)^{1/2}f_i(z)$, where z is a function of x and h_i is a normalization factor, then there exists an orthogonal transformation between the representations of DVR basis functions at N Gaussian quadrature points $\{x_\alpha\}$ and the primitive basis (FBR = finite basis representation) [144]. If the matrix elements of the potential energy operator are calculated with the appropriate N-point Gaussian quadrature, then for a one-dimensional case [144]

$$(\mathbf{T}^T\mathbf{V}^{(\mathbf{FBR})}\mathbf{T})_{\alpha\beta} = (\mathbf{V}^{(\mathbf{DVR})})_{\alpha\beta} = V(x_\alpha)\delta_{\alpha\beta} \tag{4.23}$$

where $\mathbf{V}^{(\mathbf{FBR})}$ and $\mathbf{V}^{(\mathbf{DVR})}$ are potential energy matrices in the primitive and discrete variable representations, respectively. The $\mathbf{T}$ matrix elements are given by

$$T_{i\alpha} = \sqrt{\frac{W_\alpha}{h_i}}\, \phi_i(x_\alpha) \tag{4.24}$$

where W_α and x_α are Gaussian weights and integration points, respectively. Note that in the discrete variable representation the potential energy matrix is in the diagonal form. As an example, a one-dimensional anharmonic oscillator Hamiltonian expressed in terms of the dimensionless normal coordinate q and its conjugate momentum operator p could be considered. Using the standard notation (see Section II), the Hamiltonian is

$$
\begin{aligned}
\frac{H}{hc_0} &= \tfrac{1}{2}\omega(p^2 + q^2) + \tfrac{1}{6}\phi_3\, q^3 + \tfrac{1}{24}\phi_4\, q^4 \\
&= \tfrac{1}{2}\omega(p^2 + q^2) + V_{\mathrm{anh}}(q)
\end{aligned} \tag{4.25}
$$

The Hamiltonian matrix element in the DVR basis is then

$$(\mathbf{H}^{(\mathbf{DVR})})_{\alpha\beta} = (\mathbf{T}^T\mathbf{E}\mathbf{T})_{\alpha\beta} + V_{\mathrm{anh}}(q_\alpha)\delta_{\alpha\beta} \tag{4.26}$$

where the $\mathbf{E}$ matrix is given in the harmonic oscillator basis (primitive basis) and it contains nonzero matrix elements $E_{nn} = \omega(n + \tfrac{1}{2})$ only on the diagonal. The $\mathbf{T}$ matrix elements can be calculated with the help of Eq. (4.24) using appropriate Gauss–Hermite integration points and weights. Alternatively, the $\mathbf{T}$ matrix elements could be obtained as eigenvector elements from a calculation where the coordinate operator matrix expressed in the harmonic oscillator basis (primitive basis) is diagonalized. The latter method is particularly useful for cases where orthogonal polynomials with common weight factors are nonexisting. Morse oscillator eigenfunctions provide an example of this kind of situation [138].

The Hamiltonian given in Eq. (4.17) contains a kinetic energy operator which requires more attention when the DVR method is applied to find eigenvalues. This problem has been tackled in Ref. [138]. The water molecule may be chosen as an example. In Ref. [138] it is observed that all kinetic energy terms including the pseudopotential term $V'(\mathbf{r})$ are such that they are of the form $f_1(r_1)f_2(r_2)f_3(\theta)$, where, for example, $f_1(r_1)$ is a function of r_1 and p_{r_1}. In the case of the term

$$p_\theta \frac{\cos(\theta + \alpha_e)}{m_O(r_1 + R_e)(r_2 + R_e)} p_\theta = \frac{1}{r_1 + R_e} \frac{1}{r_2 + R_e} \left(\frac{1}{m_O} p_\theta \cos(\theta + \alpha_e) p_\theta \right)$$

$$= f_1(r_1) f_2(r_2) f_3(\theta)$$

the matrix element in the standard-type basis set representation is

$$h^{(FBR)}_{ijk, i'j'k'} = \langle \phi_i(r_1) | f_1(r_1) | \phi_{i'}(r_1) \rangle \langle \varphi_j(r_2) | f_2(r_2) | \varphi_{j'}(r_2) \rangle$$

$$\times \langle \chi_k(\theta) | f_3(\theta) | \chi_{k'}(\theta) \rangle$$

$$= (\mathbf{f_1})_{ii'} (\mathbf{f_2})_{jj'} (\mathbf{f_3})_{kk'} \qquad (4.27)$$

where $\phi_i(r_1)$, $\varphi_j(r_2)$, and $\chi_k(\theta)$ are basis functions for the r_1, r_2, and θ degrees of freedom, respectively. Here, R_e and α_e are the equilibrium bond length and valence angle, respectively. The corresponding matrix is obtained in the DVR representation by employing the orthogonality property of the $\mathbf{T}$ matrix as

$$h^{DVR}_{\alpha\beta\gamma, \alpha'\beta'\gamma'} = (\mathbf{T}^T_{r_1} \mathbf{f_1} \mathbf{T}_{r_1})_{\alpha\alpha'} (\mathbf{T}^T_{r_2} \mathbf{f_2} \mathbf{T}_{r_2})_{\beta\beta'} (\mathbf{T}^T_\theta \mathbf{f_3} \mathbf{T}_\theta)_{\gamma\gamma'} \qquad (4.28)$$

where

$$(\mathbf{T}^T_{r_1} \mathbf{f_1} \mathbf{T}_{r_1})_{\alpha\alpha'} = \left(\frac{1}{r_{1\alpha}} \right) \delta_{\alpha\alpha'}$$

$$(\mathbf{T}^T_{r_2} \mathbf{f_2} \mathbf{T}_{r_2})_{\beta\beta'} = \left(\frac{1}{r_{2\beta}} \right) \delta_{\beta\beta'} \qquad (4.29)$$

$$(\mathbf{T}^T_\theta \mathbf{f_3} \mathbf{T}_\theta)_{\gamma\gamma'} = \sum_{\gamma''} (\mathbf{T}^T_\theta \mathbf{p}_\theta \mathbf{T}_\theta)_{\gamma\gamma''} \left(\frac{1}{m_O} \cos(\theta_{\gamma''} + \alpha_e) \right) (\mathbf{T}^T_\theta \mathbf{p}_\theta \mathbf{T}_\theta)_{\gamma''\gamma'}$$

It is observed that although the calculation of the potential energy matrix is straightforward in the DVR representation, the kinetic energy requires setting up momentum operators in the primitive basis and making orthogonality transformations to the DVR representation. Due to the approximations made, it may happen that this procedure does not lead to Hermitian Hamiltonian matrices (see Ref. [145] for more details). Finally, note that although the DVR technique provides an accurate way to obtain rovibrational eigenvalues, it is strictly speaking not a variational method because it does not give an upper bound for the energy levels.

TABLE XXI
Some Morse Oscillator Matrix Elements[a]

$$\langle v_r | (\tfrac{1}{2} g_{rr} p_r^2 + D_e y^2) | v_r \rangle = \omega_r (v_r + \tfrac{1}{2}) + x_{rr}(v_r + \tfrac{1}{2})^2$$

$$(v_r + j | p_r | v_r) = \tfrac{1}{2} i\hbar (-1)^{j+1} a_r$$
$$\times \left\{ \frac{(k - 2v_r - 1)(k - 2v_r - 2j - 1)(v_r + 1)(v_r + 2) \cdots (v_r + j)}{(k - v_r - 1)(k - v_r - 2) \cdots (k - v_r - j)} \right\}^{1/2}$$

$$\langle v_r | p_r | v_r \rangle = 0$$

$$\langle v_r + j | y | v_r \rangle = -i \left(\frac{g_{rr}}{2D_e} \right)^{1/2} \langle v_r + j | p_r | v_r \rangle$$

$$\langle v_r | y | v_r \rangle = \langle v_r | y^2 | v_r \rangle = \frac{2v_r + 1}{k}$$

$$\langle v_r + j | y^2 | v_r \rangle = \left[\frac{j(2v_r + j + 1)}{k} - (j - 1) \right] \langle v_r + j | y | v_r \rangle$$

[a] Expression for ω_r and x_{rr} are found in Section II. Integer $j \geq 1$; $k = -\omega_r / x_{rr} = (2/\hbar a_r)(2D_e/g_{rr})^{1/2}$. Matrix elements of y^3 and y^4 can be found in Ref. [80].

There exist examples of the use of variational and DVR methods to solve eigenvalues of both vibrational and rovibrational Hamiltonians of three [138,145,146] and four [147] atomic molecules on the ground electronic state using Hamiltonians in the internal coordinate or some other representation. The MORBID approach has also been applied to a number of triatomic problem [131,148]. The full normal-coordinate-based Hamiltonians are nowadays less frequently used in variational (or DVR) calculations, although one of the first attempts in this respect is the work by Whitehead and Handy on water [149].

C. Local Mode and Fermi Resonance Models and Variational Calculations

The simple local mode Hamiltonian for stretching vibrations in well-bent XY_2 molecules given in Eq. (2.1) is obtained from the exact internal coordinate vibrational Hamiltonian in Eq. (4.17) by constraining the bending displacement coordinate to its equilibrium value. This choice ensures that the momentum operator conjugate to bending coordinate disappears. The mass-dependent pseudopotential term V' does not disappear, but as mentioned earlier, its effect is small and it is neglected. This approximation is not so good in the case of highly excited bending states as V' becomes singular at the linear configuration. The eigenvalues of the stretching vibrational Hamiltonian can be obtained variationally using symmetrized Morse oscillator basis functions as given in Eqs. (2.7). When the basis functions are

TABLE XXII

Variational Calculations in the Stretching-Only Models for $^{28}SiH_4$, $^{74}GeH_4$, and $^{116}SnH_4$ [a]

$v_{r_1}v_{r_2}v_{r_3}v_{r_4}$	v_1v_3	$^{28}SiH_4$ (cm^{-1})		$^{74}GeH_4$ (cm^{-1})		$^{116}SnH_4$ (cm^{-1})	
			Observed		Observed		Observed
		v_{obs}	Calculated	v_{obs}	Calculated	v_{obs}	Calculated
$1000A_1$	10	2186.87	0.27	2110.71	0.13	1908.10	-0.05
$1000F_2$	01	2189.19	0.38	2111.14	0.46	1905.83	-0.17
$2000A_1$	20	4308.38	-0.70	4153.55	-0.26	3752.75	-0.22
$2000F_2$	11	4309.35	0.18	4153.83	0.02	3752.66	-0.28
$1100A_1$	02	4374.56	-0.5				
$1100F_2$	02	4378.40	1.11				
$1100E$	02	4380.28	1.8				
$3000A_1$	30	6361.98	-0.75	6128.58	-0.9	5539.07	-0.21
$3000F_2$	21	6362.08	-0.65	6128.58	-0.9	5539.04	-0.23
$2100A_1$	12	6496.13	-0.53				
$2100F_2$	12	6497.48	-0.07	6263.67	1.3		
$2100E$	12	6500.30	1.86				
$2100F_2$	03	6500.60	0.85				
$2100F_1$	03	6502.88	2.32				
$4000A_1$	40	8347.87	-1.10				
$4000F_2$	31	8347.87	-1.10				
$5000A_1$	50	10267.11	-0.76	9875.78	-2.7		
$5000F_2$	41	10267.11	-0.76	9875.78	-2.7		
$6000A_1$	60	—	—	11647.23	-4.2	10538.18	0.62
$6000F_2$	51	12118.3	-1.12	11647.23	-4.2	10538.16	0.60
$7000A_1$		—	—	13352.66	-4.4	12083.78	0.35
$7000F_2$	61	13905.18	1.54	13352.66	-4.4	12083.78	0.35
$8000A_1$		—	—	—	—	13568.63	-0.56
$8000F_2$	71	15625.4	4.86	15000	4.8	13568.63	-0.56
$9000F_2$	81	17266.6	-3.57	16574	8.2		

[a] Calculated numbers are from Refs. [49] (SiH$_4$), [50] (GeH$_4$), and [155] (SnH$_4$) and experimental data are from Refs. [49] (SiH$_4$), [50] (GeH$_4$) (where references to original experimental work are given), [153] (SiH$_4$), [154] (GeH$_4$), and [155] (SnH$_4$).

chosen to be consistent with the Morse parameters D_e and a_r appearing in Eq. (2.1), the analytic matrix elements given in Table XXI can be used [80]. This choice makes the calculations fast and accurate. The nonlinear least-squares method can be used to optimize the potential energy parameters D_e, a_r, and $f_{rr'}$ [49].

The approach described above can be easily extended to a larger number of coupled Morse oscillators. A series of papers describes these extensions to stretching vibrations in C_2H_2, C_2D_2, and CHCD [46]; AsH$_3$ and SbH$_3$ [150]; CH$_3$D, CHD$_3$, SiH$_3$D, and SiHD$_3$ [52]; CH$_4$, CD$_4$, SiH$_4$, SiD$_4$, GeH$_4$, GeD$_4$, SiF$_4$, and SnH$_4$ [45,48–50,54,151,152]; and SF$_6$, WF$_6$, and

UF_6 [47]. These models are successful in many of the examples given, although at least in methane and its isotopomers, more sophisticated approaches are needed due to Fermi resonances between CH (CD) stretching and bending levels. The calculations on silane (SiH_4), germane (GeH_4), and stannane (SnH_4) provide a convincing example of the success of these simple models where only stretching vibrations have been included. Table XXII gives experimental and calculated stretching vibrational term values for $^{28}SiH_4$, $^{74}GeH_4$, and $^{116}SnH_4$ [49,50,153–155]. The calculated values have been obtained variationally with symmetrized Morse oscillator bases where all basis functions up to the total stretching quantum number 12 have been included. Table XXIII gives potential energy parameters used to obtain the results in Table XXII. An anharmonic force field exists for silane [156]. A comparison between the stretching parts of both surfaces shows a good agreement. In any case, there must be an awareness of the effective nature of the potential energy function in the coupled Morse oscillator model where the Hamiltonian has been averaged over the bending degrees of freedom. This procedure may introduce a mass dependence on the effective potential surface. One way to improve the model is to include bilinear harmonic coupling terms between stretching and bending degrees of freedom. This can be done using second-order perturbation theory [see, e.g., Eqs. (3.3) for XY_2 molecules].

The variational method can also be applied to the internal coordinate local mode–Fermi resonance Hamiltonians discussed in Section III. So far it has only been done in some symmetrical bent triatomic molecules such as H_2O, H_2S, H_2Se, SO_2, and H_2O^+ [12,80,83,157,158]. The same kind of a model has also been applied to CH stretching vibrations and CH_2 scissoring vibrations in CH_2Cl_2 [118]. A somewhat simpler model has been used for the CH stretching and the doubly degenerate CH bending vibrations in CHD_3, CHF_3, $CHCl_3$, $CHBr_3$, and $CH(CF_3)_3$ [159,160]. Table

TABLE XXIII

Stretching Vibrational Potential Energy Parameters for $^{28}SiH_4$, $^{74}GeH_4$, and $^{116}SnH_4$ [a]

| Parameter | SiH_4 | | GeH_4 | SnH_4 |
	Local	Anharmonic force field	Local	Local
f_{rr}, aJ Å^{-2}	2.914	2.948	2.778	2.276
f_{rrr}, aJ Å^{-3}	−12.19	−13.41	−11.76	−9.114
f_{rrrr}, aJ Å^{-4}	39.6	56.4	38.7	28.4
$f_{rr'}$, aJ Å^{-2}	0.033	0.029	0.013	0.008

[a] Force constants are taken from Refs. [49] (Local, SiH_4), [156] (anharmonic force field), [50] (GeH_4), and [155] (SnH_4).

TABLE XXIV

Potential Energy Parameters for Hydrogen Sulfide and Selenide[a]

Parameter	H$_2$S				H$_2$Se			
	Fermi	Anharmonic force field	Ab initio	Morbid	Fermi	Anharmonic force field	Ab initio	Morbid
f_{rr}, aJ Å^{-2}	4.285	4.284	4.299	4.284	3.501	3.507	3.485	3.506
$f_{\theta\theta}$, aJ	0.763	0.758	0.765	0.759	0.700	0.710	0.760	0.705
$f_{rr'}$, aJ Å^{-2}	−0.020	−0.015	−0.014	−0.018	−0.022	−0.024	−0.013	−0.045
$f_{r\theta}$, aJ Å^{-1}	0.121	0.054	0.076	0.108	0.080	0.130	0.075	0.136
f_{rrr}, aJ Å^{-3}	−22.58	−23.4	−22.5	−21.31	−17.50	−16.7	−17.39	−17.52
$f_{\theta\theta\theta}$, aJ	−0.222	−0.1	−0.28	−0.257	−0.161	−0.70	−0.315	−0.148
$f_{rrr'}$, aJ Å^{-3}	0.0	−0.1	−0.02	−0.048	0.0	0.00	−0.019	−0.027
$f_{rr\theta}$, aJ Å^{-2}	0.0	−0.4	−0.04	−0.179	−0.068	−0.3	−0.110	−0.340
$f_{rr'\theta}$, aJ Å^{-2}	−0.207	−0.2	−0.19	0.0	−0.332	0.1	−0.220	−1.611
$f_{r\theta\theta}$, aJ Å^{-1}	−0.267	−0.2	−0.35	−0.435	−0.275	−0.2	−0.355	−1.048
f_{rrrr}, aJ Å^{-4}	104.34	120	103.4	82.48	76.95	63	75.76	76.81
$f_{\theta\theta\theta\theta}$, aJ	−0.770	−0.9	−0.04	0.457	0.0	−0.09	−1.074	−0.467
$f_{rrrr'}$, aJ Å^{-4}	0.0	0.7	−0.12	0.339	0.0	0.3	−0.133	0.672
$f_{rrr'r'}$, aJ Å^{-4}	0.0	0.1	0.23	0.209	0.0	0.3	0.356	0.549
$f_{rrr\theta}$, aJ Å^{-3}	0.0	0.0	−0.01	0.297	0.0	0.0	−0.077	0.755
$f_{rrr'\theta}$, aJ Å^{-3}	0.0	0.0	0.06	0.0	0.0	0.0	0.148	5.369
$f_{rr\theta\theta}$, aJ Å^{-2}	−1.25	−1.6	−0.46	0.035	−0.756	−1.8	−0.646	−3.170
$f_{rr'\theta\theta}$, aJ Å^{-2}	0.0	−0.2	0.49	0.399	0.0	0.0	0.324	4.162
$f_{r\theta\theta\theta}$, aJ Å^{-1}	0.0	0.0	0.53	−0.058	0.0	0.0	0.184	−0.205

[a] Force fields are from Refs. [80] (Fermi), [19] (anharmonic force field), [73,163] (ab initio), [161], and [162] (Morbid).

XXIV contains potential energy parameters obtained for hydrogen sulfide and selenide using the local mode–Fermi resonance model [80]. The parameters were optimized by the nonlinear least-squares method. The data included in the calculation contained vibrational term values and α parameters (see Section V) of H_2X and D_2X isotopomers [80]. Table XXIV also contains a comparison with three other surfaces: MORBID [161,162], ab initio [163], and anharmonic force field surfaces [19]. It is pleasing that these rather different kind of models give similar results.

The advantage in using full variational and DVR methods compared with block diagonal models is the explicit inclusion of interblock couplings. In the block diagonal models, these can be taken into account approximately by perturbation theory. In cases where the block diagonal approach is poor, that is, there are significant couplings between different blocks, this does not lead to a pleasing outcome, and the full variational methods provide a better solution. There is also another case where the variational method possesses advantages: intensity calculations. This is the next topic.

D. Infrared Absorption Intensities

The experimental intensities of overtone and combination bands from an important set of data which can be used to test and develop theoretical models used to describe highly excited vibrational states of polyatomic molecules. A striking feature of the high overtone infrared spectra of near-local-mode molecules is the small amount of observed bands. As an example in Fig. 6, a laser photoacoustic absorption spectrum of HCCCl in the region 9500–14,500 cm^{-1} shows the strength of the CH stretching overtone bands $3v_1$ and $4v_1$ and the simplicity of the spectrum in spite of the high density of the vibrational states, about 126 states per one wavenumber at 12,500 cm^{-1} [164]. High-resolution spectra of the individual HCCCl bands indicate that it is better to speak about, for example, $3v_1$ and $4v_1$ band systems which consist of many bands borrowing intensity from the $3v_1$ and $4v_1$ bands by vibration–rotation interactions. As another example, the historical case of benzene could be mentioned, where the CH stretching vibrational band systems carry most of the intensity. Within these band systems, the lowest members in each overtone stretching vibrational manifold are by far the strongest ones [8,10,165,166]. In $v_{\text{stretch}} = 6$, where v_{stretch} is the total CH stretching quantum number, there are 75 upper states which span the symmetry species E_{1u}. Transitions from the ground vibrational state A_{1g} to these excited upper states are infrared active, but as mentioned, only the transitions to the lowest E_{1u} component carry significant intensity.

The local mode model and the bond dipole approximation offer a natural explanation to the intensity patterns described in benzene (and in HCCCl). The dipole moment function $\mu(\mathbf{R})$ is assumed to be a sum of bond

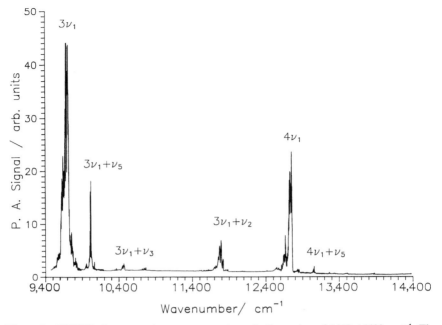

Figure 6. HCCCl photoacoustic overtone spectrum in the region of 9400–14400 cm^{-1}. The normal mode labeling scheme is such that v_1, v_2, v_3, and v_5 are the CH stretch, CC stretch, CCl stretch, and CCCl bend, respectively. (From Ref. [164] with permission.)

dipole contributions. Thus,

$$\mu(\mathbf{R}) = \sum_i \mu_i(R_i)\mathbf{e}_i \qquad (4.30)$$

where R_i is the instantaneous bond length of the ith bond and $\mathbf{e}_i$ is a unit vector along the ith bond. The stretching vibrations of bent triatomic molecules are taken as an example case. To simplify the treatment given here, the ground vibrational state is approximated as a pure local mode product state $|00\rangle = |0\rangle|0\rangle$. On the other hand, the upper state $|f\rangle$ (the final state) is expressed as a linear combination of local mode product states

$$|v_{r_1}v_{r_2}\rangle = |v_{r_1}\rangle|v_{r_2}\rangle$$

that is,

$$|f\rangle = \sum_{v_{r1}v_{r2}} c^{(f)}_{v_{r1}v_{r2}}|v_{r_1}v_{r_2}\rangle \qquad (4.31)$$

The transition moment integral $\mu_{0f} = \langle 0 | \mu(\mathbf{R}) | f \rangle$ becomes [3]

$$\mu_{0f} = \sum_{v_{r1}} c^{(f)}_{v_{r1}0} \langle 0 | \mu_1(R_1) | v_{r1} \rangle + \sum_{v_{r2}} c^{(f)}_{0v_{r2}} \langle 0 | \mu_2(R_2) | v_{r2} \rangle \qquad (4.32)$$

due to orthonormality of the bond oscillator basis functions. Besides the bond dipole function matrix elements, μ_{0f} depends on $c^{(f)}_{v_{r1}0}$ and $c^{(f)}_{0v_{r2}}$, which are coefficients of the bond oscillator functions $|v_{r1}0\rangle$ and $|0v_{r2}\rangle$ in the expansion given in Eq. (4.31). These bond oscillator wave functions become almost the eigenfunctions of the lowest local mode pair of states in each overtone manifold as energy increases. The expansion coefficient $c^{(f)}_{v_{r1}0}$ and $c^{(f)}_{0v_{r2}}$ become significantly different from zero when the upper state $|f\rangle$ is either of the local mode pair of states within each of the overtone manifold.

The obvious question concerns the form of the bond dipole function. A Taylor series expansion as a function of the instantaneous bond lengths would be one possibility [165,167–168]. In practice, it is truncated at some suitable point. An alternative choice is an analytical function such as the commonly used Mecke dipole moment function [169,170]

$$\mu(R) = \mu_0 R^m e^{-R/R^*} \qquad (4.33)$$

where μ_0 and R^* are adjustable parameters, R^* giving the bond length of the maximum dipole. Often m is taken to be an integer ≥ 1. An advantage in using the Mecke function is a realistic asymptotic limit as $R \to 0$ or $R \to \infty$. Analytic Morse oscillator matrix elements can be derived when m is an integer ≥ 1. The $m > 1$ matrix elements are complicated, and it is probably easiest to calculate them numerically using the Gauss–Laguerre quadrature [160]. For $m = 1$ analytic matrix elements are available [170].

The infrared absorption intensities for transitions from the ground state $|0\rangle$ to the vibrationally excited state $|f\rangle$ are obtained as [17]

$$I_{f0} = Cv \sum_{\xi = x, y, z} \langle f | \mu(\mathbf{R}) \cdot \mathbf{e}_\xi | 0 \rangle^2 \qquad (4.34)$$

where C is a constant, v is the transition wavenumber, and $\mathbf{e}_\xi$ is a unit vector along the molecule fixed, x, y, or z axes. In accurate calculations, both the ground and the excited state wave functions are expressed similar to the expansions of the type given in Eq. (4.31). The calculated vibrational intensities depend on different choices of molecule-fixed axes. In order to ensure the maximum separation of vibration and rotation, Eckart conditions [128] should be used to tie body-fixed axes to the molecule [171,172]. The effects of different choices of molecular fixed axes on absorption intensities have been investigated for bent XY_2-type molecules [172]. Particularly it is of interest to compare the results from Eckart axes system with bisector axes system where the XY_2 molecule is in the xz plane and the z axis bisects the instantaneous valence angle. The analytic results given in Ref. [171] allow making conclusions near the hypothetical strict local mode limit, where $m_X \gg m_Y$ and the equilibrium valence angle $\alpha_e \to \frac{1}{2}\pi$. In this limit, the two axis systems give identical results. Therefore, it is to be expected that as a good approximation this should hold for the pure stretching vibrational problem in molecules such as H_2S, H_2Se, and H_2Te where the equilibrium valence angle is close to the hypothetical value. Numerical results for H_2S have shown that the symmetrical stretching fundamental intensity (and the bending fundamental and the first bending overtone intensities) agrees almost perfectly in both axes systems. In the symmetrical stretching fundamental, the difference is about 25% [172]. In Ref. [119], similar comparisons have been made for overtones in water. It is concluded that taking into account the uncertainties in measuring intensities, the differences between the results from the two different axis systems "do not seem to be very important".

The bond dipole approach with the local mode model has been applied to several small symmetrical molecules such as AsH_3, SbH_3, CH_4, SiH_4, GeH_4, and SnH_4 [45,49,50,150]. The vibrational models in these calculations have included only stretching degrees of freedom. The principal axis system at the equilibrium configuration has been used. Some results for AsH_3 and SbH_3 are given in Table XXV. A more sophisticated approach has been adopted in CHD_3, CHF_3, $CHCl_3$, $CHBr_3$, and $CH(CF_3)_3$ in the sense that the above-described Fermi resonance model has been used to obtain the vibrational eigenfunction. In the intensity calculation, the CH stretching vibration carries all transition moments [159,160]. It is worth noticing that in some cases a comparison between observed and calculated intensities within different vibrational polyads (defined by the polyad

TABLE XXV
Overtone Intensity Calculation for AsH_3 and SbH_3 [a]

	AsH_3		SbH_3	
	I_{obs}	I_{calc}	I_{obs}	I_{calc}
$100A_1/E$	1.0^b	1.0^b	1.0^b	1.0^b
$200A_1/E$	0.021	0.022	0.022	0.021
$300A_1/E$	0.00032	0.00049	0.00032	0.00047
$400A_1/E$	0.000015	0.0000071	0.000014	0.0000078

[a] Observations and calculated values are from Ref. [150], where more details are found. In the Mecke function $m = 1$, $R^* = 0.720$ Å, and $R_e = 1.51106$ Å for arsine and $R^* = 0.813$ Å and $R_e = 1.7000$ Å for stibine.
[b] Scaled value.

quantum number $v = v_{stretch} + \frac{1}{2}v_{bend}$) has helped to choose between different potential energy surfaces obtained using the nonlinear least-squares optimization of potential energy parameters and employing experimental vibrational term values as data.

It is clear from some of the above-mentioned intensity calculations that more sophisticated intensity models are needed. One obvious improvement in the theory would be the inclusion of cross terms (i.e., terms of the form $R_i R_j$, $i \neq j$) in the dipole moment function. This idea can be developed further by expanding dipole moment functions in terms of Mecke functions. This ensures realistic asymptotic behavior of the dipole surface. One problem is the possible difficulty in obtaining values from experimental data for the parameters of these cross terms. Ab initio methods can often provide additional information on dipole moment surfaces. A series of intensity calculations have been published where simple block diagonal models and Taylor series expansions have been employed for the dipole moment function with coefficients obtained by ab initio methods [119,173]. For more discussion on the intensity calculations refer to the original papers.

V. MOLECULAR ROTATIONS AND LOCAL MODES

A. Introduction

In the normal mode picture, the two most striking results from the local mode model of stretching vibrational states in polyatomic molecules are the unusual vibrational energy level patterns near the local mode limit and the simple x–K relations between vibrational anharmonicity parameters. This poses a natural question about the nature of the rotational levels of these vibrational levels. Not surprisingly, it has been found that in the normal mode picture there are unusual features involved. In the first systematic study of this problem, it was shown that simple relations exist between vibration–rotation parameters of XH_2-, XH_3-, and XH_4-type molecules in the strict local mode limit [28]. This limit is defined by assuming that there is no kinetic or potential energy coupling between the bond oscillators, there is no coupling between the stretches and the bends, the mass of the central atom is infinite, the valence angles in XH_2 and XH_3 molecules are equal to $90°$, and the bending fundamental wavenumbers are much smaller than the stretching fundamental wavenumbers. These are strong assumptions which are not exactly valid for real molecules, but numerical calculations have shown that interesting results are obtained in wavenumber lowest states of each overtone manifold (e.g., the local mode pair of states, $[v,0\pm]$ in XH_2-type molecules). The above-mentioned relations between

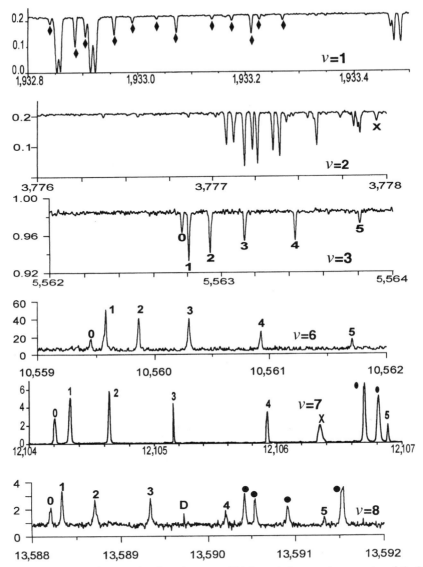

Figure 7. The $R(5)$ (i.e., $J = 6 \leftarrow 5$) peak group of high-resolution overtone spectra of the $[v_r$ $000A_1/F_2]$ band systems $(v_r = v)$ of $^{116}SnH_4$ stretching vibrational spectra. In all spectrographs the horizontal axis is given in wavenumber units and the vertical axis is given as arbitrary units of transmittance for $v = 1, 2, 3$ and as arbitrary units of absorbance for $v = 6, 7,$ 8. The numbers below/above the peaks are symmetric top K quantum numbers. ($\bullet$) Other $R(J)$ peaks. ($\blacklozenge$) Hot band transition, X refers to H_2O and D to a noise to peak. (From Ref. [155] with permission.)

the effective parameters of the local mode pair of states are more closely obeyed as the stretching energy increases [28,174,175]. The most striking result is the prediction of the dynamic symmetry change of the symmetrical molecules; for example, T_d becomes C_{3v} in XH_4. Similarly C_{3v} becomes C_s in XH_3 and C_{2v} becomes C_s in XH_2-type molecules, as discussed in theoretical contributions by Lehmann [35]. Near the local mode limit, overtone spectra of the local mode pair of states $[v,000A_1/F_2]$ in spherical top molecules should look like parallel bands of symmetric tops. Experimental work on GeH_4, SiH_4, and SnH_4 has verified this prediction and confirmed the existence of the rotational local mode relations [151,155,176,177]. In Fig. 7, high-resolution SnH_4 spectra show the change of the spectral structure in going from $v = 1$ to $v = 8$ [155]. So-called symmetric top K structure is observed for $v \geq 3$. In SbH_3 as well, it has been shown experimentally that the predictions of the special local mode relations are valid and that the SbH_3 overtone spectra of the local mode pair of states $[v,00A_1/E]$ become those of an asymmetric rotor as vibrational energy increases [57,178]. Theoretical work in this area has been carried out by Ovchinnikova on ammonia [179], which was extended by Lehmann [35], Child and Zhu [180] on the rotational structure of local mode states in spherical top molecules, Michelot [181] on the algebraic approach on local modes, and Lukka and Halonen [36], who extended Lehmann's work. Wang and Zhu have considered the effects of anharmonicity [182].

The treatment of the rotational motion as presented in this chapter is largely based on the conventional vibration–rotation theory where normal coordinates are employed. Eckart conditions are used in order to minimize interactions between vibrations and rotations [128]. This enables the use of perturbation theory to obtain theoretical expressions for rotational parameters in terms of potential energy surface parameters. The approach adopted is complicated in detail, and it requires some knowledge of the customary vibration–rotation theory formulated in terms of normal coordinates and their conjugate momenta. See Refs. [22], [23], and [183] for extensive discussion and for references to the original work. Only a brief account of the central results is given below.

The customary treatment starts from the vibration–rotation Hamiltonian given in Eq. (4.19) by expanding the $\mu_{\alpha\beta}$ tensor elements and the potential energy surface $V(\mathbf{q})$ as a Taylor series around the equilibrium configuration of the molecule in question. The result of the procedure is given in Table XXVI. The various terms obtained are arranged in order of magnitude and in powers of dimensionless normal coordinates and their conjugate momenta (q_s, p_s) and of the molecule-fixed Cartesian angular momentum components $\hat{J}_\alpha$ ($\alpha = x, y, z$) of the total angular momentum $\hat{\mathbf{J}}$ according to Oka [135]. The Born–Oppenheimer expansion parameter κ in

TABLE XXVI

Expanded Form of Rovibrational Hamiltonian Expressed in Terms of Dimensionless Normal Coordinates and Their Conjugate Momenta[a]

	Operator Form			Symbolic Form		
	$\hat{J}^0$	$\hat{J}^1$	$\hat{J}^2$	$\hat{J}^0$	$\hat{J}^1$	$\hat{J}^2$
$\kappa^0 \nu_{\text{vib}}$	$\dfrac{1}{2}\sum_r \omega_r (p_r^2 + q_r^2)$			h_{20}		
$\kappa^1 \nu_{\text{vib}}$	$+\sum_{rst} \dfrac{1}{6}\phi_{rst}\, q_r q_s q_t$			h_{30}		
$\kappa^2 \nu_{\text{vib}}$	$+\sum_{rstu} \dfrac{1}{24}\phi_{rstu}\, q_r q_s q_t q_u + \sum_\alpha B_e^{(\alpha)}[\pi_\alpha^2]$	$-2\pi_\alpha \hat{J}_\alpha$	$+\hat{J}_\alpha^2]$	h_{40}	h_{21}	h_{02}
$\kappa^3 \nu_{\text{vib}}$	$+\sum_{rstu} \dfrac{1}{5!}\phi_{rstuv}\, q_r q_s q_t q_u q_v + \sum_{\alpha\beta r} \dfrac{h^2}{2hc_0}\mu_{\alpha\beta}^{(r)} q_r[\pi_\alpha \pi_\beta]$	$-(\pi_\alpha \hat{J}_\beta + \pi_\beta \hat{J}_\alpha)$	$+\hat{J}_\alpha \hat{J}_\beta]$	h_{50}	h_{31}	h_{12}
$\kappa^4 \nu_{\text{vib}}$	$+\sum_{rstuvw} \dfrac{1}{6!}\phi_{rstuvw}\, q_r q_s q_t q_u q_v q_w + \sum_{\alpha\beta rs} \dfrac{h^2}{2hc_0}\mu_{\alpha\beta}^{(rs)} q_r q_s[\pi_\alpha \pi_\beta]$	$-(\pi_\alpha \hat{J}_\beta + \pi_\beta \hat{J}_\alpha)$	$+\hat{J}_\alpha \hat{J}_\beta]$	h_{60}	h_{41}	h_{22}

[a] The presentation is taken from Ref. [22]. The column on the left-hand side gives the order of magnitude of the Hamiltonian operators; $\kappa \cong \frac{1}{10}$ and ν_{vib} is a typical vibrational wavenumber. Note that r is this table is a summation index over normal modes and not an internal coordinate variable.

Table XXVI is defined as $(m_e/m_p)^{1/2} \approx \frac{1}{10}$, where m_e and m_p are the masses of an electron and a typical nucleus, respectively. The parameter v_{vib} is a typical vibrational wavenumber. The vibrational angular momentum π_α is defined in Eq. (4.20). The equilibrium rotational constant $B_e^{(\alpha)}$ is given as

$$B_e^{(\alpha)} = \frac{\hbar^2}{2hc_0} \mu_{\alpha\alpha}^{(e)} = \frac{\hbar^2}{2hc_0 \, I_\alpha^{(e)}} \tag{5.1}$$

where $I_\alpha^{(e)} = I_{\alpha\alpha}^{(e)}$ is the equilibrium principal moment of inertia around the α axis. The derivatives of the μ tensor elements are [22,134]

$$\mu_{\alpha\beta}^{(s)} = \left(\frac{\partial \mu_{\alpha\beta}}{\partial q_s}\right)_e = \frac{-a_s^{(\alpha\beta)}}{I_\alpha^{(e)} I_\beta^{(e)}} \, \gamma_s^{-1/2} \tag{5.2}$$

$$\mu_{\alpha\beta}^{(st)} = \left(\frac{\partial^2 \mu_{\alpha\beta}}{\partial q_s \, \partial q_t}\right)_e = \sum_\xi \frac{3(a_s^{(\alpha\xi)} a_t^{(\beta\xi)} + a_s^{(\beta\xi)} a_t^{(\alpha\xi)})}{8 I_\alpha^{(e)} I_\beta^{(e)} I_\xi^{(e)}} \, (\gamma_s \gamma_t)^{-1/2} \tag{5.3}$$

where

$$a_s^{(\alpha\beta)} = \left(\frac{\partial I_{\alpha\beta}}{\partial q_s}\right)_e \gamma_s^{1/2} \tag{5.4}$$

with $\gamma_s = 2\pi c_0 \, \omega_s/\hbar$, ω_s the harmonic wavenumber of the sth mode, and $I_{\alpha\beta}$ an element of the inertia tensor. The expanded form of the vibration–rotation Hamiltonian is also given in Table XXVI in symbolic form where the notation h_{mn} represents a term in the Hamiltonian involving an operator $p_s^k q_t^l \hat{J}_\alpha^n$ with $k + l = m$.

The standard way of proceeding is to apply perturbation theory or contact transformations to remove vibrationally off-diagonal terms in the expanded vibration–rotation Hamiltonian in order to produce an effective Hamiltonian for each vibrational state. This customary approach is explained in depth in review articles [22,183] and in a book [23]. In the case of strong resonances, that is, when at least two vibrational states are close, resulting in possible vibrational and/or rovibrational resonances, the operators for resonance interactions cannot be removed by these approximate methods. Instead, the resonance part of the vibration–rotation problem is treated by setting up Hamiltonian matrices for the interacting states. The eigenvalues are obtained by diagonalizing the matrices. This approach has already been explained in vibrational problems with Fermi resonances between stretching and bending vibrational states or Darling–Dennison resonances between stretching states. These two cases are examples of vibrational anharmonic resonances, but there are also vibration–rotation resonances between rotational states of different vibrational states. They can be significant in near-degeneracy problems.

B. Effective Rotational Hamiltonians for Bent XH_2-Type Molecules

A bent XH_2 triatomic molecule with well-separated vibrational levels is taken as an example. The rotational states of each vibrational state give rise to an isolated energy level system which can be modeled with the usual rigid asymmetric top Hamiltonian

$$\frac{H_{rot}}{hc_0} = B_v^{(x)}\hat{J}_x^2 + B_v^{(y)}\hat{J}_y^2 + B_v^{(z)}\hat{J}_z^2 \qquad (5.5)$$

where the centrifugal distortion effects have been neglected for the sake of simplicity. The operators $\hat{J}_x$, $\hat{J}_y$, and $\hat{J}_z$ are molecule-fixed Cartesian components of the total angular momentum $\hat{\mathbf{J}}$, and the $B_v^{(\alpha)}$ parameters are rotational constants of the vibrational state v. The eigenvalues of the asymmetric rotor Hamiltonian are obtained by expanding the eigenfunctions in terms of symmetric top rotational wave functions. The procedure leads to a separate matrix for each total angular momentum quantum number J. The matrices can be factorized to four blocks for each J value when symmetrized symmetric rotor wave functions (i.e., Wang combinations) are employed [141]. The eigenvalues are obtained in a closed form only for the lowest J values, but with modern computers this is not a limitation. In the symmetric top or linear molecules, the Hamiltonians corresponding to Eq. (5.5) possess eigenvalues which are expressed in closed forms [141]. The rotational constants appearing in Eq. (5.5) are usually given as a series expansion which describes the vibrational dependence of the constants; that is, the rotational constants of different vibrational states have slightly different values. The expansion is written as [9,22]

$$B_v^{(\xi)} = B_e^{(\xi)} - a_1^{(\xi)}(v_1 + \tfrac{1}{2}) - \alpha_2^{(\xi)}(v_2 + \tfrac{1}{2}) - \alpha_3^{(\xi)}(v_3 + \tfrac{1}{2}) + \cdots \qquad (5.6)$$

where ξ is x, y, or z and $B_e^{(\xi)}$ is the equilibrium rotational constant as defined in Eq. (5.1). Equation (5.6) represents an infinite series expansion, but it often converges rapidly. It remains to be seen how Eq. (5.6) [or better the Hamiltonian in Eq. (5.5) with coefficients given in Eq. (5.6)] can be obtained from the more general expanded Hamiltonian in Table XXVI. The terms involving equilibrium rotational constants, $\sum_\xi B_e^{(\xi)} \hat{J}_\xi^2$, are readily identified in Table XXVI. It is also obvious that $\alpha_s^{(\xi)}$ is the coefficient of the operator $-\tfrac{1}{2}(p_s^2 + q_s^2)\hat{J}_\xi^2$ in the expanded Hamiltonian because, according to the harmonic oscillator formulas, $\langle v_s | \tfrac{1}{2}(p_s^2 + q_s^2) | v_s \rangle = v_s + \tfrac{1}{2}$. The operator h_{22} in Table XXVI is of this form; that is, it is a two-power operator in both vibrational and rotational operators. However, this is not the full answer as there are other contributions. When second-order perturbation theory with harmonic oscillator matrix element formulas are applied to Hamiltonian terms h_{21} and h_{21} and to h_{30} and h_{12}, effective Hamiltonian terms are obtained which are of the type $(v_s + \frac{1}{2})\hat{J}_\xi^2$. This second-order treatment also contributes to the α coefficients [22]. As a sum of these three contributions, an expression for the $\alpha_s^{(\xi)}$ coefficients in Eq. (5.6) is obtained as [22]

$$
\alpha_s^{(\xi)} = -2\frac{[B_e^{(\xi)}]^2}{\omega_s}\left\{\sum_\gamma \frac{3[a_s^{(\xi\gamma)}]^2}{4I_\gamma^{(e)}} + \sum_{s'}[\zeta_{ss'}^{(\xi)}]^2 \frac{3\omega_s^2 + \omega_{s'}^2}{\omega_s^2 - \omega_{s'}^2}\right.
$$
$$
\left. + \pi\left(\frac{c_0}{h}\right)^{1/2}\sum_{s'} \frac{\phi_{sss'} a_s^{(\xi\xi)}\omega_s}{\omega_{s'}^{3/2}}\right\} \tag{5.7}
$$

A theoretical basis has been established for the coefficients of the rotational Hamiltonian.

So far only isolated vibrational states have been considered. In molecules close to the local mode limit, there are almost degenerate stretching states, as is seen from the correlation diagram given in Fig. 4. Therefore, the simple picture of the rotational levels described by the rigid-rotor Hamiltonian in Eq. (5.5) might fail. There are at least two vibrationally off-diagonal vibration–rotation operators which are important: the Coriolis operator H_{21}, which is a two-power operator in vibrational variables and one-power operator in rotational variables, and the α resonance operator H_{22}, which is a two-power operator in both vibrational and rotational variables. In the following, the molecule-fixed Cartesian axes are chosen to be the principal axes in such a way that the x axis is the C_2 axis and the y axis is perpendicular to the molecular plane. In water, for example, the y and the z axes correspond to the largest and smallest moments of inertia. The appropriate vibration–rotation Hamiltonian is [28,174]

$$
\frac{H}{hc_0} = G_v + B_e^{(x)}\hat{J}_x^2 + B_e^{(y)}\hat{J}_y^2 + B_e^{(z)}\hat{J}_z^2
$$
$$
- \bar{q}_1^2(\alpha_1^{(x)}\hat{J}_x^2 + \alpha_1^{(y)}\hat{J}_y^2 + \alpha_1^{(z)}\hat{J}_z^2)
$$
$$
- \bar{q}_3^2(\alpha_3^{(x)}\hat{J}_x^2 + \alpha_3^{(y)}\hat{J}_y^2 + \alpha_3^{(z)}\hat{J}_z^2)
$$
$$
+ d_{13}\bar{q}_1\bar{q}_3[\hat{J}_x, \hat{J}_z]_+ - 2B_e^{(y)}\zeta_{13}^{(y)}\left(\sqrt{\frac{\omega_3}{\omega_1}}q_1 p_3 - \sqrt{\frac{\omega_1}{\omega_3}}p_1 q_3\right)\hat{J}_y
$$
$$
\tag{5.8}
$$

where $[\ ]_+$ denotes an anticommutator. The vibrational operators ($i = 1, 3$ for the symmetric and antisymmetric stretch, respectively) are defined as

$$\bar{q}_i^2 = \tfrac{1}{2}(q_i^2 + p_i^2) \qquad \bar{q}_1\bar{q}_3 = \tfrac{1}{2}(q_1q_3 + p_1p_3) \tag{5.9}$$

and G_v is the vibrational term value. As discussed in Section II, it takes the same value for the local mode pair of states in the local mode limit [3]. Here, d_{13} and the Coriolis constant $\zeta_{13}^{(y)}$ are the coefficients of the vibrationally off-diagonal H_{22} and H_{21} terms, respectively.

The Coriolis operator is obtainable from the operator h_{21} in Table XXVI. The Coriolis coupling coefficient $\zeta_{ss'}^{(\xi)}$ is defined as a cross product of the displacement of atoms,

$$\zeta_{ss'}^{(\xi)} = \sum_k (\mathbf{l}_{k,s} \times \mathbf{l}_{k,s'})_\xi \tag{5.10}$$

where $\xi \doteq x$, y, z, the summation is over all atoms, and the $\mathbf{l}$ matrix is defined as $(l)_{\xi k, s} = (\partial \xi_k / \partial Q_s)_e$, with ξ_k being the mass-adjusted Cartesian coordinate and Q_s the usual normal coordinate. The Coriolis coefficients can be calculated from the structure and the harmonic part of the potential energy surface of the molecule in question [184].

The H_{22} resonance operator (the α resonance operator) is more difficult to derive than its Coriolis counterpart. Besides the h_{22} term given in Table XXVI, there are other contributions which are obtained by using second-order perturbation theory (or equivalently contact transformations). The procedure is analogous with the derivation of the coefficients of the Hamiltonian given in Eq. (5.6). The coefficient of the α resonance operator d_{13} becomes

$$d_{13} = \frac{4B_e^{(x)}B_e^{(z)}}{(\omega_1\omega_3)^{1/2}} \left[\frac{3}{8} \left(\frac{a_1^{(xx)}a_3^{(xz)}}{I_x^{(e)}} + \frac{a_1^{(zz)}a_3^{(xz)}}{I_z^{(e)}} \right) \right.$$
$$\left. + \pi \left(\frac{c_0}{h} \right)^{1/2} \phi_{133}\, a_3^{(xz)} \, \frac{\omega_1^{1/2}\omega_3}{\omega_3^2 - (\omega_1 - \omega_3)^2} \right] \tag{5.11}$$

It follows from the theoretical expressions for the α and d_{13} coefficients that these parameters depend on five different sets of fundamental spectroscopic parameters: harmonic vibrational wavenumbers ω_s, equilibrium rotational constants $B_e^{(\xi)}$ (or equivalently equilibrium moments of inertia $I_{\xi\xi}^{(e)}$), Coriolis coupling coefficients $\zeta_{su}^{(\xi)}$, Wilson's $a_s^{(\alpha\beta)}$ constants [see Eq. (5.4)], and cubic anharmonic force constants ϕ_{stu}. The equilibrium rotational constants are calculated from the equilibrium structure. The parameters ω_s, $\zeta_{su}^{(\xi)}$, and $a_s^{(\alpha\beta)}$

can be calculated from the harmonic force field of the molecule in question [17,134,184].

The effects of the Coriolis and α resonances on line positions and intensities have been discussed in the literature [136,185]. The Coriolis resonances often cause large global effects where whole vibration–rotation bands are affected. The α resonance operator, being a higher order term than the Coriolis counterpart, is often responsible for localized effects. In the case of stretching vibrational states of molecules near the local mode limit, the situation is different, as is discussed below.

By assuming that the bending vibration is completely decoupled from the stretches, the Hamiltonian given above can be transformed to an internal coordinate representation using the simple scaling given in Eq. (2.27), where the symmetry coordinates S_1 and S_3 are defined in Eq. (2.24). The transformed Hamiltonian is

$$
\begin{aligned}
\frac{H}{hc_0} = {} & G_v + B_e^{(x)}\hat{J}_x^2 + B_e^{(y)}\hat{J}_y^2 + B_e^{(z)}\hat{J}_z^2 \\
& - \tfrac{1}{2}(\bar{r}_1^2 + 2\bar{r}_1\bar{r}_2 + \bar{r}_2^2)(\alpha_1^{(x)}\hat{J}_x^2 + \alpha_1^{(y)}\hat{J}_y^2 + \alpha_1^{(z)}\hat{J}_z^2) \\
& - \tfrac{1}{2}(\bar{r}_1^2 - 2\bar{r}_1\bar{r}_2 + \bar{r}_2^2)(\alpha_3^{(x)}\hat{J}_x^2 + \alpha_3^{(y)}\hat{J}_y^2 + \alpha_3^{(z)}\hat{J}_z^2) \\
& + \tfrac{1}{2}d_{13}(\bar{r}_1^2 - \bar{r}_2^2)[\hat{J}_x, \hat{J}_z]_+ \\
& - \hbar^{-1}B_e^{(y)}\zeta_{13}^{(y)}\left[\sqrt{\frac{\omega_3}{\omega_1}}\,(r_1 p_1 + r_2 p_1 - r_1 p_2 - r_2 p_2) \right. \\
& \left. - \sqrt{\frac{\omega_1}{\omega_3}}\,(r_1 p_1 - r_2 p_1 + r_1 p_2 - r_2 p_2) \right]\hat{J}_y
\end{aligned}
\tag{5.12}
$$

In the equation above ($i = 1, 2$ and $i < j = 2$),

$$
\bar{r}_i^2 = \tfrac{1}{2}(\alpha r_i^2 + \hbar^{-2}\alpha^{-1}p_{ri}^2) \qquad \bar{r}_i\bar{r}_j = \tfrac{1}{2}(\alpha r_i r_j + \hbar^{-2}\alpha^{-1}p_{ri}p_{rj}) \tag{5.13}
$$

A single scaling factor α is used instead of two different scaling factors α_1 and α_3 [see Eq. (5.12)] because in the zeroth-order model $\omega = \omega_1 = \omega_3$ [21].

In the local mode basis, the terms containing vibrational operators of the type $\bar{r}_1\bar{r}_2$, $r_1 p_2$, and $r_2 p_1$ couple rotational states of different vibrational states. The case with no coupling between the local mode states is called the vibrational local mode limit. If the intuitive physical picture that rotation does not couple the different localized states is assumed, all rovibrational terms containing a vibrational part of the above-mentioned type should disappear. This is called the rovibrational local mode limit [36]. In the case of the H_{22} coefficients, this leads to the relationships [35,36]

$$\alpha_1^{(x)} = \alpha_3^{(x)} \qquad \alpha_1^{(y)} = \alpha_3^{(y)} \qquad \alpha_1^{(z)} = \alpha_3^{(z)} \tag{5.14}$$

In the local mode limit, $\omega_1 = \omega_3$ and the Coriolis coefficient disappears, that is,

$$\zeta_{13}^{(y)} = 0. \tag{5.15}$$

The results given in Eqs. (5.14) and (5.15) are called α relations, in analogy with x–K relations in the vibrational theory. The message is such that in the rovibrational local mode limit the rotational constants of the local mode pair of states $|v_r, 0+\rangle$ and $|v_r, 0-\rangle$ are equal and there is no Coriolis coupling between the rotational states of the degenerate vibrational states. On the other hand, the α resonance given by the Hamiltonian operator in d_{13} may remain significant depending on the magnitude of the d_{13} coefficient. The argument can be developed further by retaining the rovibrational local mode limit assumption; that is, Eqs. (5.14) and (5.15), together with the constraint $\omega_1 = \omega_3$, are assumed to be obeyed. By averaging the internal coordinate Hamiltonian in Eq. (5.12) with the unsymmetrized local mode wave function $|v, 0\rangle$, an effective rotational Hamiltonian is obtained for this state as [180]

$$\langle v_r 0 | \frac{H}{hc_0} | v_r 0 \rangle = G_v + B_v^{(x)} \hat{J}_x^2 + B_v^{(y)} \hat{J}_y^2 + B_v^{(z)} \hat{J}_z^2$$

$$+ \tfrac{1}{2} v_r d_{13} [\hat{J}_x, \hat{J}_z]_+ \tag{5.16}$$

where harmonic oscillator matrix elements have been used. Here, $B_v^{(\xi)} = B_e^{(\xi)} - (v_r + 1)\alpha^{(\xi)}$, $\alpha^{(\xi)} = \alpha_1^{(\xi)} = \alpha_3^{(\xi)}$, and the bend has been omitted. The rotational part in the matrix representation is

$$\frac{H_{\text{rot}}}{hc_0} = (\hat{J}_x \quad \hat{J}_y \quad \hat{J}_z) \begin{pmatrix} B_v^{(x)} & 0 & \tfrac{1}{2} v_r d_{13} \\ 0 & B_v^{(y)} & 0 \\ \tfrac{1}{2} v_r d_{13} & 0 & B_v^{(z)} \end{pmatrix} \begin{pmatrix} \hat{J}_x \\ \hat{J}_y \\ \hat{J}_z \end{pmatrix} \tag{5.17}$$

Diagonalization of the 3×3 matrix on the right-hand side provides a transformation to a form where the rotational Hamiltonian does not contain rotational cross terms. In the present problem, the solution is

$$\frac{H_{\text{rot}}}{hc_0} = B_v^{(x')} \hat{J}_{x'}^2 + B_v^{(y')} \hat{J}_{y'}^2 + B_v^{(z')} \hat{J}_{z'}^2. \tag{5.18}$$

where

$$B_v^{(x')} = \tfrac{1}{2}(B_v^{(x)} + B_v^{(z)}) - \tfrac{1}{2}[(B_v^{(x)} - B_v^{(z)})^2 + v_r^2 d_{13}^2]^{1/2}$$
$$B_v^{(y')} = B_v^{(y)} \tag{5.19}$$
$$B_v^{(z')} = \tfrac{1}{2}(B_v^{(x)} + B_v^{(z)}) + \tfrac{1}{2}[(B_v^{(x)} - B_v^{(z)})^2 + v_r^2 d_{13}^2]^{1/2}$$

and

$$\hat{J}_{x'} = \cos \theta\, \hat{J}_x + \sin \theta\, \hat{J}_z \qquad \hat{J}_{y'} = \hat{J}_y \qquad \hat{J}_{z'} = -\sin \theta\, \hat{J}_x + \cos \theta\, \hat{J}_z$$
$$\tag{5.20}$$

with

$$\theta = \tfrac{1}{2} \arctan \frac{v_r d_{13}}{B_v^{(x)} - B_v^{(z)}} \quad \text{and} \quad -\tfrac{1}{4}\pi < \theta < \tfrac{1}{4}\pi$$

A single asymmetric top rotational Hamiltonian is obtained, with rotational constants related to the parameters of the coupled problems, as given above. Thus, in this special case (in the rovibrational local mode limit), the rotational energy level structure of the degenerate local mode states $|v, 0+\rangle$ and $|v, 0-\rangle$ becomes that of an asymmetric rotor. Note that the result is not limited to the use of harmonic oscillator matrix elements, but it is necessary to assume that Eqs. (5.14) and (5.15) are obeyed. Otherwise even in the case of pure local mode states, there remains rovibrational coupling between different states.

In Ref. [28] it was seen for the first time that there exist simple relations between vibration–rotation parameters of near XH_2-type local mode molecules. The results were based on the strict local mode limit concept defined earlier in this chapter. Using perturbation theory formulas for the vibration–rotation parameters, simple relations such as the α relations are obtained. This results in the disappearance of the Coriolis coupling coefficient, but for the H_{22} coefficients the relations obtained are more strict than the rovibrational local mode limit results.

Equations (5.14) and (5.18) have been published for the first time in Ref. [35], where a different kind of approach was used. The starting point was an empirical local mode internal coordinate vibration–rotation Hamiltonian. A transformation was made to the normal coordinate Hamiltonian. This procedure is mathematically equivalent to the one presented for the rovibrational local mode limit case.

The local mode relations derived are valid only within a hypothetical limit where rotation does not introduce any coupling between different local mode states. In real molecules, this kind of coupling is not necessarily absent. For molecules near the rovibrational local mode limit, it is useful to define a new set of α constants such as $\alpha_+^{(\xi)} = \frac{1}{2}(\alpha_1^{(\xi)} + \alpha_3^{(\xi)})$ and $\alpha_-^{(\xi)} = \alpha_1^{(\xi)} - \alpha_3^{(\xi)}$, where $\xi = x, y, z$. The plus and minus subscripts should not be confused with the local mode symmetry labels. The Hamiltonian operator expressed in terms of these parameters is

$$\frac{H_{22}}{hc_0} = -(\bar{r}_1^2 + \bar{r}_2^2) \sum_{\xi = x, y, z} \alpha_+^{(\xi)} \hat{J}_\xi^2 - \bar{r}_1 \bar{r}_2 \sum_{\xi = x, y, z} \alpha_-^{(\xi)} \hat{J}_\xi^2 \qquad (5.21)$$

The first term is diagonal in the local mode basis and contributes to the vibrational dependence of the rotational constants of the localized states. The second term connects the rotational states of the different local mode states. Near the rovibrational local mode limit, $\alpha_-^{(\xi)} = \alpha_1^{(\xi)} - \alpha_3^{(\xi)}$ is small and the vibration–rotation coupling is weak between the rotational states of different local mode states. Outside the rovibrational local mode limit, the derived relationships in Eqs. (5.14) and (5.15) are invalid (and so $\alpha_- \neq 0$), and it is necessary to add the last correction term in Eq. (5.21) into the model of Ref. [35]. Physically, it would describe the rotational dependence of the interbond coupling parameter λ and it is a term of the type $H_{2n, 2}$, where n is the stretching vibrational quantum number [35]. The term in α_+ is an H_{22} operator, similar to the terms involving the standard α parameters.

C.   Effective Vibration–Rotation Parameters in XH$_2$

As discussed in Section II, according to the customary vibration–rotation theory formulated in terms of rectilinear normal coordinates, the analysis of the stretching overtones of XH$_2$-type bent molecules requires the inclusion of the quartic Darling–Dennison anharmonic resonance term in the Hamiltonian. This resonance interaction is strong in near local mode molecules with close-lying stretching overtone energy levels. This causes the overtone wave functions in the normal mode representation to be strongly mixed combinations of the basis functions. The overtone spectra are normally analyzed with the standard theory based on normal coordinates, that is, with equations similar to Eq. (5.8). Due to the above-mentioned Darling–Dennison resonance, the parameters obtained in the case of overtone states are effective. In order to understand the vibrational dependence of these effective constants, it is necessary to use a vibrational model which includes the Darling–Dennison resonance terms. If no coupling between the rotational states of local mode states $|v_r, 0+\rangle$ and $|v_r, 0-\rangle$ exists, the α relations given before in Eqs. (5.14) and (5.15) hold both for the effective constants and for the constants obtained from an analysis of the stretching fundamentals. However, so far no answer has been provided to the question of how well the relations between vibration–rotation parameters are obeyed in practice. For the parameters obtained from the stretching fundamentals of real molecules, the α relations are not exactly obeyed (see Ref. [28] for examples in some specific molecules). On the other hand, for the effective vibration–rotation parameters obtained from the analysis of stretching overtones, the situation can be different because the vibrational matrix elements also make contributions to the effective parameters. First it was found both experimentally [151,155,176–178] and numerically [28,174,175, 186] that in the near-local-mode molecules the effective vibration–rotation parameters obey simple relations more accurately as the stretching energy increases. Later this was proved to be the case using simple arguments as are discussed below [36].

The Hamiltonian given in Eq. (5.8) for XH_2-type molecules can be used to calculate rotational constants for any excited stretching vibrational state. In the following, only the local mode pair of states $|v_r 0+\rangle = |v_r 0A_1\rangle$ and $|v_r 0-\rangle = |v_r 0B_1\rangle$ are considered. As discussed before for near-local-mode molecules, these states become quickly degenerate as the stretching energy increases. The reason for this restriction is the observation that near the vibrational local mode limit, transitions from the ground state to these states are stronger than transitions to any other states within the overtone manifold in question [3]. Most of the experimental data available are these kinds of strong transitions. It follows from Eq. (5.8) that the rotational constants of the local mode pair of states can be calculated as (at this stage the effects of bending vibrations have been disregarded, i.e., in this equation they can be thought to be part of $B_e^{(\xi)}$)

$$B_{\pm, \text{ eff}}^{(\xi)} = B_e^{(\xi)} - \alpha_1^{(\xi)}\langle v_r 0 \pm | \bar{q}_1^2 | v_r 0 \pm \rangle - \alpha_3^{(\xi)}\langle v_r 0 \pm | \bar{q}_3^2 | v_r 0 \pm \rangle \quad (5.22)$$

where $\xi = x, y, z$ and where normal mode wave functions are used, although the more suitable local mode notation for the wave function is employed. When Darling–Dennison resonance between the stretching vibrational states within the overtone manifolds is unimportant, Eq. (5.22) reduces to the standard expression

$$B_{i, \text{ eff}}^{(\xi)} = B_e^{(\xi)} - \alpha_1^{(\xi)}(v_1 + \tfrac{1}{2}) - \alpha_3^{(\xi)}(v_3 + \tfrac{1}{2}) \quad (5.23)$$

where $i = +, -$ (v_1 and v_3 quantum numbers differ for the + and − states). When Darling–Dennison resonance cannot be neglected as in near-local-mode molecules, the expectation values must be calculated with the

help of eigenvectors obtained by diagonalizing Darling–Dennison resonance Hamiltonian matrices. Calculations of this type have been performed
in the past [174].

The rotational constants of the local mode pair of states can also be
calculated as

$$B^{(\xi)}_{\pm,\,\text{eff}} = B^{(\xi)}_e - \tfrac{1}{2}\alpha^{(\xi)}_1 \langle v_r, 0\pm \,|\, (\bar{r}_1^2 + 2\bar{r}_1\bar{r}_2 + \bar{r}_2^2)\,|\, v_r, 0\pm \rangle$$
$$- \tfrac{1}{2}\alpha^{(\xi)}_3 \langle v_r, 0\pm \,|\, (\bar{r}_1^2 - 2\bar{r}_1\bar{r}_2 + \bar{r}_2^2)\,|\, v_r, 0\pm \rangle \tag{5.24}$$

where the wave functions are local mode eigenfunctions and they are
obtained, for example, from the HCAO model [13]. Both the Darling–
Dennison resonance model and the HCAO model have been earlier shown
to produce identical eigenvalues. At the high-energy limit, the wave functions appearing in Eq. (5.24) become eigenfunctions of pure local mode
states for near-local-mode limit molecules: $|v_r 0+\rangle = (1/\sqrt{2})(|v_r, 0\rangle$
$+ |0v_r\rangle)$ and $|v_r, 0-\rangle = (1/\sqrt{2})(|v_r, 0\rangle - |0v_r\rangle)$. In this case, within the
HCAO model, the cross terms $\bar{r}_1\bar{r}_2$ in Eq. (5.24) possess zero matrix elements for $v_r > 1$. Both of the integrals in Eq. (5.24) are identical for the local
mode pair of states, and the rotational constants become equal. It follows
that the local mode relations given between the effective α constants in Eqs.
(5.14) become more exact as the stretching energy increases. Thus, if the
effective α constants are defined as

$$\alpha^{(\xi)}_{A1,\,\text{eff}} = B^{(\xi)}_0 - B^{(\xi)}_{A1,\,\text{eff}} = B^{(\xi)}_0 - B^{(\xi)}_{+,\,\text{eff}}$$
$$\alpha^{(\xi)}_{B1,\,\text{eff}} = B^{(\xi)}_0 - B^{(\xi)}_{B1,\,\text{eff}} = B^{(\xi)}_0 - B^{(\xi)}_{-,\,\text{eff}} \tag{5.25}$$

where $B^{(\xi)}_0$ is the rotational constant of the ground vibrational state, then

$$\alpha^{(\xi)}_{A1,\,\text{eff}} = \alpha^{(\xi)}_{B1,\,\text{eff}} \tag{5.26}$$

for $\xi = x,\, y,\, z$. This is equivalent to the statement that the rotational con-
stants of the local mode pair of states $|v_r, 0+\rangle$ and $|v_r, 0-\rangle$ become equal as
v_r increases. The conclusion is not restricted to harmonic oscillator matrix
elements. The use of Morse oscillator matrix elements leads to the same
result, although there are some additional points worth stressing. First, the
diagonal matrix elements of the type $\langle v_r | r | v_r \rangle$ do not disappear as v_r
increases. In spite of this, the conclusion regarding the equivalence of the
rotational constants at high excitations is still valid [182]. Second, as v_r
increases, the local mode limit relations are reached more slowly than in the
HCAO model. This is due to the slower disappearance of the matrix ele-
ments of the type $\langle v_r | r | 0 \rangle$. Third, in the anharmonic model there are only

numerical proofs of the decoupling of the bond oscillators as energy increases. Finally, when the states $|v_r 0\pm\rangle$ become pure local mode states, Eq. (5.24) and harmonic oscillator matrix elements can be used to obtain a closed form for the effective rotational constants. The result is [187]

$$B_{\pm,\text{ eff}}^{(\xi)} = B_0^{(\xi)} - \tfrac{1}{2}(\alpha_1^{(\xi)} + \alpha_3^{(\xi)})v_r \tag{5.27}$$

The behavior of the effective Coriolis coefficients in the XH_2-type molecules can be derived in a similar way. By observing that both the operators $q_1 p_3$ and $-p_1 q_3$ have identical matrix elements, the vibrational matrix element of the Coriolis Hamiltonian is obtained from Eq. (5.8) as

$$\langle v_r 0- | \frac{H_{\text{Cor}}}{hc_0} | v_r 0+\rangle$$

$$= -2B_e^{(y)}\zeta_{13}^{(y)}\Omega_{13}\langle v_r 0- |(q_1 p_3 - p_1 q_3)|v_r 0+\rangle \hat{J}_y$$

$$= -2\hbar^{-1}B_e^{(y)}\zeta_{13}^{(y)}\Omega_{13}\langle v_r 0- |(p_{r_1}r_2 - r_1 p_{r_2})|v_r 0+\rangle \hat{J}_y \tag{5.28}$$

where $v_r > 1$, $\Omega_{13} = \tfrac{1}{2}(\sqrt{(\omega_1/\omega_3)} + \sqrt{(\omega_3/\omega_1)})$ and the wave functions are eigenfunctions obtained from the HCAO model. For $v_r > 1$, the integrals disappear for pure local mode wave functions. Thus, Coriolis coupling disappears at the high energy limit. For the stretching fundamentals, the vibrational matrix element becomes

$$\langle 1, 0| \frac{H_{\text{Cor}}}{hc_0} |0, 1\rangle = i2B_e^{(y)}\zeta_{13}^{(y)}\Omega_{13}\hat{J}_y \tag{5.29}$$

By comparing Eqs. (5.28) and (5.29), the effective Coriolis coefficient for the excited local mode pair of states is

$$\zeta_{\text{eff}} = i\zeta_{13}^{(y)}\langle v_r 0- |(q_1 q_3 - p_1 q_3)|v_r 0+\rangle \tag{5.30}$$

It is concluded that the effective Coriolis coefficient between the local mode pair of states disappears for high overtone states.

Unlike Coriolis coupling, α resonance is important for the rotational states of near-local-mode vibrational states. The α resonance matrix element is obtained from Eq. (5.8) for the stretching fundamentals as

$$\langle 1, 0| \frac{H_\alpha}{hc_0} |0, 1\rangle = \tfrac{1}{2}d_{13}[\hat{J}_x, \hat{J}_z]_+ \tag{5.31}$$

and for the local mode pair of states as

$$\langle v_r 0 - | \frac{H_\alpha}{hc_0} | v_r 0 + \rangle = \tfrac{1}{2} d_{13} \langle v_r 0 - | \bar{q}_1 \bar{q}_3 | v_r 0 + \rangle [\hat{J}_x, \hat{J}_z]_+$$

$$= \tfrac{1}{2} d_{13} \langle v_r 0 - | (\bar{r}_1^2 - \bar{r}_2^2) | v_r 0 + \rangle [\hat{J}_x, \hat{J}_z]_+ \quad (5.32)$$

The effective d coefficient is defined as

$$d_{\mathrm{eff}} = d_{13} \langle v_r 0 - | \bar{q}_1 \bar{q}_3 | v_r 0 + \rangle = d_{13} \langle v_r 0 - | (\bar{r}_1^2 - \bar{r}_2^2) | v_r 0 + \rangle \quad (5.33)$$

By assuming that the states $| v_r 0 \pm \rangle$ are pure local mode states, the previous equation with harmonic oscillator matrix elements leads to the result [187]

$$d_{\mathrm{eff}} = v_r d_{13} \quad (5.34)$$

As the stretching energy increases, the effective rovibrational Hamiltonian of the local mode pair of states becomes that of a single asymmetric top Hamiltonian given in Eq. (5.18). When harmonic approximation is used to evaluate vibrational matrix elements, the rotational constants in Eq. (5.18) are replaced by the effective constants given in Eq. (5.27) and $v_r d_{13}$ by d_{eff}, as shown in Eq. (5.34). The final conclusion of the validity of Eq. (5.18) for the local mode pair of states is not restricted to the use of harmonic oscillator matrix elements.

As an example, the theory presented is applied to the rotational constants of bent XY_2 molecules in the $2v_1$ and $2v_3$ states, which, according to the standard vibration–rotation theory, are coupled by the Darling–Dennison resonance term as shown by Eqs. (2.10). In the local mode notation, the states are labeled as $[20+]$ and $[11+]$. The standard approach is used below, although the same results are obtainable by the local mode formulation. The a, b, c, and d coefficients are defined by

$$| 2v_1 \rangle = a | 2, 0 \rangle + b | 0, 2 \rangle \qquad | 2v_3 \rangle = c | 2, 0 \rangle + d | 0, 2 \rangle \quad (5.35)$$

and are given in Table XXVII for SO_2, H_2O, and H_2Se, of which SO_2 is near the normal mode limit, H_2Se is near the local mode limit, and H_2O is an intermediate case. As before, $| 2, 0 \rangle = | 2 \rangle | 0 \rangle$ and $| 0, 2 \rangle = | 0 \rangle | 2 \rangle$ are products of harmonic oscillator wave functions in the normal coordinate representation. The rotational constants for these states are calculated with Eq. (5.22), where the expectation values are evaluated using the wave function expansion of the type given in Eq. (5.35) for the $2v_1/2v_3$ pair. The results in Table XXVII indicate that, for SO_2, $a, d \sim 1$ and $b, c \sim 0$. Thus, the vibrational dependence of the rotational constants of the stretching

TABLE XXVII

Eigenvalues and Eigenvectors for $2\nu_1/2\nu_3$ Hamiltonian Matrix of SO_2, H_2O, and H_2Se[a]

	SO_2		H_2O		H_2Se	
E, cm^{-1}	2296.6	2712.3	7190.8	7424.9	4614.5	4703.3
$\lvert 2, 0 \rangle$	0.99985	-0.017	0.934	-0.356	0.810	-0.587
$\lvert 0, 2 \rangle$	0.017	0.99985	0.356	0.934	0.587	0.810

[a] In the case of H_2O and SO_2 the model parameters ω'_r, x_{rr}, and λ_r are taken from Table III, and in the case of $H_2\,^{80}Se$ they are $\omega'_r = 2434.96$ cm^{-1}, $x_{rr} = -42.20$ cm^{-1}, and $\lambda_r = -6.91$ cm^{-1}.

vibrational states of SO_2 is well understood using Eq. (5.23). On the other hand, for H_2Se, a, $d \sim 0.810$, $b \sim -0.587$, and $c \sim 0.587$. Equation (5.23) becomes a poor representation of the problem for H_2Se. For the $2\nu_1/2\nu_3$ interacting pair in the local mode limit [Eq. (5.27)],

$$B^{(\xi)}_{2\nu_1} = B^{(\xi)}_{2\nu_3} = B^{(\xi)}_0 - \alpha^{(\xi)}_1 - \alpha^{(\xi)}_3 \qquad (5.36)$$

compared with

$$B^{(\xi)}_{2\nu_1} = B^{(\xi)}_0 - 2\alpha^{(\xi)}_1 \qquad B^{(\xi)}_{2\nu_3} = B^{(\xi)}_0 - 2\alpha^{(\xi)}_3 \qquad (5.37)$$

in the normal mode limit [Eq. (5.23)]. In the equations above, $B^{(\xi)}_0 = B^{(\xi)}_e - \frac{1}{2}(\alpha^{(\xi)}_1 + \alpha^{(\xi)}_3)$ is the rotational constant of the ground vibrational state (the effects of bending vibrations are included in $B^{(\xi)}_e$). The difference between Eqs. (5.36) and (5.37) is immediately anticipated.

Table XXVIII contains observed and calculated vibrationally diagonal α parameters $\alpha^{(A)}$, $\alpha^{(B)}$, and $\alpha^{(C)}$ for the observed stretching states of $H_2\,^{32}S$ and $H_2\,^{80}Se$ [91,188,189], and Table XXIX contains observed and calculated d and ζ parameters between the $\lvert v_r, 0 + \rangle$ and $\lvert v_r, 0 - \rangle$ local mode pair of states. All calculated values have been obtained using the eigenvectors from the Darling–Dennison resonance model. It is clear from the results that, on the whole, this approach produces experimental observations well. In the case of the Coriolis coefficients, there are deviations between observed and calculated values. This might be due to the small magnitude of the Coriolis coefficients.

D. Dynamic Aspects of Rotations and Local Modes in XH_2

In the vibrational local mode limit, the bond oscillators are independent in XH_2-type molecules. In the unsymmetrized local mode state $\lvert v_r, 0 \rangle$, the first bond being longer, the principal axis system is rotated from the ground vibrational state position in such a way that the A ($= B^{(z')}_v$) axis is rotated toward this excited bond oscillator. The rotational energy level structure of

TABLE XXVIII

Observed and Calculated Vibrational Term Values and Vibrationally Diagonal α Parameters (cm^{-1}) of Stretching States in $H_2{}^{32}S$ and $H_2{}^{80}Se^a$

v_1v_3	$v_{r_1}v_{r_2} \pm$	O/C^b	$H_2{}^{32}S$				$H_2{}^{80}Se$			
			ν	$\alpha^{(A)}$	$\alpha^{(B)}$	$\alpha^{(C)}$	ν	$\alpha^{(A)}$	$\alpha^{(B)}$	$\alpha^{(C)}$
10	10+	O	2614.41	0.1596	0.1237	0.0698	2344.36	0.1201	0.1051	0.0519
		C	2613.35	—	—	—	2343.66	—	—	—
01	10−	O	2628.46	0.2178	0.0789	0.0544	2357.66	0.1589	0.0696	0.0485
		C	2627.15	—	—	—	2357.48	—	—	—
20	20+	O	5144.99	0.3592	0.2144	0.1289	4615.32	0.2640	0.1873	0.1006
		C	5143.37	0.3612	0.2151	0.1285	4614.53	0.2668	0.1858	0.1014
11	20−	O	5147.22	0.3765	0.2012	0.1235	4617.39	0.2778	0.1754	0.1009
		C	5145.33	0.3774	0.2026	0.1242	4616.74	0.2789	0.1747	0.1004
02	11+	O	5243.10	0.3868	0.1956	0.1183	4702.54	0.2855	0.1690	0.1009
		C	5242.46	0.3936	0.1901	0.1199	4703.34	0.2910	0.1636	0.0993
30	30+	O	—	—	—	—	6798.10	0.3866	0.2891	0.1534
		C	—	—	—	—	6797.60	0.3849	0.2927	0.1535
21	30−	O	—	—	—	—	6798.21	0.4522	0.2307	0.1558
		C	—	—	—	—	6797.74	0.4142	0.2659	0.1509
12	21+	O	—	—	—	—	6953.46	0.3610	0.3142	0.1533
		C	—	—	—	—	6954.40	0.4130	0.2670	0.1510
40	40+	O	9911.02	0.7453	0.4028	0.2542	—	—	—	—
		C	9909.28	0.7491	0.4095	0.2499	—	—	—	—
31	40−	O	9911.02	0.7453	0.4028	0.2542	—	—	—	—
		C	9909.29	0.7491	0.4095	0.2499	—	—	—	—
50	50+	O	12149.46	0.8852	0.5462	0.3196	—	—	—	—
		C	12148.88	0.9382	0.5105	0.3119	—	—	—	—
41	50−	O	12149.46	0.8852	0.5462	0.3196	—	—	—	—
		C	12148.88	0.9382	0.5105	0.3119	—	—	—	—
32	41+	O	12524.60	0.8937	0.5386	0.3154	—	—	—	—
		C	12528.61	0.9291	0.5176	0.3143	—	—	—	—
23	41−	O	12525.20	0.9002	0.5331	0.3131	—	—	—	—
		C	12529.04	0.9345	0.5134	0.3129	—	—	—	—

[a] The vibrational model parameters used for $H_2{}^{80}Se$ are taken from Table XXVII. Observed and calculated data for $H_2{}^{32}S$ are from Ref. [91], and observed data for $H_2{}^{80}Se$ are taken from Refs. [188].

[b] O = observed; C = calculated.

the state is described by the asymmetric rotor Hamiltonian given in Eq. (5.18). If, on the other hand, the unsymmetrized local mode state $|0v_r\rangle$ is excited, the principal axis system is rotated by the same amount as in the previous case but in the opposite direction. Again, the rotation energy level structure is described by an asymmetric rotor Hamiltonian with the same rotational constants as in the previous example. Splitting of the vibrational levels occurs if we allow for coupling between the bond oscillators. This splitting is related to the time scale for tunneling to a wave function for which the excitation has swapped bonds. If the rotational period is much shorter than the tunneling time, then the rotational Hamiltonian should reflect the reduced dynamic symmetry; that is, instead of the coupled two-

TABLE XXIX

Observed and Calculated d Parameters (cm^{-1}) and Coriolis Coefficients ζ (Dimensionless) between Local Mode Pair of States in H$_2$ ^{32}S and H$_2$ ^{80}Se^a

		H$_2$ ^{32}S		H$_2$ ^{80}Se	
$v_{r_1}v_{r_2}$	O/C	d	ζ	d	ζ
10	O	0.2961	0.0092	0.2299	0.0363
20	O	0.5713	0.0002	0.4448	0.0587
	C	0.5863	0.0026	0.4542	0.0121
30	O	—	—	0.6747	0.0962
	C	—	—	0.6875	0.0011
40	O	1.1393	0.0	—	—
	C	1.1830	0.0	—	—
50	O	1.3601	0.0	—	—
	C	1.4796	0.0	—	—
41	O	0.8191	0.0	—	—
	C	0.8829	0.0	—	—

^{a} See the footnote to Table XXVIII. ζ constants for H$_2$ ^{32}S are obtained from the references given in Ref. [91].

state Hamiltonian, a single asymmetric rotor Hamiltonian provides a good description of the problem. Now, interestingly in cases with non zero coupling between the bond oscillators, it is found that the tunneling rates of the rovibrational states are reduced from the corresponding vibrational rates. This has been interpreted to arise from the fact that in addition to transferring the vibrational action in the rotating molecule, the angular momenta in the body-fixed frame must also be reoriented to obtain an equivalent state. This discussion is treated mathematically in Refs. [35]. At this point it is also worth mentioning the usual clustering of rotational energy levels at high rotational excitation (fourfold clustering) both in the ground vibrational state and in the v_1/v_3 pair of states [35,190]. This can be thought to arise as follows: The molecule rotates around its two bonds in a clockwise or an anticlockwise manner. The two choices for the bond and the two choices for the sense of the rotation provide a total of four equivalent situations corresponding to a four-fold energy clustering. For more details see a review in Ref. [190].

Finally, relations between the rovibrational eigenstates of the local mode pair of eigenstates have been considered in H$_2$S [191]. This is useful in considering rovibrational intensity simulations using single asymmetric rotor Hamiltonians in the local mode limit.

E.   Rotations and Local Modes in XH$_3$ and XH$_4$

Extensions of the approach to pyramidal XH$_3$ and tetrahedral XH$_4$ is straightforward. Discussion of the appropriate rotational Hamiltonians is found in Refs. [35], [36], [174], [180], and [182]. The α relations obtained for the vibration–rotation parameters in the rovibrational local mode limit and for effective parameters of the local mode pair of states at high stretching energies are given in Tables XXX, and XXXI, respectively. The general conclusion obtained is the same as before; that is, the Coriolis effects are not important but the vibration–rotation couplings due to vibrationally off-diagonal H_{22} coefficients cannot be ignored. The appropriate rotational constants of the local mode pair of states are equal. The vibrational dependence of effective Hamiltonian parameters is well produced with vibrational models similar to the ones discussed above. If harmonic oscillator formulas are used, then the vibrational dependencies of the parameters at the high energy limit can be expressed in simple closed forms as given in Table XXXI. The readers are referred to the SbH_3 results given in Refs. [57] and [178] to AsH_3 results in Ref. [100], and to SnH_4 results in Refs. [151] and [155]. See also results for SiH_4 and GeH_4 in Ref. [36].

TABLE XXX
α Relations for Bent XH_2, Pyramidal XH_3, and Tetrahedral XH_4 Molecules[a]

α Relations	Redefined Molecular Parameters
XH_2	
$\alpha_1^{(x)} = \alpha_3^{(x)},\ \alpha_1^{(y)} = \alpha_3^{(y)},\ \alpha_1^{(z)} = \alpha_3^{(z)}$	$\alpha_+^{(\xi)} = \frac{1}{2}(\alpha_1^{(\xi)} + \alpha_3^{(\xi)})$
$\zeta_{13}^{(y)} = 0$	$\alpha_-^{(\xi)} = \alpha_1^{(\xi)} - \alpha_3^{(\xi)}\ (\xi = x, y, z)$
XH_3	
$\alpha_1^{(x)} = \alpha_3^{(x)},\ \alpha_1^{(z)} = \alpha_3^{(z)}$	$\alpha_+^{(\xi)} = \alpha_1^{(\xi)} + 2\alpha_3^{(\xi)}$
$-\sqrt{2}\,q_3 = \alpha_{13}^{(xx)},\ 4\sqrt{2}\,r_3 = \alpha_{13}^{(zx)}$	$\alpha_-^{(\xi)} = \alpha_1^{(\xi)} - \alpha_3^{(\xi)}\ (\xi = x, z)$
$\zeta_{1,\,3a}^{(y)} = \zeta_{3a,\,3b}^{(z)} = 0$	$q_+ = q_3 - \sqrt{2}\,\alpha_{13}^{(xx)}$
	$q_- = q_3 + \dfrac{1}{\sqrt{2}}\,\alpha_{13}^{(xx)}$
	$r_+ = -4r_3 - \sqrt{2}\,\alpha_{13}^{(zx)}$
	$r_- = 4r_3 - \dfrac{1}{\sqrt{2}}\,\alpha_{13}^{(zx)}$
XH_4	
$\alpha_1^{(B)} = \alpha_3^{(B)}$	$d_\pm = d_{13} \pm \alpha_{220} \mp 8\alpha_{224}\ \ a = \alpha_{220} + 12\alpha_{224}$
$d_{13} = \frac{5}{3}\alpha_{220} = -20\alpha_{224}$	
$\zeta = 0$	

[a] The parameters with a minus sign as a subscript and the a parameter disappear for near-local-mode molecules as stretching vibrational energy increases. For XH_3 and XH_4 results given see Refs. [28], [35], [36], and [174] for the definition of the parameters and the appropriate Hamiltonian operators.

TABLE XXXI

α Relations and Vibrational Dependence of the Effective Vibration–Rotation Parameters for XH_2, XH_3, and XH_4 Molecules[a]

XH_2

$$\alpha_{\text{eff}}^{(\xi)} = \alpha_{A_1, \text{eff}}^{(\xi)} = \alpha_{B_1, \text{eff}}^{(\xi)} = B_0^{(\xi)} - B_{+, \text{eff}}^{(\xi)} = B_0^{(\xi)} - B_{-, \text{eff}}^{(\xi)} = \tfrac{1}{2} v_r(\alpha_1^{(\xi)} + \alpha_3^{(\xi)}) \ (\xi = x, y, z)$$
$$d_{\text{eff}} = v_r d_{13}$$
$$\zeta_{\text{eff}} = 0$$

XH_3

$$\alpha_{\text{eff}}^{(\xi)} = \alpha_{A_1, \text{eff}}^{(\xi)} = \alpha_{E, \text{eff}}^{(\xi)} = B_0^{(\xi)} - B_{A_1, \text{eff}}^{(\xi)} = B_0^{(\xi)} - B_{E, \text{eff}}^{(\xi)} = \tfrac{1}{3} v_r(\alpha_1^{(\xi)} + 2\alpha_3^{(\xi)}) \ (\xi = x, z)$$

$$q_{\text{eff}} = -\frac{1}{\sqrt{2}} \alpha_{\text{eff}}^{(xx)} = \frac{v_r}{3}(q_3 - \sqrt{2}\,\alpha_{13}^{(xx)})$$

$$r_{\text{eff}} = \frac{1}{4\sqrt{2}} \alpha_{\text{eff}}^{(zx)} = \frac{v_r}{6}\left(2r_3 + \frac{1}{\sqrt{2}} \alpha_{13}^{(zx)}\right)$$

$$(\zeta_{1, 3a}^{(y)})_{\text{eff}} = (\zeta_{3a, 3b}^{(z)})_{\text{eff}} = 0$$

XH_4

$$\alpha_{\text{eff}}^{(B)} = \alpha_{A_1, \text{eff}}^{(B)} = \alpha_{F_2, \text{eff}}^{(B)} = B_0 - B_{A_1, \text{eff}} = B_0 - B_{F_2, \text{eff}} = \frac{v_r}{4}(\alpha_1^{(B)} + 3\alpha_3^{(B)})$$

$$d_{\text{eff}} = \tfrac{5}{3}\alpha_{220, \text{eff}} = -20\alpha_{224, \text{eff}} = \frac{v_r}{2}(d_{13} + \alpha_{220} - 8\alpha_{224})$$

$$\zeta_{\text{eff}} = 0$$

[a] The α relations given apply only at high stretching energies to effective molecular parameters of the local mode pair of states, i.e., for states labelled as $[v_r 0 \pm] = [v_r, 0 A_1/B_1]$ in XH_2, $[v_r, 0 0 A_1/E]$ in XH_3 and $[v_r, 0 0 0 A_1/F_2]$ in XH_4. For XH_3 and XH_4 results given see Refs. [28], [36], and [174] for the definition of the parameters and for the appropriate Hamiltonian operators. For similar results of the $[(v_r - 1)1 \pm]$ states in XH_2 see Ref. [91].

The behavior of the vibration–rotation Hamiltonian of the local mode pair of states in the rovibrational local mode limit is obtained as in the XH_2 case by averaging over single local mode states $|v_r 00\rangle$ and $|v_r 000\rangle$ in XH_3 and XH_4, respectively. In XH_3, the problem is formulated in terms of an asymmetric top Hamiltonian given in Eq. (5.18). When the z axis is chosen to be the symmetry axis and harmonic oscillator matrix elements are used, the rotational constants become

$$B_v^{(x')} = \tfrac{1}{2}\{B_v^{(z)} + B_v^{(x)} + E - [(B_v^{(z)} - B_v^{(x)} - E)^2 + 4G^2]^{1/2}\}$$
$$B_v^{(z')} = \tfrac{1}{2}\{B_v^{(z)} + B_v^{(x)} + E + [(B_v^{(z)} - B_v^{(x)} - E)^2 + 4G^2]^{1/2}\} \quad (5.38)$$
$$B_v^{(y')} = B_v^{(x)} - E$$

where $B_v^{(\xi)} = B_e^{(\xi)} - \alpha^{(\xi)}(v_r + \frac{3}{2})$, $\alpha^{(\xi)} = \alpha_1^{(\xi)} = \alpha_3^{(\xi)}$, $\xi = x$, z, $E = -v_r q_3$, and $G = 4v_r r_3$, $B_e^{(\xi)}$ the equilibrium rotational constant. The spectroscopic parameters $\alpha_1^{(\xi)}$, $\alpha_3^{(\xi)}$, q_3, and r_3 are defined in Refs. [28] and [36] (see also Table XXX). The same result holds for the local mode pair of states $[v_r 00A_1/E]$ at high stretching energy when the rotational constants $B_v^{(\xi)}$ are replaced by $B_{eff}^{(\xi)} = B_{A1, eff}^{(\xi)} = B_{E, eff}^{(\xi)}$ and when $E = -q_{eff}$ and $G = 4r_{eff}$ (see Table XXXI). In XH_4 tetrahedra in the rovibrational local mode limit, the effective rotational Hamiltonian becomes that of a symmetric top, that is,

$$\frac{H}{hc_0} = B_v' \hat{J}^2 + (A_v' - B_v')\hat{J}_z^2 \tag{5.39}$$

where the rotational constants B_v' and A_v' are

$$B_v' = B_v - \tfrac{1}{2}v_r d_{13} \qquad A_v' = B_v + v_r d_{13} \tag{5.40}$$

and $B_v = B_e - \alpha^{(B)}(v_r + 2)$, $\alpha^{(B)} = \alpha_1^{(B)} = \alpha_3^{(B)}$, and B_e is the equilibrium rotational constant. The spectroscopic coefficients $\alpha_1^{(B)}$, $\alpha_3^{(B)}$, and d_{13} are defined in Refs. [28], [36], and [174]. As before, in XH_2 and XH_3 molecules, the same result holds for the local mode pair of states $[v_r 000A_1/F_2]$ at high v_r values when the rotational constant B_v is replaced by $B_{eff} = B_{A1, eff} = B_{F2, eff}$ and $v_r d_{13} = d_{eff}$ (see Table XXXI). The conclusion is such that at high stretching excitations the dynamic symmetry of the local mode pair of states changes from C_{3v} to C_s in XH_3 and from T_d to C_{3v} in XH_4. Visually, most drastically this happens in high-resolution spherical top spectra of SnH_4 (see Fig. 7) [151,155]. Recently rotational local mode effects have also been studied in H_3XY-type symmetric top molecules such as H_3 ^{74}Ge ^{79}Br [192], H_3 ^{74}GeI [192], H_3 ^{70}GeD [187,193], and H_3SiD [194]. In many cases results similar to XH_3 systems are obtained; that is, the dynamic symmetry changes in the local mode pair of states.

VI. CONCLUSION

This chapter has presented a molecular vibrational model which is based on curvilinear internal valence coordinates. Both stretching and bending vibrations are included. The model has mainly been applied to XH_2-, XH_3-, and XH_4-type molecules, but there does not seem to be any reason why these ideas cannot be applied to larger molecules. This extension would mainly be of a technical nature. The rapid development of modern computers means that in the near future these calculations will become feasible.

The curvilinear internal coordinate models have been shown to be equivalent to traditional normal coordinate models. This might imply that there is little gained in the internal coordinate formulation. However, this is not the case for several reasons. First, the physical picture of the internal coordinate model is superior. Second, the internal coordinate parameters are directly related to Born–Oppenheimer potential energy parameters, which by definition are isotope independent. Third, the wave functions in the internal coordinate representation for molecules near the local mode limit are often close to simple zeroth-order basis functions. This is important in overtone spectroscopy, where the strongest bands belong to hydrogen stretching vibrations, which behave in a near-local-limit manner.

The rotational motion with stretching vibrations only has been discussed. The inclusion of bending vibrations to this analysis has not been covered. It seems that for the sake of locallike stretching vibrations a rather complete picture is presented with dynamic symmetry changes in symmetrical molecules and with special relations between rovibrational molecular parameters. Another aspect, the clustering of rotational levels at high rotational excitations, is, apart from XH_2-type molecules, probably less well characterized.

Finally, it should be realized that there exist other ways to model overtone problems. This review has not discussed the so-called algebraic models which are used widely in the literature [195,196]. This approach is based on the use of symmetry, and it is not directly related to Born–Oppenheimer potential energy surfaces. However, it has been successful in interpreting overtone spectra, including both stretching and bending overtone states in a variety of different polyatomic molecules. The algebraic models often produce simple block diagonal vibrational Hamiltonian matrices which are similar to those presented here. It remains a future task to relate in detail the algebraic models and the conventional models presented in this chapter with each other.

ACKNOWLEDGMENTS

The author is grateful to Marjo Halonen, Matthias Horn, and Juha Lummila for helpful comments. The author thanks the Rector of the University of Helsinki and the European Commission (TMR network contract number FMRX-CT96-0088) for financial support.

REFERENCES

1. B. R. Henry, in *Vibrational Spectra and Structure*, J. R. Durig, Ed., Elsevier, New York, 1981, Vol. 10; B. R. Henry, *Acc. Chem. Res.*, **10**, 207 (1977), **20**, 429 (1987). B. R. Henry, H. G. Kjaergaard, B. Niefer, B. J. Schattka, and D. M. Turnbull, *Can. J. Appl. Spectrosc.*, **38**, 42 (1993).

2. M. L. Sage and J. Jortner, *Adv. Chem. Phys.*, **47**, Pt. 1, 293 (1981).

3. M. S. Child and L. Halonen, *Adv. Chem. Phys.*, **57**, 1 (1984); M. S. Child, *Acc. Chem. Res.*, **18**, 45 (1985).

4. *Faraday Trans.* 2, **84**, part 9 (1988); *Comp. Physics Commun.*, **51** 1, 2 (1988); *Chem. Phys.*, **190** 2, 3 (1995); K. K. Lehmann, G. Scoles, and B. H. Pate, *Ann. Rev. Phys. Chem.*, **45**, 241 (1994); M. Kellman, in *Advanced Series in Physical Chemistry, Molecular Dynamics and Spectroscopy by Stimulated Emission Pumping*, H.-L. Dai and R. W. Field, Eds., World Scientific, Singapore, 1995, Vol. 4.

5. L. Halonen, *J. Phys. Chem.*, **93**, 3386 (1989).

6. K. K. Lehmann, G. J. Scherer, and W. Klemperer, *J. Chem. Phys.*, **77**, 2853 (1982); H. L. Fang and D. A. C. Compton, *Appl. Opt.*, **21**, 55 (1982); R. F. Menefee, R. R. Hall, and M. J. Berry, *Appl. Phys. B*, **28**, 121 (1982); O. M. Sarkisov, E. A. Sviridenkov, and A. F. Suchkov, *Sov. J. Chem. Phys.*, **1**, 1993 (1984); J. W. Perry, D. J. Moll, A. Kuppermann, and A. H. Zewail, *J. Chem. Phys.*, **82**, 1195 (1984); T. R. Rizzo, C. C. Hayden, and F. F. Crim, *J. Chem. Phys.*, **81**, 4501 (1984); X. Luo and T. R. Rizzo, *J. Chem. Phys.*, **93**, 8620 (1990); R. D. F. Settle and T. R. Rizzo, *J. Chem. Phys.*, **97**, 2823 (1992); O. V. Boyarkin and T. R. Rizzo, *J. Chem. Phys.*, **103**, 1985 (1995); A. M. de Souza, D. Kaur, and D. S. Perry, *J. Chem. Phys.*, **88**, 4569 (1988); R. H. Page, Y. R. Shen, and Y. T. Lee, *J. Chem. Phys.*, **88**, 4621, 5362 (1988); C. Douketis and J. P. Reilly, *J. Chem. Phys.*, **91**, 5239 (1989); *J. Chem. Phys.*, **96**, 3431 (1992); J. Davidsson, J. H. Gutow, and R. N. Zare, *J. Phys. Chem.*, **94**, 4069 (1990); J. K. Holland, D. A. Newnham, and I. M. Mills, *Mol. Phys.*, **70**, 319 (1990); D. A. Newnham, Ph.D. Thesis, University of Reading, Reading, England (1990); X. Zhan, O. Vaittinen, E. Kauppi, and L. Halonen, *Chem. Phys. Lett.*, **180**, 310 (1991); D. Newnham, X. Zhan, O. Vaittinen, E. Kauppi, and L. Halonen, *Chem. Phys. Lett.*, **189**, 205 (1992); X. Zhan, E. Kauppi, and L. Halonen, *Rev. Sci. Instrum.*, **63**, 5546 (1992); A Campargue, F. Stoeckel, and M. Chenevier, *Spectrochim. Acta Rev.* 13, 69 (1990); A. Campargue, M. Chenevier, and F. Stoeckel, *Chem. Phys. Lett.*, **183**, 153 (1991); A. Campargue and D. Permogorov, *Chem. Phys.*, **182**, 281 (1994); A. Campargue, D. Permogorov, and R. Jost, *J. Chem. Phys.*, **102**, 5910 (1995); T. Yoshida and H. Sasada, *J. Mol. Spectrosc.*, **153**, 208 (1992); S. Hassoon and D. L. Snavely, *J. Chem. Phys.*, **99**, 2511 (1993); D. Romanini and K. K. Lehmann, *J. Chem. Phys.*, **99**, 6287 (1993); J. E. Gambogi, E. R. T. Kerstel, K. K. Lehmann, and G. Scoles, *J. Chem. Phys.*, **100**, 2612 (1994); J. E. Gambogi, R. Z. Pearson, X. Yang, K. K. Lehmann, and G. Scoles *Chem. Phys.*, **190**, 191 (1995); A. Callegari, H. K. Srivastava, U. Merker, K. K. Lehmann, G. Scoles, and M. J. Davis, *J. Chem. Phys.*, **106**, 432 (1997); M. Hippler and M. Quack, *Ber. Bunseges. Phys. Chem.*, **99**, 417 (1995); J. A. Barnes, T. E. Gough, and M. Stoer, *Chem. Phys. Lett.*, **237**, 437 (1995); D. Romanini, A. A. Kachanov, N. Sadeghi, and F. Stoeckel, *Chem. Phys. Lett.*, **264**, 316 (1997); P. Jungner and L. Halonen, *J. Chem. Phys.*, **107**, 1680 (1997); O. Votava, J. R. Fair, D. F. Plusquellic, E. Riedle, and D. J. Nesbitt, *J. Chem. Phys.*, **107**, 8854 (1997); J. R. Fair, O. Votava, and D. J. Nesbitt, *J. Chem. Phys.*, **108**, 72 (1998); H.-C. Chang and W. Klemperer, *J. Chem. Phys.*, **98**, 2497 (1993); *J. Chem. Phys.*, **98**, 9266 (1993); A. P. Milce and B. J. Orr, *J. Chem. Phys.*, **104**, 6423 (1996); *J. Chem. Phys.*, **106**, 3592 (1997).

7. X. Zhan, *Ann. Acad. Scientiarum Fennicae A*, **252**, 1 (1993).

8. T. E. Martin and A. H. Kalantar, *J. Chem. Phys.*, **49**, 235 (1968); R. L. Swofford, M. E. Long, and A. C. Albrecht, *J. Chem. Phys.*, **65**, 179 (1976); B. R. Henry and W. Siebrand, *J. Chem. Phys.*, **49**, 5369 (1968); R. J. Hayward, B. R. Henry, and W. Siebrand, *J. Mol. Spectrosc.*, **46**, 207 (1973); R. G. Bray and M. J. Berry, *J. Chem. Phys.*, **71**, 4909 (1979).

9. G. Herzberg, *Infrared and Raman Spectra*, Van Nostrand, New York, 1945.

10. J. W. Ellis, *Phys. Rev.*, **32**, 906 (1928); *Phys. Rev.*, **33**, 27 (1929); *Trans. Faraday Soc.*, **25**, 888 (1929); R. Z. Mecke, *Z. Phys. Chem. B*, **17**, 1 (1932); *Z. Phys.*, **81**, 311 (1933); *Z. Phys.*,

99, 217 (1936); B. Timm and R. Mecke, *Z. Phys.*, **98**, 363 (1936); K. Rumf and R. Mecke, *Z. Phys. Chem. B*, **44**, 299 (1939).

11. P. Jensen, S. A. Tashkun, and Vl. G. Tyuterev, *J. Mol. Spectrosc.*, **168**, 271 (1994).

12. E. Kauppi and L. Halonen, *J. Chem. Phys.*, **96**, 2933 (1992); E. Kauppi, Ph.D. Thesis, University of Helsinki, Helsinki, Finland (1992).

13. M. S. Child and R. T. Lawton, *Faraday Discuss. Chem. Soc.*, **71**, 273 (1981).

14. P. M. Morse, *Phys. Rev.*, **34**, 57 (1929).

15. M. L. Sage and J. A. Williams III, *J. Chem. Phys.*, **78**, 1348 (1983); M. L. Sage, *Chem. Phys.*, **35**, 375 (1978).

16. O. S. Mortensen, B. R. Henry, and M. A. Mohammadi, *J. Chem. Phys.*, **75**, 4800 (1981).

17. E. B. Wilson, J. C. Decius, and P. C. Cross, *Molecular Vibrations*, McGraw-Hill, New York, 1955.

18. H. W. Kroto, *Molecular Rotation Spectra*, Dover, New York, 1992.

19. I. M. Mills, in *A Specialist Periodical Report, Theoretical Chemistry*, R. N. Dixon, Ed., Chemical Society, London, 1974, Vol. 1.

20. B. T. Darling and D. M. Dennison, *Phys. Rev.*, **57**, 128 (1940).

21. I. M. Mills and A. G. Robiette, *Mol. Phys.*, **56**, 743 (1985).

22. I. M. Mills, in *Molecular Spectroscopy: Modern Research*, K. N. Rao and C. W. Mathews, Eds., Academic, New York, 1972, Vol. I.

23. D. Papousek and M. R. Aliev, *Molecular Vibrational-Rotational Spectra*, Elsevier, Amsterdam, 1982.

24. K. K. Lehmann, *J. Chem. Phys.*, **79**, 1098 (1983).

25. K. K. Lehmann, *J. Chem. Phys.*, **84**, 6524 (1986).

26. R. G. Della Valle, *Mol. Phys.*, **63**, 611 (1988).

27. A. Messiah, *Quantum Mechanics*, Wiley, New York, 1958.

28. L. Halonen and A. G. Robiette, *J. Chem. Phys.*, **84**, 6861 (1986).

29. M. Abbouti Temsamani and M. Herman, *J. Chem. Phys.*, **102**, 6371 (1995).

30. A. Campargue, M. Abbouti Temsamani, and M. Herman, *Mol. Phys.*, **90**, 787 (1997); S.-F. Yang, L. Binnier, A. Campargue, M. Abbouti Temsamani, and M. Herman, *Mol. Phys.*, **90**, 807 (1997).

31. R. R. Hall, Ph.D. Thesis, Rice University, Houston, (1984).

32. M. Herman and A. Pisarchik, *J. Mol. Spectrosc.*, **164**, 210 (1994).

33. C. Camy-Peyret, J.-M. Flaud, J.-Y. Mandin, J.-P. Chevillard, J. Brault, D. A. Ramsay, M. Vervloet, and J. Chauville, *J. Mol. Spectrosc.*, **113**, 208 (1985).

34. J. E. Baggott, *Mol. Phys.*, **65**, 739 (1988).

35. K. K. Lehmann, *J. Chem. Phys.*, **95**, 2361 (1991); *J. Chem. Phys.*, **96**, 7402 (1992).

36. T. Lukka and L. Halonen, *J. Chem. Phys.*, **101**, 8380 (1994).

37. H. C. Allen, Jr., E. D. Tidwell, and E. K. Plyler, *J. Res. Nat. Bur. Standards*, **57**, 213 (1956).

38. H. C. Allen and E. K. Plyler, *J. Chem. Phys.*, **25**, 1132 (1956).

39. R. A. Hill and T. H. Edwards, *J. Chem. Phys.*, **42**, 1391 (1965).

40. I. M. Mills and F. J. Mompean, *Chem. Phys. Lett.*, **124**, 425 (1986).

41. R. J. Hayward and B. R. Henry, *J. Mol. Spectrosc.*, **50**, 58 (1974); *J. Mol. Spectrosc.*, **57**, 221 (1975).

42. R. Wallace, *Chem. Phys.*, **11**, 189 (1975).
43. I. A. Watson, B. R. Henry, and I. G. Ross, *Spectrochim. Acta*, **37A**, 857 (1981).
44. R. T. Lawton and M. S. Child, *Mol. Phys.*, **40**, 773 (1980); M. S. Child and R. T. Lawton, *Chem. Phys. Lett.*, **87**, 217 (1982).
45. L. Halonen and M. S. Child, *Mol. Phys.*, **46**, 239 (1982).
46. L. Halonen, M. S. Child, and S. Carter, *Mol. Phys.*, **47**, 1097 (1982); L. Halonen, D. W. Noid, and M. S. Child, *J. Chem. Phys.*, **78**, 2803 (1983).
47. L. Halonen and M. S. Child, *J. Chem. Phys.*, **79**, 559 (1983).
48. L. Halonen, *J. Mol. Spectrosc.*, **120**, 175 (1986).
49. L. Halonen and M. S. Child, *Comp. Phys. Comm.*, **51**, 173 (1988).
50. L. Halonen, *J. Phys. Chem.*, **93**, 631 (1989).
51. J. S. Griffith, *The Theory of Transition Metal Ions*, Cambridge University Press, Cambridge, England, 1961.
52. L. Halonen and M. S. Child, *J. Chem. Phys.*, **79**, 4355 (1983).
53. B. R. Henry, A. W. Tarr, O. S. Mortensen, W. F. Murphy, and D. A. C. Compton, *J. Chem. Phys.*, **79**, 2583 (1983).
54. M. Halonen, L. Halonen, H. Bürger, and S. Sommer, *J. Phys. Chem.*, **94**, 5222 (1990).
55. J. L. Duncan, G. D. Nivellini, F. Tullini, and L. Fusina, *Chem. Phys. Lett.*, **165**, 362 (1990).
56. T. Lukka, E. Kauppi, and L. Halonen, *J. Chem. Phys.*, **102**, 5200 (1995).
57. J. Lummila, T. Lukka, L. Halonen, H. Bürger, and O. Polanz, *J. Chem. Phys.*, **104**, 488 (1996).
58. G. J. Scherer, K. K. Lehmann, and W. Klemperer, *J. Chem. Phys.*, **78**, 2817 (1983).
59. A. Baldacci, S. Ghersetti, S. C. Hurlock, and K. N. Rao, *J. Mol. Spectrosc.*, **59**, 116 (1976).
60. J. E. Baggott, D. A. Newnham, J. L. Duncan, D. C. McKean, and A. Brown, *Mol. Phys.*, **70**, 715 (1990); J. E. Baggott, D. W. Law, and I. M. Mills, *Mol. Phys.*, **61**, 1309 (1987).
61. L. Lubich, O. L. Boyarkin, D. F. Settle, D. S. Perry, and T. R. Rizzo, *Faraday Discuss. Chem. Soc.*, **102**, 167 (1995).
62. K. K. Lehmann, *Mol. Phys.*, **66**, 1129 (1989); *Mol. Phys.*, **75**, 739 (1992).
63. M. M. Law, Ph.D. Thesis, University of Aberdeen, Aberdeen, Scotland (1992).
64. L. Halonen, *J. Chem. Phys.*, **106**, 7931 (1997).
65. J. H. Van Vleck, *Phys. Rev.*, **33**, 467 (1929).
66. C. E. Blom, L. P. Otto, and C. Altona, *Mol. Phys.*, **32**, 1137 (1976).
67. I. M. Mills, *Faraday Discuss. Chem. Soc.*, **102**, 244 (1995).
68. H. C. Allen, Jr., L. R. Blaine, and E. K. Plyler, *J. Res. Nat. Bur. Standards*, **56**, 279 (1956).
69. W. S. Benedict, *J. Chem. Phys.*, **24**, 1139 (1956).
70. S. L. Coy and K. K. Lehmann, *Spectrochim. Acta*, **45A**, 47 (1989).
71. D. L. Gray and A. G. Robiette, *Mol. Phys.*, **37**, 1901 (1979).
72. N. C. Handy, J. F. Gaw, and E. D. Simandiras, *J. Chem. Soc., Faraday Trans. 2*, **83**, 1577 (1987).
73. J. Senekowitsch, S. Carter, A. Zilch, H.-J. Werner, N. C. Handy, and P. Rosmus, *J. Chem. Phys.*, **90**, 783 (1989).
74. J. M. L. Martin, T. J. Lee, and P. R. Taylor, *J. Chem. Phys.*, **97**, 8361 (1992).
75. T. J. Lee, J. M. L. Martin, and P. R. Taylor, *J. Chem. Phys.*, **102**, 254 (1995).

76. M. M. Law and J. L. Duncan, *Mol. Phys.*, **83**, 757 (1994).

77. J. E. Baggott, *Mol. Phys.*, **62**, 1019 (1987).

78. I. L. Cooper, *Chem. Phys.*, **112**, 67 (1987).

79. T. Carrington, Jr., *J. Chem. Phys.*, **86**, 2207 (1987).

80. E. Kauppi and L. Halonen, *J. Phys. Chem.*, **94**, 5779 (1990).

81. H. M. Pickett, *J. Chem. Phys.*, **56**, 1715 (1972).

82. R. Meyer and Hs. H. Günthard, *J. Chem. Phys.* **49**, 1510 (1968).

83. L. Halonen and T. Carrington, Jr., *J. Chem. Phys.*, **88**, 4171 (1988).

84. R. Wallace, *Chem. Phys.*, **11**, 189 (1975).

85. L. A. Gribov, *Opt. Spectrosc.*, **31**, 456 (1981).

86. C. R. Quade, *J. Chem. Phys.*, **64**, 2783 (1976); *J. Chem. Phys.*, **79**, 4989 (1983).

87. E. L. Sibert III, J. T. Hynes, and W. P. Reinhardt, *J. Phys. Chem.*, **87**, 2032 (1983).

88. A. R. Hoy, I. M. Mills, and G. Strey, *Mol. Phys.*, **24**, 1265 (1972).

89. A. D. Bykov, O. V. Naumenko, M. A. Smirnov, L. N. Sinitsa, L. R. Brown, J. Crisp, and D. Crisp, *Can. J. Phys.*, **72**, 989 (1994).

90. J.-M. Flaud, R. Grosskloss, S. B. Rai, R. Stuber, W. Demtröder, D. A. Date, L.-G. Wang, and T. H. Gallagher, *J. Mol. Spectrosc.*, **172**, 275 (1995).

91. O. Vaittinen, L. Biennier, A. Campargue, J.-M. Flaud, and L. Halonen, *J. Mol. Spectrosc.*, **184**, 288 (1997).

92. A. Baldacci, V. M. Devi, and K. N. Rao, *J. Mol. Spectrosc.*, **81**, 179 (1980).

93. G. Tarrago, M. Dang-Nhu, and A. Goldman, *J. Mol. Spectrosc.*, **88**, 311 (1981).

94. G. Di Lonardo, L. Fusina, and J. W. C. Johns, *J. Mol. Spectrosc.*, **104**, 282 (1984).

95. Wm. B. Olsen, A. G. Maki, and R. L. Sams, *J. Mol. Spectrosc.*, **55**, 252 (1975).

96. L. Fusina (private communication).

97. J. Pesonen and L. Halonen (unpublished results).

98. J. Breidung and W. Thiel, *J. Mol. Spectrosc.*, **169**, 166 (1995); W. Thiel (private communication, 1996).

99. K. K. Lehmann, *J. Chem. Phys.*, **96**, 8117 (1992).

100. O. N. Ulenikov, F. Sun, X. Wang, and Q. Zhu, *J. Chem. Phys.*, **105**, 7310 (1996).

101. E. Kauppi and L. Halonen, *J. Chem. Phys.*, **103**, 6861 (1995).

102. K. K. Lehmann and S. L. Coy, *J. Chem. Soc. Faraday Trans. 2*, **84**, 1389 (1988).

103. L. Halonen, *J. Chem. Phys.*, **106**, 831 (1997).

104. T. Simanouti, *J. Chem. Phys.*, **17**, 245 (1949).

105. S. Califano, *Vibrational States*, Wiley, London, 1976.

106. W. T. Raynes, P. Lazzeretti, R. Zanasi, A. J. Sadlej, and P. W. Fowler, *Mol. Phys.*, **60**, 509 (1987); K. Kuchitsu and L. S. Bartell, *J. Chem. Phys.*, **36**, 2470 (1962); D. Bermejo and S. Montero, *J. Chem. Phys.*, **81**, 3835 (1984).

107. J. K. Holland, D. Newnham, I. M. Mills, and M. Herman, *J. Mol. Spectrosc.*, **151**, 346 (1992); J. K. Holland, W. D. Lawrance, and I. M. Mills, *J. Mol. Spectrosc.*, **151**, 369 (1992); A. F. Borro and I. M. Mills, *J. Mol. Struct.*, **320**, 237 (1995); O. Vaittinen, M. Saarinen, L. Halonen, and I. M. Mills, *J. Chem. Phys.*, **99**, 3277 (1993); X. Zhan, O. Vaittinen, and L. Halonen, *J. Mol. Spectrosc.*, **160**, 172 (1993); X. Zhan and L. Halonen, *J. Mol. Spectrosc.*, **160**, 464 (1993); O. Vaittinen, L. Halonen, H. Bürger, and O. Polanz, *J. Mol. Spectrosc.*, **167**, 55 (1994); M. Halonen, *J. Mol. Spectrosc.*, **167**, 225 (1994); O. Vait-

tinen, T. Lukka, L. Halonen, H. Bürger, and O. Polanz, *J. Mol. Spectrosc.*, **172**, 503 (1995); O. Vaittinen, *Chem. Phys. Lett.* **238**, 319 (1995); J. Lummila, O. Vaittinen, P. Jungner, L. Halonen, and A.-M. Tolonen, *J. Mol. Spectrosc.*, **185**, 296 (1997); O. Vaittinen, M. Hämäläinen, P. Jungner, K. Pulkkinen, L. Halonen, H. Bürger, and O. Polanz, *J. Mol. Spectrosc.*, **187**, 193 (1998); E. Venuti, G. Di Lonardo, P. Ferracuti, L. Fusina, and I. M. Mills, *Chem. Phys.*, **190**, 279 (1995); A. F. Borro, I. M. Mills, and A. Mose, *Chem. Phys.*, **190**, 363 (1995); A. F. Borro, I. M. Mills, and E. Venuti, *J. Chem. Phys.*, **190**, 3938 (1995); M. Herman, T. R. Huet, and V. Vervloet, *Mol. Phys.*, **66**, 333 (1989); Q. Kou, G. Guelachvili, M. Abbouti Temsamani, and M. Herman, *Can. J. Chem.*, **72**, 1241 (1994); J. Lievin, M. Abbouti Temsamani, P. Gaspard, and M. Herman, *Chem. Phys.*, **190**, 419 (1995); M. Abbouti Temsamani and M. Herman, *J. Chem. Phys.*, **105**, 1355 (1996); D. McNaughton, D. McGilvery, and F. Shanks, *J. Mol. Spectrosc.*, **149**, 458 (1991); D. McNaughton and M. Shallard, *J. Mol. Spectrosc.*, **165**, 185 (1994); A.-M. Tolonen, S. Alanko, M. Koivusaari, R. Paso, and V.-M. Horneman, *J. Mol. Spectrosc.*, **165**, 249 (1994); A.-M. Tolonen, S. Alanko, R. Paso, V.-M. Horneman, and B. Nelander, *Mol. Phys.*, **83**, 1233 (1994).

108. J. L. Duncan, *Spectrochim. Acta*, **47A**, 1 (1991).

109. M. Quack, *Ann. Rev. Phys. Chem.*, **41**, 839 (1990).

110. J. E. Baggott, H. J. Clase, and I. M. Mills, *Spectrochim. Acta*, **42A**, 319 (1986); J. E. Baggott, D. W. Law, P. D. Lightfoot, and I. M. Mills, *J. Chem. Phys.*, **85**, 5414 (1986).

111. J. L. Duncan and I. M. Mills, *Chem. Phys. Lett.*, **145**, 347 (1988); L. J. Duncan and A. M. Ferguson, *J. Chem. Phys.*, **89**, 4216 (1988); J. L. Duncan, D. C. McKean, I. Torto, A. Brown, and A. M. Ferguson, *J. Chem. Soc. Faraday Trans. 2*, **84**, 1423 (1988); J. L. Duncan, A. M. Ferguson, and S. T. Goodlad, *Spectrochim. Acta*, **49A**, 149 (1993).

112. J. L. Duncan and M. M. Law, *J. Mol. Spectrosc.*, **140**, 13 (1990).

113. J. L. Duncan, *Spectrochim. Acta*, **45A**, 1067 (1989).

114. J. L. Duncan, A. M. Ferguson, and S. Mathews, *J. Chem. Phys.*, **91**, 783 (1989).

115. F. Tullini, G. Nivellini, and L. Fusina, *J. Mol. Structr.*, **320**, 81 (1994).

116. J. L. Duncan, C. A. New, and B. Leavitt, *J. Chem. Phys.*, **102**, 4012 (1995).

117. L. Ricard-Lespade, G. Longhi, and S. Abbate, *Chem. Phys.*, **142**, 245 (1990); L. Lespade, S. Rodin, D. Cavagnat, and S. Abbate, *J. Phys. Chem.*, **97**, 6134 (1993); S. Rodin-Bercion, D. Cavagnat, L. Lespade, and P. Maraval, *J. Phys. Chem.*, **99**, 3005 (1995); D. Cavagnat, L. Lespade, and C. Lapouge, *Ber. Bunsenges. Phys. Chem.*, **99**, 544 (1995); D. Cavagnat, L. Lespade, and C. Lapouge, *J. Chem. Phys.*, **103**, 10502 (1995); D. Cavagnat and L. Lespade, *J. Chem. Phys.*, **106**, 7946 (1997).

118. L. Halonen, *J. Chem. Phys.*, **88**, 7599 (1988).

119. H. G. Kjaergaard, B. R. Henry, H. Wei, S. Lefebvre, T. Carrington, Jr., O. S. Mortensen, and M. L. Sage, *J. Chem. Phys.*, **100**, 6228 (1994).

120. Z. Bacic and J. C. Light, *Ann. Rev. Phys. Chem.*, **40**, 469 (1989).

121. H. Margenau and G. M. Murphy, *The Mathematics of Physics and Chemistry*, Van Nostrand, New York, 1956.

122. R. Meyer, Intramolekulare Bewegung, lecture notes 1985/1986. Laboratorium für Physikalische Chemie, ETH, Zürich.

123. B. Podolsky, *Phys. Rev.*, **32**, 812 (1928).

124. E. C. Kemble, *Fundamental Principles of Quantum Mechanics*, Dover, New York, 1958.

125. S. Carter and N. C. Handy, *Comp. Phys. Rep.*, **5**, 117 (1986).

126. N. C. Handy, *Mol. Phys.*, **61**, 207 (1987); A. G. Csaszar and N. C. Handy, *Mol. Phys.*, **86**,

959 (1995); *J. Chem. Phys.*, **102**, 3962 (1995).

127. S. Carter, N. C. Handy, and B. T. Sutcliffe, *Mol. Phys.*, **49**, 745 (1983).

128. C. Eckart, *Phys. Rev.*, **47**, 552 (1935); P. R. Bunker, *Molecular Symmetry and Spectroscopy*, Academic, New York, 1979.

129. T. J. Lukka, *J. Chem. Phys.*, **102**, 3945 (1995); T. J. Lukka, Ph.D. Thesis, University of Helsinki, Helsinki, Finland (1995).

130. K. L. Mardis and E. L. Sibert III, *J. Chem. Phys.*, **106**, 6618 (1997); H. Wei and T. Carrington, Jr., *J. Chem. Phys.*, **107**, 2813 (1997); *J. Chem. Phys.*, **107**, 9493 (1997); T. J. Lukka and E. Kauppi, *J. Chem. Phys.*, **103**, 6586 (1995).

131. P. Jensen, *J. Mol. Spectrosc.*, **128**, 478 (1988); *J. Mol. Spectrosc.*, **133**, 438 (1989).

132. J. T. Hougen, P. R. Bunker, and J. W. C. Johns, *J. Mol. Spectrosc.*, **34**, 136 (1970).

133. J. W. Cooley, *Math. Comp.*, **15**, 363 (1961).

134. J. K. G. Watson, *Mol. Phys.*, **15**, 479 (1968).

135. T. Oka, *J. Chem. Phys.*, **47**, 5410 (1967).

136. C. Di Lauro and I. M. Mills, *J. Mol. Spectrosc.*, **21**, 386 (1966).

137. J. Tennyson and B. T. Sutcliffe, *J. Chem. Phys.*, **77**, 4061 (1982).

138. H. Wei and T. Carrington, Jr., *J. Chem. Phys.*, **97**, 3029 (1992).

139. S. Carter and N. C. Handy, *J. Chem. Phys.*, **87**, 4294 (1987).

140. B. R. Johnson and W. P. Reinhardt, *J. Chem. Phys.*, **85**, 4538 (1986).

141. W. Gordy and R. L. Cook, *Microwave Molecular Spectra*, Interscience, New York, 1970.

142. R. N. Zare, *Angular Momentum*, Wiley, New York, 1988.

143. I. P. Hamilton and J. C. Light, *J. Chem. Phys.*, **84**, 306 (1986).

144. A. S. Dickinson and P. R. Certain, *J. Chem. Phys.*, **49**, 4209 (1968).

145. H. Wei and T. Carrington, Jr., *J. Chem. Phys.*, **101**, 1343 (1994).

146. M. J. Bramley and T. Carrington, Jr., *J. Chem. Phys.*, **99**, 8519 (1993); *J. Chem. Phys.*, **101**, 8494 (1994).

147. N. M. Poulin, M. J. Bramley, T. Carrington, Jr., H. G. Kjaergaad, and B. R. Henry, *J. Chem. Phys.*, **104**, 7807 (1996).

148. P. Jensen, *J. Chem. Soc. Faraday Trans. 2*, **84**, 1315 (1988); I. N. Kozin and P. Jensen, *J. Mol. Spectrosc.*, **161**, 186 (1993); *J. Mol. Spectrosc.*, **163**, 483 (1994); P. Jensen, Y. Li, G. Hirsch, R. J. Bunker, T. J. Lee, and I. N. Kozin, *Chem. Phys.*, **190**, 179 (1995); J. Gräf and P. Jensen, *J. Mol. Spectrosc.*, **159**, 175 (1993).

149. R. J. Whitehead and N. C. Handy, *J. Mol. Spectrosc.*, **55**, 356 (1975).

150. M. Halonen, L. Halonen, H. Bürger, and P. Moritz, *J. Phys. Chem.*, **96**, 4225 (1992).

151. M. Halonen, L. Halonen, H. Bürger, and S. Sommer, *J. Chem. Phys.*, **93**, 1607 (1990); M. Halonen, L. Halonen, and H. Bürger, *Chem. Phys. Lett.*, **205**, 380 (1993); M. Halonen and X. Zhan, *J. Chem. Phys.*, **101**, 950 (1994).

152. D. Permogorov and A. Campargue, *Mol. Phys.*, **92**, 117 (1997).

153. M. Chevalier and A. De Martino, *J. Chem. Phys.*, **90**, 2077 (1989); Q. Zhu, H. Qian, H. Ma, and L. Halonen, *Chem. Phys. Lett.*, **177**, 261 (1991); Q. Zhu, H. Ma, B. Zhang, Y. Ma, and H. Qian, *Spectrochim. Acta*, **46A**, 1217, 1323 (1990); Q. Zhu, A. Campargue, and F. Stoeckel, *Spectrochim. Acta*, **50A**, 663 (1994).

154. Q. Zhu, H. Qian, and B. A. Thrush, *Chem. Phys. Lett.*, **186**, 436 (1991); G. Sun, X. Wang, Q. Zhu, C. Pierre, and G. Pierre, *Chem. Phys. Lett.*, **239**, 373 (1995); A. Campargue, J. Vetterhöffer, and M. Chenevier, *Chem. Phys. Lett.*, **192**, 353 (1992); F. Sun, X. Wang, J.

Liao, and Q. Zhu, *J. Mol. Spectrosc.*, **184**, 12 (1997).

155. X. Zhan, M. Halonen, L. Halonen, H. Bürger, and O. Polanz, *J. Chem. Phys.*, **102**, 3911 (1995).

156. D. L. Gray, Ph.D. Thesis, University of Reading, Reading, England (1978).

157. L. Halonen, *J. Chem. Soc., Faraday Trans. 2*, **84**, 1573 (1988).

158. E. Kauppi and L. Halonen, *Chem. Phys. Lett.*, **169**, 393 (1990).

159. T. Carrington, Jr., L. Halonen, and M. Quack, *Chem. Phys. Lett.*, **140**, 512 (1987); L. Halonen, T. Carrington, Jr., and M. Quack, *J. Chem. Soc., Faraday Trans. 2*, **84**, 1371 (1988).

160. E. Kauppi and L. Halonen, *J. Chem. Phys.*, **90**, 6980 (1989); L. Halonen and E. Kauppi, *J. Chem. Phys.*, **92**, 3278 (1990); E. Kauppi, *J. Mol. Spectrosc.*, **167**, 314 (1994); E. Kauppi, *J. Chem. Phys.*, **101**, 6470 (1994).

161. I. N. Kozin and P. Jensen, *J. Mol. Spectrosc.*, **163**, 483 (1994).

162. P. Jensen and I. N. Kozin, *J. Mol. Spectrosc.*, **160**, 39 (1993).

163. J. Senekowitsch, A. Zilch, S. Carter, H.-J. Werner, P. Rosmus, and P. Botschwina, *Chem. Phys.*, **122**, 375 (1988).

164. M. Saarinen, L. Halonen, and O. Polanz, *Chem. Phys. Lett.*, **219**, 181 (1994).

165. M. S. Burberry and A. C. Albrecht, *J. Chem. Phys.*, **71**, 4631 (1979).

166. L. Halonen, *Chem. Phys. Lett.*, **87**, 221 (1982).

167. P. R. Stannart, M. L. Elert, and W. M. Gelbart, *J. Chem. Phys.*, **74**, 6050 (1981).

168. O. S. Mortensen, M. K. Ahmed, B. R. Henry, and A. W. Tarr, *J. Chem. Phys.*, **82**, 3903 (1985); A. W. Tarr, D. J. Swanton, and B. R. Henry, *J. Chem. Phys.*, **85**, 3463 (1986).

169. R. Mecke, *Z. Electrochem.*, **54**, 38 (1950).

170. I. Schek, J. Jortner, and M. L. Sage, *Chem. Phys. Lett.*, **64**, 209 (1979).

171. R. T. Lawton and M. S. Child, *Mol. Phys.*, **40**, 773 (1980).

172. C. R. Le Sueur, S. Miller, J. Tennyson, and B. T. Sutcliffe, *Mol. Phys.*, **76**, 1147 (1992).

173. H. G. Kjaergaard, C. D. Daub, and B. R. Henry, *Mol. Phys.*, **90**, 201 (1997); D. M. Turnbull, H. G. Kjaergaard, and B. R. Henry, *Chem. Phys.*, **195**, 129 (1995); H. G. Kjaergaard and B. R. Henry, *J. Phys. Chem.*, **99**, 899 (1995); H. G. Kjaergaard, D. M. Turnbull, and B. R. Henry, *J. Chem. Phys.*, **99**, 9438 (1993); H. G. Kjaergaard, J. D. Goddard, and B. R. Henry, *J. Chem. Phys.*, **95**, 5556 (1991); H. G. Kjaergaard, B. R. Henry, and A. W. Tarr, *J. Chem. Phys.*, **94**, 5844 (1991); H. G. Kjaergaard, H. Yu, B. J. Schattka, B. R. Henry, and A. W. Tarr, *Chem. Phys.*, **93**, 6239 (1990).

174. L. Halonen, *J. Chem. Phys.*, **86**, 588 (1987).

175. L. Halonen, *J. Chem. Phys.*, **86**, 3115 (1987).

176. Q. Zhu, B. A. Thrush, and A. G. Robiette, *Chem. Phys. Lett.*, **150**, 181 (1988).

177. Q. Zhu and B. A. Thrush, *J. Chem. Phys.*, **92**, 2691 (1990); Q. Zhu, A. Campargue, J. Vetterhöffer, D. Permogorov, and F. Stoeckel, *J. Chem. Phys.*, **99**, 2359 (1993).

178. M. Halonen, L. Halonen, H. Bürger, and P. Moritz, *J. Chem. Phys.*, **95**, 7099 (1991); *Chem. Phys. Lett.*, **203**, 157 (1993); M. Halonen, *Ann. Acad. Scientiarum Fennicae*, **A256**, 1 (1994).

179. M. Ya. Ovchinnikova, *Chem. Phys.*, **120**, 249 (1988).

180. M. S. Child and Q. Zhu, *Chem. Phys. Lett.*, **184**, 41 (1991); Q. Zhu, *Spectrochim. Acta*, **48A**, 193 (1992).

181. F. Michelot, J. Moret-Bailly, and A. De Martino, *Chem. Phys. Lett.*, **148**, 52 (1988).

182. X. Wang and Q. Zhu, *J. Chem. Phys.*, **105**, 8011 (1996).

183. M. R. Aliev and J. K. G. Watson, in *Molecular Spectroscopy: Modern Research*, K. N. Rao, Ed., Academic, New York, 1985, Vol. III.

184. J. H. Meal and S. R. Polo, *J. Chem. Phys.*, **24**, 1119 (1956); *J. Chem. Phys.*, **24**, 1126 (1956).

185. G. L. Caldow, L. O. Halonen, and J. Kauppinen, *Chem. Phys. Lett.*, **101**, 100 (1983); L. Halonen, J. Kauppinen, and G. L. Caldow, *J. Chem. Phys.*, **81**, 2257 (1984); *J. Chem. Phys.*, **86**, 5888 (1987); F. W. Birss, *Mol. Phys.*, **31**, 491 (1976).

186. M. S. Child and Q. Zhu, *Chem. Phys. Lett.*, **207**, 116 (1993); Q. Zhu, H. Li, and X. Wang, *Chem. Phys. Lett.*, **212**, 403 (1993).

187. H. Bürger, H. Ruland, and L. Halonen, *J. Mol. Spectrosc.*, **182**, 195 (1997).

188. J.-M. Flaud, C. Camy-Peyret, H. Bürger, and H. Willner, *J. Mol. Spectrosc.*, **161**, 157 (1993); J.-M. Flaud, C. Camy-Peyret, PH, Arcas, H. Bürger, and H. Willner, *J. Mol. Spectrosc.*, **165**, 124 (1994); *J. Mol. Spectrosc.*, **168**, 556 (1994); O. N. Ulenikov, A. B. Malikova, H. Li, H. Qian, Q. Zhu, and B. A. Thrush, *J. Chem. Soc. Faraday Trans. 2*, **91**, 13 (1995).

189. C. Camy-Peyret and J. W. C. Johns, *Can. J. Phys.*, **61**, 1462 (1983); L. Lechuga-Fossat, J.-M. Flaud, C. Camy-Peyret, and J. W. C. Johns, *Can. J. Phys.*, **62**, 1889 (1984); L. Brown, D. Crisp, J. Crisp, A. Bykov, O. Naumenko, and L. Sinitsa, *SPIE*, **2205**, 238 (1993).

190. P. Jensen, G. Osmann, and I. N. Kozin, in *Advanced Series in Physical Chemistry: Vibration-Rotational Spectroscopy and Molecular Dynamics*, D. Papousek, Ed., World Publishing, Vol. 9.

191. M. S. Child, O. V. Naumenko, M. A. Smirnov, and L. R. Brown, *Mol. Phys.*, **92**, 885 (1997).

192. H. Bürger and G. Graner, *J. Mol. Spectrosc.*, **149**, 491 (1991).

193. H. Bürger and L. Halonen, *Mol. Phys.*, **87**, 227 (1996); H. Bürger, A. Campargue, L. Halonen, and H. Ruland, *J. Mol. Spectrosc.*, **183**, 99 (1997).

194. G. Graner, O. Polanz, H. Bürger, and P. Pracna, *J. Mol. Spectrosc.*, **188**, 115 (1998).

195. F. Iachello and R. D. Levine, *Algebraic Theory of Molecules*, Oxford University Press, Oxford, 1995.

196. S. Oss, *Adv. Chem. Phys.*, **XCIII**, 455 (1996).

WIDEBAND MEASUREMENT AND ANALYSIS TECHNIQUES FOR THE DETERMINATION OF THE FREQUENCY-DEPENDENT, COMPLEX SUSCEPTIBILITY OF MAGNETIC FLUIDS

P. C. FANNIN

Department of Electronic and Electrical Engineering, Trinity College, Dublin 2, Ireland

CONTENTS

Advances in Chemical Physics, Volume 104, Edited by I. Prigogine and Stuart A. Rice.
ISBN 0-471-29338-5   © 1998 John Wiley & Sons, Inc.

Symbols

A	Area, m^2; coaxial line parameter
a	Radius, m
B	Magnetic flux density, T
b	Radius, m
$b(t)$	Aftereffect function
D	Einstein's coefficient for rotational diffusion, s^{-1}
$D(x)$	Dawsons integral
d	Sample depth, m
E	Electric field, $V\,m^{-1}$
$E(b)$	Energy barrier, J
$Ei(x)$	Integral exponential function
$F(t)$	Aftereffect function
$F(\omega)$	Fourier transform of $f(t)$
f	Frequency, Hz
f_{max}	Frequency at which $\chi''(\omega)$ is a maximum, Hz
f_{res}	Resonant frequency, Hz
f_0	Frequency of free precession, Hz
f_o^{-1}	Néel prefactor, s
$f(t)$	Function of time t
$f(\tau)$	Distribution function
$G(\ln \tau)$	Distribution function
g	Landé splitting factor

H	Constant magnetic field, A m^{-1}
H_A	Equivalent anisotropy field, A m^{-1}
H_c	Coercivity
$H(f)$	Fourier transform of $h(t)$
h	Alternating magnetic field, A m^{-1}
$h(t)$	Impulse response
$h[K]$	$n \times n$ matrix
i	imaginary number, $= \sqrt{-1}$
K	Anisotropy constant, J m^{-3}
K_β	Bessel function
L	Inductance, H
L_0	Inductance of empty coil, H
L_1	Mean circumference, m
L_2	Gap thickness, m
$L(x)$	Langevin function
ΔL	Incremental change in inductance, H
M	Magnetization, T
$M_r(t)$	Magnetic viscosity
M_s	Saturation magnetization, T
m	Magnetic moment, Wb m
N	Number of particles per unit volume, m^{-3}
N_a, N_b, N_c	General ellipsoid demagnetizing factors
$P(t)$	Delta impulse response function
R_i	Parameters
R_W	Resistance, Ω
ΔR	Incremental change in resistance, Ω
r	Radius, m
T	Absolute temperature, K
t	Time, s
S_{11}	Scattering parameter
U_A	Anisotropy energy of a particle
U_d	Demagnetizing energy
V	Hydrodynamic volume of a particle, m^3
v	Magnetic volume of a particle, m^3
W	Scaling parameter in the Nakagami distribution
X	Reactance, Ω
$X(f)$	Fourier transform of $x(t)$
$\hat{X}(f)$	Fourier transform of $\hat{x}(t)$
x	Distance, m
$\hat{x}(t)$	Hilbert transform of $x(t)$
Z	Intrinsic impedance, Ω
Z_0	Characteristic impedance of a transmission line, Ω

Z_R	Load impedance, Ω
Z_1	Characteristic impedance of sample-filled coaxial line
Ω	Ohms, unit of resistance
Z_{in}	Input impedance of a transmission line, Ω
Z_R	Load impedance, Ω
Z_{ce}	Impedance, Ω
Z_{cf}	Impedance, Ω
α	Damping parameter; attenuation coefficient; parameter in Cole–Cole equation
α_0	Attenuation constant, Np m^{-1}
α_1	Attenuation constant of sample material, Np m^{-1}
β	Width parameter for Nakagami distribution; phase change coefficient, rad
β_0	Parameter in Cole–Davidson equation
β_1	Phase change coefficient of sample material, rad
γ	Magnetomechanical ratio, s^{-1} A^{-1} m
γ_g	L_1/L_2
γ_0	Propagation constant
γ_1	Propagation constant of sample material
ε	Permittivity, F m^{-1}
$\varepsilon(\omega)$	Complex, frequency-dependent, relative permittivity, $= \varepsilon'(\omega) - i\varepsilon''(\omega)$
ε_0	Absolute permittivity of free space
ε_r	Relative permittivity
ε_1	Absolute permittivity of sample material
η	Carrier liquid viscosity, Ns m^{-2}
η_m	Magnetic viscosity of a particle, Ns m^{-2}
θ	Polar angle, rad
λ	Wavelength, m
$\mu(\omega)$	Complex, frequency-dependent, relative permeability, $= \mu'(\omega) - i\mu''(\omega)$
μ_{eff}	Effective permeability
μ_0	Absolute permeability of free space
μ_r	Relative permeability
μ_1	Absolute permeability of sample material, H m^{-1}
ξ	Ratio of magnetic energy to thermal energy
ρ	Particle density, kg m^{-3}
σ	Ratio of anisotropy energy to thermal energy
τ	Relaxation time, s
τ_B	Brownian relaxation time, s
τ_D	Diffusion time, s
τ_N	Néel relaxation time, s

τ_{eff}	Effective relaxation time, s
τ_o	Decay time of precession, s
τ_o	Debye relaxation time
τ_{od}	Cole–Davidson time parameter
$\tau_{\parallel}$	Parallel relaxation time, s
$\tau_{\perp}$	Transverse relaxation time, s
τ_1, τ_2	Relaxation time
ϕ	Azimuthal angle, rad
φ	Superparamagnetic fraction
χ	Magnetic susceptibility
χ_0	Static susceptibility
χ'	Real component of complex magnetic susceptibility
$\chi(\omega)$	Complex, frequency-dependent, relative susceptibility, $= \chi'(\omega) - i\chi''(\omega)$
$\chi_{\parallel}(\omega)$	Parallel component of magnetic susceptibility
$\chi_{\parallel}(0)$	Static component of parallel susceptibility
$\chi_{\perp}(\omega)$	Perpendicular component of magnetic susceptibility
$\chi_{\perp}(0)$	Static component of perpendicular susceptibility
$\hat{\chi}'(\omega)$	Hilbert transform of $\chi'(\omega)$
χ_0	Static susceptibility
χ_{∞}	Susceptibility at very high frequencies
$\chi(\omega, H)$	Field-dependent susceptibility
ω	Angular frequency, rad s^{-1}
ω_{max}	Angular frequency at which $\chi''(\omega)$ is a maximum, rad s^{-1}
ω_{res}	Resonant angular frequency, rad s^{-1}
ω_0	Angular frequency of free precession, rad s^{-1}
ζ	Drag coefficient, Ns m

I. INTRODUCTION

Since the mid-1960s technological interest in magnetic fluids has grown rapidly, with a significant increase in theoretical research on their properties and applications in areas as diverse as the treatment of cancer [1] and loudspeaker damping [2]. However, it is interesting to note that almost 200 years previously attempts were being made to make a "magnetic liquid". In fact, the first reported attempt at producing a ferromagnetic liquid was by Gowan Knight [3] in 1779. Knight mixed iron filings in water but the dispersion did not have long-term stability. In 1932 Bitter [4] produced a colloid, for magnetic domain studies, which consisted of a suspension of γ-Fe_2O_3 in ethyl acetate; the particles had radii of approximately 500 nm. However, the first stable ferromagnetic liquid was manufactured by Elmore in 1938 [5] at the Massachusetts Institute of Technology. He successfully produced a water-based colloidal suspension of magnetite (Fe_3O_4) stabilized by soap. Elmore did not pursue his investigations, and interest in ferrofluids lay dormant until the mid-1960s when Papell [6] and Rosensweig [7,8] reported on practical applications of ferrofluids.

This work is concerned with the measurement and analysis of the frequency-dependent, complex susceptibility $\chi(\omega)$ of magnetic fluids. Accurate data on $\chi(\omega)$ is vital for an understanding of the dynamic behavior of such colloidal suspensions. It provides information on particle size r, anisotropy constant K, and internal field H_A; it also enables relaxation mechanisms, both Brownian and Néel, as well as ferromagnetic resonance to be identified.

Maiorov [9] appears to be the first to have reported on frequency-dependent susceptibility measurements. Others, including Soffge and Schmidbauer [10], although making measurements of susceptibility at discrete frequencies, had concentrated on measurement of the variation of susceptibility as a function of temperature. Maiorov's method of measurement was based on the variation of inductance of a coil when a magnetic fluid was inserted in its field. His measurement range was restricted to the approximate frequency range of 30 Hz–100 kHz, with a combination of an alternating current bridge and a Q-meter being used to perform the measurements. Here a combination of computer-controlled, automatic bridges and a network analyzer is used with the toroidal technique [11], and an adaptation of the short-circuited transmission line technique [12] to perform measurements over the approximate frequency range 10 Hz–6 GHz.

The data and analysis are presented in the form of case studies that are representative of the author's work over the past twelve years; ten case studies covering more than twenty magnetic fluids are presented. Fitting techniques are used for the determination of data both within and outside the measurement frequency range.

A. Single-Domain Particles

A magnetic fluid is a colloidal suspension of single-domain [13] ferri- or ferromagnetic particles dispersed in a carrier liquid and which may be stabilized by a suitable organic surfactant. The surfactant coating creates an entropic repulsion between particles such that thermal agitation alone is sufficient to prevent aggregation. The particles have radii ranging from approximately 2 to 10 nm, and when in suspension, their magnetic properties can be described by the Langevin theory of paramagnetism suitably modified to take account of a distribution of particle sizes. Being single domain, the particles are considered to be in a state of uniform magnetization with magnetic moment m given by

$$m = M_s v \tag{1.1}$$

where M_s (in webers per square meter) denotes saturation magnetization and v is the volume of the particle. In some magnetic fluids, depending upon the magnetic anisotropy and the size and shape of the particles, the magnetic moment may be regarded as being fixed with respect to a set of axes fixed within a particle during the duration of a measurement. In such particles we describe the magnetic moment as being blocked. The direction of the magnetic moment is referred to as the axis of easy magnetization.

Frenkel and Dorfman [13] were the first to predict the possible existence of single-domain particles. Kittel [14] estimated the critical sizes for single-domain particles in the form of rods, cubes, and spheres. His calculations were based on the minimization of the total energy of a particle. A comparison of the magnetostatic energy of a single-domain particle with the anisotropy and domain wall energy of the particle in a multidomain state led to a critical diameter in the region of 20 nm for a spherical iron particle. Raĭkher and Shliomis [15] calculated the critical size for mono domainness for a typical ferromagnet by comparing the stray field energy of a uniformly magnetized particle with the energy of a single-domain wall and obtained a critical diameter of 30 nm. Aharoni and Shtrikman [16] calculated the critical size for cobalt to be 60 nm.

Consider a single-domain particle whose magnetic moment m is directed at an angle θ to the applied field H. If there are no additional anisotropy terms in the energy, the energy of the particle is given as

$$U = -mH \cos \theta \tag{1.2}$$

For the assembly of particles at a temperature T that has achieved a state of thermal equilibrium with the field, there will be a Boltzmann distribution of θ's over the particle assembly. The fraction of the magnetization that has been aligned by the field is calculated by averaging $\cos \theta$ over the Boltzmann distribution, which yields the Langevin function

$$L(x) = \coth x - \frac{1}{x} \tag{1.3}$$

This is similar to the behavior of an assembly of paramagnetic atoms which are also described by the Langevin function. Therefore an assembly of such particles behaves like paramagnetic material with no hysteresis in their magnetization curve.

The Langevin equation also describes the magnetization resulting from an assembly of particles whose magnetic moments undergo an internal diffusion process. However, in general, particles have some anisotropy and the magnetization resulting from internal diffusion cannot be described by a simple Langevin equation. When a magnetic field is applied to an assembly of say uniaxial particles, the energy of one such particle will be

$$U = Kv \sin^2 \theta - mH \cos \theta \qquad (1.4)$$

where K is the anisotropy coefficient, also known as the anisotropy constant, and where the field H is applied along the easy axis. In this case there will be a different Boltzmann distribution of θ's in thermal equilibrium than without the anisotropy term. The magnetic moment of a single-domain particle of volume $< 10^{-18}$ cm^3 exceeds that of an individual atom by four or five orders of magnitude—hence the term *superparamagnetism* as coined by Bean in 1955 [17] and first discovered by Elmore in 1938 [5]. Furthermore an assembly of particles is called supermagnetic when its magnetization has assumed a thermodynamic equilibrium value after the time t' of a measurement following an arbitrary field change.

It is interesting to note that below a certain size the spontaneous magnetization, M_s, of a ferromagnetic particle decreases. A number of researchers have addressed this topic, including Bean and Livingston [18] and Kneller [19]. For very small particles there is a greater surface-to-volume ratio and so more of the atoms are surface atoms which have a deficiency of magnetic neighbors. This means one can expect the spontaneous magnetization to diminish below a critical particle size because there are simply not enough atoms to contribute to the exchange forces. The critical size at which a ferromagnetic particle becomes paramagnetic depends on the material in question. Raĭkher and Shliomis estimated a critical diameter of 1 nm using the fact that the magnetic ordering only exists if the exchange energy is greater than the electron kinetic energy. Another factor is that on the surface of a particle there exists a demagnetized layer due to the absence of electron partners outside the surface, and also a demagnetized layer may form by chemical reaction with the surfactant in a ferrofluid. This implies that even for particles with a diameter less than approximately 3–5 nm the spontaneous magnetization has decreased somewhat.

B. Magnetic Anisotropy

By magnetic anisotropy is meant the dependence of the internal energy on the direction of spontaneous magnetization [20]. There are several sources of anisotropy, but here we will consider magnetocrystalline and shape anisotropies.

Crystalline anisotropy arises from the crystalline structure of the material where the magnetization prefers to align along some crystallographic axis or axes in order to minimize the anisotropy energy of the particle while shape anisotropy is associated with the demagnetizing field within a particle. In general, all ferromagnetic materials are to some degree anisotropic, but some, such as cobalt, have much stronger crystalline anisotropy than others, such as nickel or magnetite, whose main source of anisotropy is from their shape.

1.   Magnetocrystalline Anisotropy

The origin of the magnetocrystalline anisotropy lies in the incomplete quenching of the orbital angular momentum of the ions in the crystal lattice. The existence of magnetocrystalline anisotropy implies that, in addition to the mutual coupling between electron spins that gives rise to ferromagnetism, there must also be some interaction between the spins and the crystal lattice.

Consider the simple case of uniaxial anisotropy where the moment tends to align along one axis, known as the easy axis, within the particle. The energy of the particle depends in part on the magnetization direction relative to this easy axis; this energy is referred to as the anisotropy energy. In the uniaxial crystal the anisotropy energy may be written as

$$U = Kv \sin^2 \theta \qquad (1.5)$$

where θ is the angle between the magnetization vector and the easy axis or principal axis of the particle. The lowest energy is attained when $\theta = 0, \pi$, that is, when the magnetization direction lies along the easy axis.

It is convenient to consider the effect of the anisotropy energy in terms of an equivalent internal magnetic field. This equivalent field within the particle is known as the anisotropy field H_A (in amperes per meter). It can be defined in terms of the torque exerted on the magnetic moment by such a field and is equal to the torque exerted by the anisotropy energy, that is,

$$|mH_A \sin \theta| = \left| \frac{\partial U}{\partial \theta} \right| \qquad (1.6)$$

which from (1.1) may be written as

$$M_s v H_A \sin \theta = \frac{\partial U}{\partial \theta} \qquad (1.7)$$

Differentiating U from Eq. (1.5) with respect to θ gives

$$\frac{\partial U}{\partial \theta} = 2Kv \sin \theta \cos \theta \qquad (1.8)$$

Finally equating Eqs. (1.7) and (1.8), we obtain

$$H_A = \frac{2K}{M_s} \cos \theta \qquad (1.9)$$

This is the equivalent anisotropy field for a uniaxial particle, and when the magnetic moment vector is pointing at an angle relative to the direction of H_A, it will experience a torque which tries to bring it into line with this field. Notice that the equivalent anisotropy field is directly proportional to the anisotropy constant and also that it depends on the angle with which the magnetization vector makes with the easy axis. When the anisotropy energy is large, the magnetic moment will be closely aligned in the direction of the easy axis so that the angle θ is generally small and $\cos \theta \approx 1$; in this case the anisotropy field becomes

$$H_A = \frac{2K}{M_s} \quad \text{for small } \theta \qquad (1.9a)$$

The expression for the equivalent anisotropy field [Eq. (1.9)] can also be described in the following manner. If a magnetic field is applied parallel to the easy axis in the sense shown in Fig. 1(a), then Eqn. (1.9) gives the value of the field required to rotate the magnetic moment from its initial position (making the angle θ with the easy axis) into the field direction. Equation (1.9a) gives the maximum value of this field that is required to reverse the magnetization from one easy direction to the other, and this maximum gives the coercivity. Therefore Eq. (1.9a) gives the coercivity of a spherical particle with uniaxial anisotropy. However, if the anisotropy energy barrier is small enough, $Kv/kT < 25$ approximately, then the magnetic moment may reverse direction in the time of measurement by thermal activation without any applied field, and in this case the coercivity of the particle is zero; however, Eqs. (1.9) and (1.9a) for the equivalent anisotropy field will still apply.

2. Shape Anisotropy

Shape anisotropy arises from the demagnetizing fields within a particle. The demagnetizing fields in turn arise from the external shape of the particles. Kittel [14] in 1947 treated shape anisotropy and its effect on ferromagnetic resonance. Consider the prolate ellipsoid shown in Fig. 1(b). If the demag-

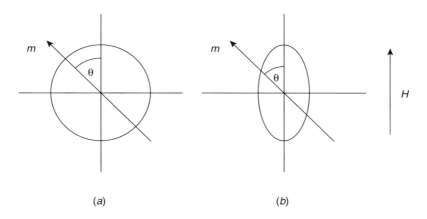

Figure 1. (a) Spherical particle with uniaxial crystalline anisotropy. (b) Prolate ellipsoid with shape anisotropy.

netizing factors are N_a and N_b in the direction of the major axis and in any direction perpendicular to the major axis, respectively, then the demagnetizing energy is

$$U_d = \frac{1}{2\mu_0} M_s^2 v(N_a \cos^2 \theta + N_b \sin^2 \theta) \qquad (1.10)$$

Thus

$$\frac{\partial U_d}{\partial \theta} = \frac{1}{\mu_0} M_s^2 v(N_b - N_a) \sin \theta \cos \theta \qquad (1.11)$$

Equating this with an equivalent magnetic field as before, one obtains

$$H_A = \frac{1}{\mu_0} M_s(N_b - N_a) \cos \theta \qquad (1.12)$$

The maximum value of this equivalent field is the coercivity H_c, where

$$H_c = \frac{1}{\mu_0} M_s(N_b - N_a) \qquad (1.13)$$

For a general ellipsoid the demagnetizing factors N_a, N_b, N_c along the three principal axes are related by the expression

$$N_a + N_b + N_c = 1 \qquad (1.14)$$

In the case of spherical particles $N_a = N_b = N_c = \frac{1}{3}$; therefore, from Eq. (1.12) spheres have no shape anisotropy. It is only those particles which have a difference in their demagnetizing factors that possess shape anisotropy. For the infinite cylinder magnetized longitudinally we have $N_b = N_c = \frac{1}{2}$ and $N_a = 0$, which results in a coercivity proportional to M_s/μ_0 and therefore possess a very large shape anisotropy. In a ferrofluid there exist particles of different shape and hence there is a distribution in the values of shape anisotropy ranging between the extremes of perfect spheres with no shape anisotropy and infinite cylinders with very large anisotropy.

C. Complex Susceptibility

The complex, magnetic susceptibility $\chi(\omega) = \chi'(\omega) - i\chi''(\omega)$ of a random assembly of single-domain particles can be described in terms of its parallel, $\chi_{\parallel}(\omega)$, and perpendicular, $\chi_{\perp}(\omega)$, components, with [15]

$$\chi(\omega) = \tfrac{1}{3}[\chi_{\parallel}(\omega) + 2\chi_{\perp}(\omega)] \tag{1.15}$$

The parallel susceptibility $\chi_{\parallel}(\omega)$ is purely relaxational in character and can be described by the Debye equation [21], with

$$\chi_{\parallel}(\omega) = \frac{\chi_{\parallel}(0)}{1 + i\omega\tau_{\parallel}} \tag{1.16}$$

where $\chi_{\parallel}(0)$ is the static parallel susceptibility and $\tau_{\parallel}$ is the effective parallel relaxation time which is composed of two relaxational processes.

1. Relaxation Mechanisms

The theory developed by Debye [13] to account for the anomalous dielectric dispersion in dipolar fluids has been used [9,11] to account for the analogous case of magnetic fluids.

Debye's theory holds for spherical particles when the magnetic dipole–dipole interaction energy is small compared to the thermal energy kT.

According to Debye's theory, the complex susceptibility $\chi(\omega)$ has a frequency dependence given by the equation

$$\chi(\omega) - \chi_\infty = \frac{\chi_0 - \chi_\infty}{1 + i\omega\tau} \tag{1.17}$$

$$= \frac{\chi_0 - \chi_\infty}{1 + \omega^2\tau^2} - \frac{i\omega\tau(\chi_0 - \chi_\infty)}{1 + \omega^2\tau^2} \tag{1.18}$$

where τ is the relaxation time and the static susceptibility

$$\chi_0 = \frac{nm^2}{3kT\mu_0} \tag{1.19}$$

Here, χ_∞ is the high-frequency susceptibility at a frequency below that of resonance, n is the particle number density, and μ_0 is the permeability of free space.

Equation (1.18) and Fig. 2 demonstrate how $\chi'(\omega)$ falls monotonically while the $\chi''(\omega)$ component has a maximum at $\omega_m\tau = 1$, giving

$$\tau = \frac{1}{\omega_m} = \frac{1}{2\pi f_m} \tag{1.20}$$

Also the theoretical maximum of $\chi''(\omega)$ is found to be

$$\chi''(\omega_m) = \tfrac{1}{2}\chi_0 \tag{1.21}$$

For the case where the relaxation times τ are distributed continuously with a weight function $f(\tau)$, $\chi(\omega)$ may also be expressed as

$$\chi(\omega) = \chi_\infty + (\chi_0 - \chi_\infty)\int_0^\infty f(\tau)\frac{d(\tau)}{1 + i\omega\tau} \tag{1.22}$$

The term τ is composed of two relaxation processes, and in the case of the first process the magnetic moment may rotate along with the particle, with the time associated with this "Brownian rotation" being the Brownian

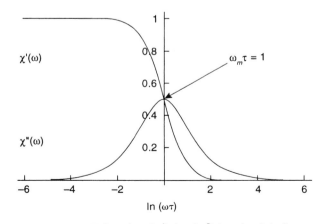

Figure 2.   Debye plot of $\chi'(\omega)$ and $\chi''(\omega)$ against $\ln(\omega\tau)$.

relaxation time τ_B [22], where

$$\tau_B = \frac{3V'\eta}{kT} \tag{1.23}$$

where V' is the hydrodynamic volume of the particle and η is the dynamic viscosity of the carrier liquid.

In the case of the second relaxation mechanism, Néel relaxation, the magnetic moment may reverse direction within the particle by overcoming an energy barrier, which for uniaxial anisotropy is given by Kv. The existence of the energy minima arises from the particle's anisotropy energy so that in the case of isotropic particles there is no energy barrier and the magnetic moment vectors are free to move within the particles by internal diffusion.

The Einstein formula gives the coefficient of internal rotational diffusion of the magnetic moment as $D = \alpha\gamma kT/M_s v$. Thus the characteristic time of orientational diffusion of the magnetic moment m is

$$\tau_D = \frac{1}{2D} = \frac{M_s v}{2\alpha\gamma kT} \tag{1.24}$$

This can be written in the more convenient form $\tau_D = 3v\eta_m/kT$, where the parameter $\eta_m = M_s/6\alpha\gamma$ is the internal "magnetic" viscosity which inhibits the movement of the magnetic moment vector. As it has the same units (newton-seconds per square meter) as viscosity, η_m plays the same role as the fluid viscosity η in Eq. 1.23.

For anisotropic particles the magnetic moments within the particles have to surmount the anisotropy energy barrier. The probability of such a transition is approximately equal to $\exp(\sigma)$, where σ is the ratio of anisotropy energy to thermal energy (Kv/kT). This reversal, or switching time, is referred to as the Néel relaxation time τ [23], where Néel estimated the relaxation time τ to be

$$\tau = f\bar{o}^1 \exp(\sigma) \tag{1.25}$$

where $f\bar{o}^1$ is a damping or extinction time having an often-quoted approximate value of between 10^{-8} and 10^{-12} s [24–26] and can be determined from the expression [27]

$$f\bar{o}^1 = \frac{M_s}{2\gamma\alpha K} \tag{1.26}$$

γ being the gyromagnetic ratio and α a damping constant.

For particles with $\sigma < 1$ the movement of the magnetic moment within the particle is barely affected by the existence of the potential barrier and the Néel relaxation time is approximately the same as the internal diffusion time τ_D for isotropic particles. However, for larger particles where $\sigma > 1$ there is an exponential dependence of τ_N on σ. This means that there is a well-defined particle size at which the transition to stable behavior occurs.

Brown realized that Néel's expressions did not take into account the fact that the magnetic moment could spend some time in directions other than at the minimum of the potential well (easy axis). He thus derived a differential equation to describe the motion of the direction of the magnetic moment during its "random walk" from one energy minimum to another. Brown realized that in the limit when the parameter σ becomes zero, the Néel relaxation time should equal the internal diffusion time for isotropic particles, τ_D. He proposed the following approximate expressions for τ_N for low and high barrier heights:

$$\tau_N = \tau_D(1 + \tfrac{2}{5}\sigma) \qquad \sigma \ll 1$$

$$\tau_N = \frac{\sqrt{\pi}}{2} \frac{\tau_D}{\sigma^{3/2}} \exp(\sigma) \qquad \sigma > 2 \tag{1.27}$$

From Eq. (1.27) one can see that τ_N does indeed become τ_D in the limit $\sigma \to 0$. An obvious weakness with Eq. (1.27) is that they do not cater for a continuous range of values in the region of $\sigma \approx 1$, and recognizing this, Bessais et al. [25] and Aharoni [28] derived empirical expressions while Coffey et al. [29,30] have subsequently derived a single expression to cater for a continuous range of σ. The equations of Coffey et al. in Ref. [30] are exact but complicated to use, unlike that of Ref. [29], which is user friendly and has the form

$$\tau_N = \frac{\tau_D(e^\sigma - 1)}{2\sigma} \left(\frac{1}{1 + 1/\sigma} \sqrt{\frac{\sigma}{\pi}} + 2^{-\sigma-1} \right)^{-1} \tag{1.28}$$

Figure 3 shows plots of τ_N/τ_D versus σ. It is quite obvious that plot 3 of Coffey et al. gives the best approximation to that of the exact equation while the plot of Brown's equation for low barrier heights is accurate up to $\sigma = 2$. Brown's equation for high barriers gives a poor approximation for high values of σ.

An estimate [31] shows that usually $\eta_m/\eta < 10^{-2}$ and therefore $\tau_B \gg \tau_D$. Hence for particles with barrier height potentials $\sigma < 1$ the relaxation is mainly due to the internal diffusion with characteristic time τ_D.

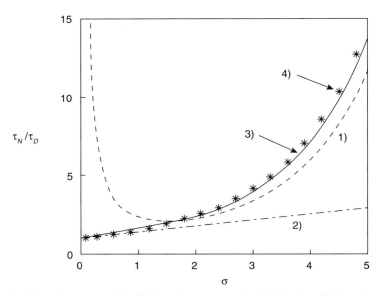

Figure 3. Plots of τ_N/τ_D vs. σ for (1) Brown's expression for high barriers, (2) Brown's expression for low barrier, (3) the Coffey et al. equation for continuous values of σ, and (4) the Coffey et al. exact solution.

The exponential growth of τ_N with particle volume means that for particles with $\sigma > 10$, $\tau_N \gg \tau_B$, so that for particles with large values of σ the relaxation of the magnetic moments is due to the Brownian mechanism; the magnetic moments are effectively frozen in the particle.

An insight into the spectrum of relaxation times likely to be encountered in a colloidal suspension of magnetite in water with particle radii ranging from 1.44 to 11.2 nm has been reported by Fannin [32] using a 23-particle fraction distribution, as shown in Fig. 4. He showed that τ_B had values ranging from approximately 10^{-7} to 10^{-5} s while τ_N had values ranging from 10^{-9} to 10^{15} s and that $\tau_N = \tau_B = 1.56 \times 10^{-6}$ s, corresponding to an approximate particle size of 5.94 nm. This demonstrates that particles which have a radius greater than 6 nm (approximately) relax via the Brownian mechanism whereas those whose radius is less than 6 nm relax via the Néel mechanism, as indicated in Fig. 5. In the case where τ_B is the dominant relaxation time, we speak of the magnetic moment being blocked.

A distribution of particle sizes implies the existence of a distribution of relaxation times, with both relaxation mechanisms contributing to the mag-

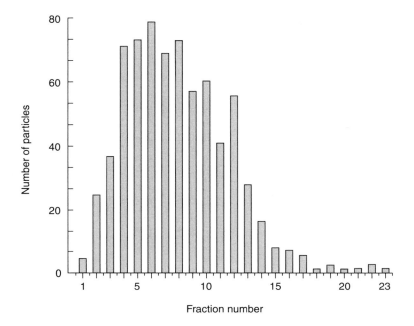

Figure 4.   Particle distribution of magnetic fluid sample.

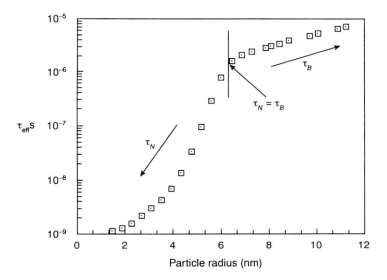

Figure 5.   Plot of t_{eff} vs. particle radius for 23-fraction sample.

netization. They do so with an effective time $\tau_{\text{eff}} = \tau_{\parallel}$, where [33]

$$\tau_{\text{eff}} = \frac{\tau_N \tau_B}{\tau_N + \tau_B} \tag{1.29}$$

the mechanism with the shortest relaxation time being dominant.

In general, one can anticipate a single loss peak in the susceptibility profile; however, circumstances can arise whereby, depending on the ratio of τ_N to τ_B, the particle distribution appears as a two-particle fraction and exhibits two loss peaks, as illustrated in Fig. 6. This situation can be represented by the approximate expression

$$\chi(\omega) - \chi_\infty = \left(\frac{A}{1 + i\omega\tau_B} + \frac{B}{1 + i\omega\tau_N} \right) \tag{1.30}$$

In the case of Fig. 6, A and B are in the ratio of $2 : 1$ while $\tau_B = 1000\tau_N$.

2. Superparamagnetic Fraction φ

In the case where the susceptibility profile indicates the presence of blocked particles (giving rise to a low-frequency loss peak) as well as superparamagnetic unblocked particles (giving rise to a high-frequency loss peak), as indicated in Fig. 6, it is possible to determine the superparamagnetic fraction φ of the sample [9].

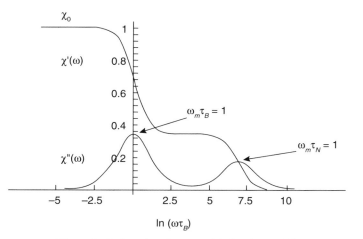

Figure 6. Debye plot of two-particle fraction model.

From [9] if the fraction of particles with relaxation times τ_N, is φ, then for $\omega\tau_N \ll 1$

$$\chi'(\omega) = \chi_0\left(\frac{1 - \varphi}{1 + \omega^2\tau_B^2} + \frac{\varphi}{1 + \omega^2\tau_N^2}\right) \qquad (1.31)$$

$$\approx \chi_0\left(\frac{1 - \varphi}{1 + \omega^2\tau_B^2} + \varphi\right) \qquad (1.32)$$

Similarly,

$$\chi''(\omega) = \chi_0\left(\frac{(1 - \varphi)\omega\tau_B}{1 + \omega^2\tau_B^2} + \frac{\varphi\omega\tau_N}{1 + \omega^2\tau_B^2}\right) \qquad (1.33)$$

$$\approx \chi_0\left(\frac{(1 - \varphi)\omega\tau_B}{1 + \omega^2\tau_B^2}\right) \qquad (1.34)$$

For $\omega\tau_B = 1$ [maximum $\chi''(\omega)$],

$$\chi'(\omega)_m = \tfrac{1}{2}\chi_0(1 + \varphi) \qquad (1.35)$$

and

$$\chi''(\omega)_m = \tfrac{1}{2}\chi_0(1 - \varphi) \qquad (1.36)$$

Therefore, for $\varphi \ll 1$

$$\chi_0 = 2\chi'(\omega)_m(1 - \varphi) \qquad (1.37)$$

and

$$\chi''(\omega)_m = \frac{2\chi'(\omega)_m(1 - \varphi)}{2(1 + \varphi)} \qquad (1.38)$$

$$= \frac{\chi'(\omega)_m(1 - \varphi)}{1 + \varphi} \qquad (1.39)$$

giving

$$\varphi = \frac{1 - \chi''(\omega)_m/\chi'(\omega)_m}{1 + \chi''(\omega)_m/\chi'(\omega)_m} \qquad (1.40)$$

Using the two-fraction model, Maiorov [9] has reported on the determination of φ for colloidal suspensions of magnetite and cobalt ferrite. The cobalt ferrite sample was found to have approximately half the number of superparamagnetic particles found in the magnetite sample. This result was attributed to the fact that the value of K for cobalt ferrite was an order of magnitude greater than that of magnetite, and so the condition $\tau_B \gg \tau_N$ held for a smaller fraction of particles in the case of the cobalt ferrite dispersion.

3. Analyzing Susceptibility Data

The major advances of recent times in the understanding of ferrofluids are in no small way due to the application of dielectric formalism in the representation of ferrofluid data. The magnetic analogues of the equations of Debye [34–37] are just a number of examples of those that have lightened the burden of workers involved in the measurement and analysis of ferrofluids.

In cases where the complex susceptibility data fits a depressed circular arc, the relation between $\chi'(\omega)$ and $\chi''(\omega)$ and their dependence on frequency, $\omega/2\pi$, can be displayed by means of the magnetic analogue of the Cole–Cole plot [34]. In the Cole–Cole case [Fig. 7(a)] the circular arc cuts the χ' axis at an angle of $\frac{1}{2}\alpha\pi$, where α is referred to as the Cole–Cole parameter and is a measure of the particle size distribution.

The magnetic analogue of the Cole–Cole circular arc is described by the equation

$$\chi(\omega) = \chi_\infty + \frac{\chi_0 - \chi_\infty}{1 + (i\omega\tau_o)^{1-\alpha}} \qquad 0 < \alpha < 1 \tag{1.41}$$

For cases where the data fits an asymmetric Cole–Cole plot the magnetic analogue to the Cole–Davidson plot is applicable.

As can be seen from Fig. 7(b) the Cole–Davidson plot cuts the χ' axis at the high-frequency side at an angle $\frac{1}{2}\pi\beta$. The bisector of this angle, when projected forward, cuts the plot at $\omega\tau_{od} = 1$.

For the Cole–Davidson equation [35], $\chi(\omega)$ is given by

$$\chi(\omega) = \chi_\infty + \frac{\chi_0 - \chi_\infty}{(1 + i\omega\tau_{od})^\beta} \qquad 0 < \beta < 1 \tag{1.42}$$

where β is the Cole–Davidson parameter.

The Havriliak–Negami equation [36] uses the Cole–Cole and Cole–Davidson parameters α and β, respectively, and its corresponding magnetic analogue is described by the empirical expression

$$\chi(\omega) = \chi_\infty + \frac{\chi_0 - \chi_\infty}{[(1 + (i\omega\tau_o)^{1-\alpha})]^\beta} \qquad 0 < \beta < 1 \tag{1.43}$$

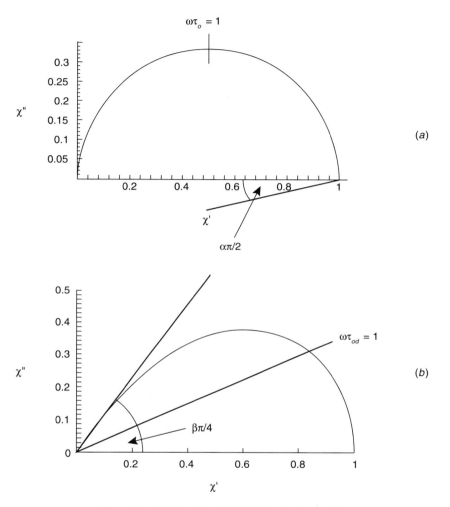

Figure 7. (*a*) Cole–Cole plot. (*b*) Cole–Davidson plot.

The term $\chi(\omega)$ may also be expressed in terms of a logarithmic distribution function $G(\ln \tau)$ in the form

$$\chi(\omega) = \chi_\infty + (\chi_0 - \chi_\infty) \int_0^\infty \frac{G(\ln \tau) \, d \ln \tau}{1 + i\omega\tau} \tag{1.44}$$

In the Cole–Cole case the distribution function [37] is described by

$$G(\ln \tau) = \frac{1}{2\pi} \frac{\sin \pi\alpha}{\cosh[(1-\alpha)\ln(\tau_o/\tau)] - \cos \pi\alpha} \tag{1.45}$$

For the Cole–Davidson case the distribution function [37] has the form

$$G(\ln \tau) = \frac{1}{\pi} \left(\frac{\tau}{\tau_{od} - \tau}\right)^{\beta} \sin(\pi\beta) \qquad \tau < \tau_{od} \tag{1.46}$$

In the case of Havriliak–Negami the distribution function [36] is given by the relation

$$G(\ln \tau) = \frac{1}{\pi} \frac{[y^{(1-\alpha)\beta}]\sin(\beta\theta)}{\{(y^{2(1-\alpha)} + [2y^{(1-\alpha)}]\cos[\pi(1-\alpha)] + 1)\}^{\beta/2}} \tag{1.47}$$

where $y = (\tau/\tau_o)$ and

$$\theta = \arctan\left(\frac{\sin[\pi(1-\alpha)]}{y^{(1-\alpha)} + \cos[\pi(1-\alpha)]}\right) \tag{1.48}$$

Typical plots of $G(\ln \tau)$ versus $\ln(\tau/\tau_o)$ are shown in Fig. 8.

The Fröhlich [38] distribution function assumes that the distribution of relaxation times is limited within a certain range from τ_1 to τ_2, with

$$f(\tau) = \begin{cases} \dfrac{1}{A\tau} & \tau_2 < \tau < \tau_1 \\ 0 & 0 < \tau < \tau_2, \tau_1 < \tau < \infty \end{cases} \tag{1.49}$$

where

$$A = \text{const} = \ln\left(\frac{\tau_1}{\tau_2}\right) \tag{1.50}$$

The latter distribution function is often written in a logarithmic form, $G(\ln \tau)$, with

$$G(\ln \tau) = \begin{cases} \dfrac{1}{\ln(\tau_1/\tau_2)} & \tau_2 < \tau < \tau_1 \\ 0 & 0 < \tau < \tau_2, \tau_1 < \tau < \infty \end{cases} \tag{1.51}$$

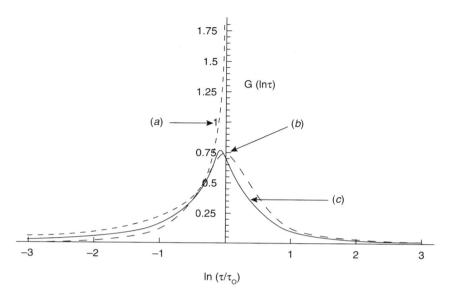

Figure 8. Plots of $G(\ln \tau)$ vs. $\ln(\tau/\tau_o)$ for (a) Havriliak–Negami, (b) Cole–Cole, and (c) Cole–Davidson distributions with $\alpha = 0.15$ and $\beta = 0.5$.

Thus

$$\chi(\omega) = \chi_\infty + (\chi_0 - \chi_\infty) \int_{\tau_2}^{\tau_1} \frac{G(\ln \tau)\, d(\ln \tau)}{1 + i\omega\tau} \tag{1.52}$$

$$= \chi_\infty + (\chi_0 - \chi_\infty)\left\{\frac{1}{\ln(\tau_1/\tau_2)}\left[\ln\left(\frac{\tau_1(1 + i\omega\tau_2)}{\tau_2(1 + i\omega\tau_1)}\right)\right]\right\} \tag{1.53}$$

where

$$\chi'(\omega) = \chi_\infty + (\chi_0 - \chi_\infty)\left\{\frac{1}{\ln(\tau_1/\tau_2)}\left[\ln\left(\frac{\tau_1\sqrt{1 + \omega^2\tau_2^2}}{\tau_2\sqrt{1 + \omega^2\tau_1^2}}\right)\right]\right\} \tag{1.54}$$

and

$$\chi''(\omega) = \chi_\infty + (\chi_0 - \chi_\infty)\frac{1}{\ln(\tau_1/\tau_2)}(\text{arctg } \omega\tau_1 - \text{arctg } \omega\tau_2) \tag{1.55}$$

The values of τ_1 and τ_2 can be determined by means of (1) fitting Eqs. (1.54) and (1.55) to the measured susceptibility data or (2) the approximate ellipse technique [39].

4. Approximate Ellipse Technique

If one accepts the Fröhlich assumption that the distribution of relaxation times is limited within a certain range from τ_1 to τ_2 as defined in Eqs. (1.49)–(1.53), then the locus of $\chi(\omega)$ in the complex plane is found [40] to be close to an ellipse and one obtains the relation

$$\frac{[\chi'(\omega) - \chi''(\omega)] - \chi_\infty}{\chi_0 - \chi_\infty} = \frac{1}{A} \ln\left(\frac{\tau_2(1 + \omega\tau_1)}{\tau_1(1 + \omega\tau_2)}\right) \tag{1.56}$$

which in the case of the loss component $\chi''(\omega_m)$ reduces to

$$\chi''(\omega_m) = (\chi_0 - \chi_\infty) \frac{\tan^{-1} c}{\tanh^{-1} c} \tag{1.57}$$

where

$$c = \tanh(\tfrac{1}{4}A) \tag{1.58}$$

and

$$\omega_m^{-1} = \tau_m = \sqrt{\tau_1 \tau_2} \tag{1.59}$$

To determine τ_1 and τ_2, it is necessary to obtain values for $\chi''(\omega_m)$, $(\chi_0 - \chi_\infty)$, and ω_m. Here, $\chi_0 - \chi_\infty$ is determined by fitting the experimental data to an ellipse whose major axis lies on the $\chi'(\omega)$ axis of the complex susceptibility plot; the major axis of the ellipse corresponds to $\chi_0 - \chi_\infty$. Similarly $\chi''(\omega_m)$ can be determined from the minor axis of the ellipse as its value is equal to half the minor-axis measurement. By inserting the values obtained for $\chi''(\omega_m)$ and $(\chi_0 - \chi_\infty)$ into Eq. (1.57), a value for c is found which, by means of Eqs. (1.50) and (1.58), enables the ratio τ_1/τ_2 to be determined. The product $\tau_1\tau_2$ is obtained from Eq. (1.59), where $\omega_m/2\pi$ is the frequency at which $\chi''(\omega)$ is a maximum.
Now

$$(\tau_1\tau_2)^{\frac{\tau_2}{\tau_1}} = \tau_2^2 \tag{1.60}$$

thus enabling τ_2 to be determined; τ_1 is then found from Eq. (1.59).

D. Aftereffect Function or Magnetization Decay

When a ferrofluid sample is placed in a steady magnetic field, the magnetic moments m of the particles tend to align in the direction of the field. If the field is suddenly removed, the magnetization decays with time, and this decay is usually referred to as the aftereffect function $M_r(t)$, or magnetic viscosity, of the system of particles. For the situation where a small steady field H_1, having been applied to a system of single-domain ferromagnetic particles at time $t = -\infty$, is suddenly switched off at $t = 0$, the resulting decay transient of the magnetization $M_r(t)$ may be written as

$$M_r(t) = (\chi_0 - \chi_\infty)H_1 F(t) \tag{1.61}$$

where $F(t)$ is the normalized aftereffect function.

According to linear response theory [41,42], the frequency-dependent susceptibility $\chi(\omega)$, arising from the application of an alternating field $H_1(t) = H_1\exp(-i\omega t)$, is given by

$$\frac{\chi(\omega) - \chi_\infty}{\chi_0 - \chi_\infty} = -\int_0^\infty \frac{d[F(t)]}{dt}\exp(-i\omega t)\,dt \tag{1.62}$$

$$= 1 - i\omega \int_0^\infty F(t)\exp(-i\omega t)\,dt \tag{1.63}$$

The component $-d[F(t)]/dt$ is the delta impulse–response function $P(t)$, and thus Eq. (1.62) identifies the relationship between the susceptibility components and the one-sided Fourier transform (FT) of $P(t)$.

The parameters $F(t)$ and $P(t)$ are related in the manner

$$F(t) = \int_0^t P(t')\,dt' \tag{1.64}$$

In the case of the Fröhlich distribution of relaxation times [Eq. (1.51)] one obtains [37]

$$P(t) = \int_0^\infty \frac{1}{\tau}\exp\left(\frac{-t}{\tau}\right)G(\ln\tau)\,d\ln\tau \tag{1.65}$$

$$= \frac{\exp(-t/\tau_1) - \exp(-t/\tau_2)}{t\,\ln(\tau_1/\tau_2)} \tag{1.66}$$

with

$$F(t) = \frac{1}{\ln(\tau_2/\tau_1)}\left[\mathrm{Ei}\left(\frac{-t}{\tau_1}\right) - \mathrm{Ei}\left(\frac{-t}{\tau_2}\right)\right] \tag{1.67}$$

where $\mathrm{Ei}(x)$ is the integral exponential function [43,44].

Similarly, for the Cole–Cole function [37]

$$
P(t) = \begin{cases}
\dfrac{1}{\tau_o} \displaystyle\sum_{n=0}^{\infty} \dfrac{-1^n}{\Gamma[(n+1)(1-\alpha)]} \left(\dfrac{t}{\tau_o}\right)^{[n(1-\alpha)-\alpha]} & t \ll \tau_o \quad (1.68) \\[3ex]
\dfrac{1}{\tau_o} \displaystyle\sum_{n=1}^{\infty} \dfrac{-1^n}{\Gamma[n(\alpha-1)]} \left(\dfrac{t}{\tau_o}\right)^{[-n(1-\alpha)-1]} & t \gg \tau_o \quad (1.69)
\end{cases}
$$

and

$$
F(t) = \begin{cases}
1 - \dfrac{1}{\tau_o} \displaystyle\sum_{n=0}^{\infty} \dfrac{-1^n}{\Gamma[(n+1)(1-\alpha)]\{[n(1-\alpha)-\alpha+1]\}} \left(\dfrac{t}{\tau_o}\right)^{[n(1-\alpha)-\alpha+1]} & \\
& t \ll \tau_o \quad (1.70) \\[2ex]
1 - \dfrac{1}{\tau_o} \displaystyle\sum_{n=0}^{\infty} \dfrac{-1^n}{\Gamma[n(\alpha-1)][-n(1-\alpha]} \left(\dfrac{t}{\tau_o}\right)^{[-n(1-\alpha)]} & \\
& t \gg \tau_o \quad (1.71)
\end{cases}
$$

For the Cole–Davidson function [37]

$$
P(t) = \dfrac{1}{\tau_{od}\,\Gamma(\beta)} \left(\dfrac{t}{\tau_{od}}\right)^{(\beta-1)} \exp\left(\dfrac{-t}{\tau_{od}}\right) \tag{1.72}
$$

and

$$
F(t) = \dfrac{2}{\Gamma(\beta)} \left(\dfrac{t}{\tau_{od}}\right)^{(\beta/2)} K_\beta[2(t/\tau_{od})^{1/2}] \tag{1.73}
$$

where K_β is a Bessel function of the second kind [45].
Similarly, for the Havriliak–Negami function [46]

$$
P(t) = \begin{cases}
\dfrac{1}{(\tau_o)\Gamma[\beta(1-\alpha)]} \left(\dfrac{t}{\tau_o}\right)^{-[1-\beta(1-\alpha)]} & t \ll \tau_o \quad (1.74) \\[3ex]
\dfrac{\beta(1-\alpha)}{\tau_o\,\Gamma[\alpha]} \left(\dfrac{t}{\tau_o}\right)^{-(2-\alpha)} & t \gg \tau_o \quad (1.75)
\end{cases}
$$

and

$$
F(t) = \begin{cases}
\dfrac{1}{\beta(1-\alpha)\{\tau_o\,\Gamma[\beta(1-\alpha)]\}} \left(\dfrac{t}{\tau_o}\right)^{\beta(1-\alpha)} & t \ll \tau_o \quad (1.76) \\[3ex]
\dfrac{\beta(1-\alpha)}{\{(\alpha-1)\tau_o\,\Gamma[\alpha]\}} \left(\dfrac{t}{\tau_o}\right)^{-(1-\alpha)} & t \gg \tau_o \quad (1.77)
\end{cases}
$$

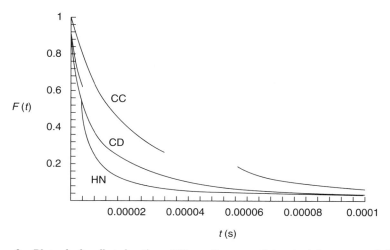

Figure 9. Plot of aftereffect functions $F(t)$ vs. time t as determined by means of Cole–Davidson (CD), Cole–Cole (CC), and Havriliak–Negami (HN) distribution functions.

Figure 9 shows the aftereffect functions obtained of a ferrofluid with arbitrary values of $f_m = 7$ kHz, $\tau_{od} = 45.5$ ms, $\alpha = 0.5$, and $\beta = 0.5$ using the Cole–Cole, Cole–Davidson, and Havriliak–Negami expressions. (Note that the values of α and β are the same as those used in determining the distribution functions of Fig. 8.)

Figure 9 highlights the effect which the different distribution functions have on $F(t)$. These results show how careful one must be in the choice of equation if one wishes to obtain realistic decay curves. Here it was found that the plots of the Cole–Davidson and Havriliak–Negami decay functions were similar, particularly at high and low times, whereas the decay function generated from the Cole–Cole equation differed substantially over the entire decay. However, the Cole–Davidson had an additional advantage over the use of the other equations in that a continuous magnetization decay curve could be generated. This was because no closed-form decay function exists in the other cases.

II.   MEASUREMENT OF FREQUENCY-DEPENDENT COMPLEX SUSCEPTIBILITY $\chi(\omega)$

It is apparent from Section I that relaxation and resonant mechanisms cover a corresponding frequency range from hertz to gigahertz, and thus measuring instrumentation and test cells must be capable of operating over these frequencies. With modern instrumentation this requires at least two instruments. The data presented here were obtained by using a combination of the Hewlett–Packard HP 4193A LF Bridge and the HP 8753C network analyzer. The approach here will be to separate measurements into two regions covering (1) Brownian and Néel relaxation and (2) mainly the frequency region of ferromagnetic resonance, with the data being presented in the form of case studies.

A.   Measurement in the Brownian and Néel Frequency Range of 10 Hz–13 MHz

The conventional method of determining the frequency-dependent complex susceptibility $\chi(\omega) = \chi'(\omega) - i\chi''(\omega)$ of a magnetic fluid is to insert the fluid into the alternating magnetic field of a coil of inductance L and resistance R and observe the changes in its inductance, ΔL, and resistance, ΔR, as the frequency is varied. The ratio $\Delta L/L$ is proportional to $\chi'(\omega)$, while $\chi''(\omega)$ is proportional to $\Delta R\ \omega L$. However, the coil technique requires the use of a large inductance which, with its associated capacitance, restricts the frequency range over which measurements can be made. Maiorov [9] used this technique and was among the first to publish measurements of $\chi(\omega)$ for magnetic fluids over the frequency range 30 Hz to kilohertz.

A more convenient and efficient method of measuring $\chi(\omega)$ is to use the "toroidal technique" of Fannin et al. [11]. This technique involves the placing of the fluid in an alternating magnetic field which is generated within a high-permeability (e.g., mu-metal or similar magnetic alloy) toroid. A narrow slit is cut in the toroid (Fig. 10) and the fluid is inserted into the slit. By means of appropriate measurements, the value of $\chi(\omega)$ can be readily determined. Over a wide frequency range it may be necessary to use a number of toroids. As an example, a toroid with the dimensions shown in Table I was found suitable for use up to a frequency of 10 MHz.

As has already been stated, $\chi(\omega)$ can be expressed in terms of its complex components, where $\chi'(\omega)$ is proportional to the increment in inductance and

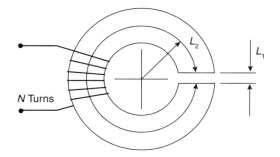

Figure 10.   Experimental toroid.

TABLE I
Details of Toroid

Outside diameter	57.20 mm
Inside diameter	38.10 mm
Axial length	9.52 mm
Mean circumference	150.00 mm
Cross-sectional area	90.70 mm^2
Mu-metal strip thickness	0.013 mm
Slit thickness	0.33 mm

$\chi''(\omega)$ is proportional to the variation in resistance due to the presence of the fluid in the slit in the toroid. Thus by measuring the inductance and resistance of the toroid when (a) the slit is empty and (b) the slit is full of ferrofluid, one can determine $\chi'(\omega)$ and $\chi''(\omega)$ for the fluid.

A full analysis of the measuring technique is given in Appendix A, and so only an explanation of the final result will be given here. Consider a test cell of relative permeability μ_r with a slit length L_1 cut in a toroid of rectangular cross section A and mean length L_2. The term μ_{eff} is the effective permeability of the cut toroid.

The impedance $Z(\omega)$ measured across N turns wound on the toroid is given by

$$Z(\omega) = R_w + i\omega\mu_{\text{eff}} L_0 \tag{2.1}$$

where R_w is the resistance of the winding and

$$L_0 = \frac{\mu_{\text{eff}} N^2 A}{L_2(1 + \gamma_g)} \qquad \gamma_g = \frac{L_1}{L_2} \tag{2.2}$$

When the slit is empty, the impedance is denoted by

$$Z_{ce} = R_w + \frac{i\omega\mu_r L_0}{1 + \mu_r \gamma_g} \tag{2.3}$$

and when it is filled with magnetic fluid of permeability μ_1, the impedance is

$$Z_{cf} = R_w + \frac{i\omega\mu_r L_0}{1 + \gamma_g \mu_r/\mu_1} \tag{2.4}$$

After some algebraic manipulation the following expression for the complex permeability of the ferrofluid, μ_1, is arrived at:

$$\mu_1' - i\mu_1'' = \frac{-\gamma_g(Z_{cf} - R_w)(Z_{ce} - R_w)}{(Z_{cf} - R_w)(i\omega L_0 + \gamma_g R_w - \gamma_g Z_{ce}) - (Z_{ce} - R_w)i\omega L_0} \quad (2.5)$$

Thus, by making measurements of Z_{cf} and Z_{ce}, one can determine μ_1' and μ_1'' for the fluid in the slit, and since

$$\mu_1' = \chi' + 1 \quad \text{and} \quad \mu_1'' = \chi''$$

the complex susceptibility components are readily obtained.

The results presented here were obtained with a test cell which consisted of a mu-metal strip toroid would with a suitable number of turns of wire, as shown in Fig. 10. The number of turns of wire used is determined by the necessity to avoid magnetic saturation of the colloidal sample and the core itself while magnetic saturation is avoided by ensuring that the condition $mH/kT < 1$ is satisfied. This then allows the use of the Langevin approximation $L(x) = \frac{1}{3}x$, where $x = mH/kT$ and H is the magnetizing force.

In order to ensure that the gap, and hence the relative positions, of the pole pieces remained fixed, the following procedure was followed:

1. A sector of the toroid was encapsulated in an epoxy resin to a depth of 20 mm.
2. A radial cut was then made in the encapsulated section by means of a diamond saw.

The effect of cutting the toroid is to reduce the permeability μ_r of the toroid to an effective value μ_{eff} such that

$$\mu_{eff} = \frac{1}{1/\mu_r + L_1/L_2} \quad (2.6)$$

Thus provided μ_r is large, μ_{eff} is equal to the ratio L_2/L_1. In deriving (2.6), the magnetic flux density is assumed to be uniform over the cross section of the toroid. Fringing effects near the slot are ignored and it is also assumed that there is negligible leakage flux.

B. Polarized Measurements

The toroidal technique is well suited for susceptibility measurements in the presence of a polarizing magnetic field H [47] since a DC biasing field H can be readily generated by (a) the addition of an auxiliary biasing winding to the arrangement of Fig. 10 or (b) using a variable-gap technique [48].

Assuming a Langevin function for the magnetization of the fluid, an expression for the field dependence of the AC susceptibility, $\chi(\omega, H)$, is

$$\chi(\omega, H) = \frac{\chi_0[1 + f(H)]}{1 + i\omega\tau_{\text{eff}}} \tag{2.7}$$

with

$$1 + f(H) = 3\left[1 + \left(\frac{kT}{mH}\right)^2 - \coth^2\left(\frac{mH}{kT}\right)\right] \tag{2.8}$$

Equation (2.8) predicts a reduction in both $\chi'(\omega)$ and $\chi''(\omega)$ with increasing bias, as will be confirmed in the results to be presented.

C. Application of the Toroidal Technique

The following examples of measurements of the frequency-dependent, complex susceptibility $\chi(\omega) = \chi'(\omega) - i\chi''(\omega)$ taken over the approximate frequency range of 10 Hz–13 MHz represent some of the work undertaken by the author and fellow workers. Five case studies involving 11 magnetic fluids are presented.

1. Case Study 1

The susceptibility plots here are chosen to demonstrate how the loss peak can lie in any part of the frequency spectrum (for $f \ll f_1$, the Lamor frequency) depending on the particle size and viscosity of the magnetic fluid sample.

An illustration of actual susceptibility data obtained for four magnetic fluid samples (samples 1, 2, 3, and 4, respectively) [32] is given in Fig. 11 for measurements taken up to a frequency of 10 MHz. The median particle radii of samples 1, 2, and 3 was 5 nm while that of sample 4 was 4.5 nm. The samples were colloidal solutions of (1) cobalt ferrite in hexadecene, (2) magnetite in water, (3) cobalt ferrite in toluene, and (4) magnetite in isopar M with corresponding viscosity values of 3.5×10^{-3}, 1×10^{-3}, 0.6×10^{-3}, and 1×10^{-3} Ns m^{-2}, respectively, being assumed. The figure is a normalized plot of $\chi'(\omega)$ and $\chi''(\omega)$ versus $\log f$ and shows four Debye-type profiles with loss peaks at 400 Hz, 2 kHz, 63 kHz, and 3.5 MHz, respectively. The corresponding values of hydrodynamic particle radii, as determined by the Debye equation (1.16), at room temperature are 33, 29.7, 10, and 2.4 nm, respectively.

One can comment on the fact that the values of radii obtained for samples 1 and 2 are indicative of aggregation having occurred. This demonstrates that the dominant relaxation mechanism in these samples was

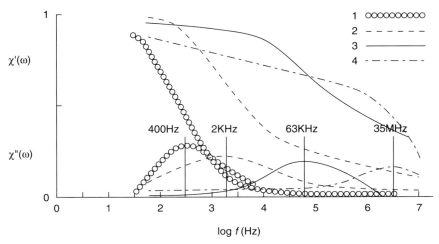

Figure 11. Plot of $\chi'(\omega)$ and $\chi''(\omega)$ vs. $\log f$ for samples 1, 2, 3, and 4.

Brownian. By using the data of Fig. 5 as a rough guide, the corresponding results for samples 3 and 4 indicate that both Néel and Brownian mechanisms made contributions to the magnetization, with the dominant mechanism in the former sample being Brownian and that of the latter sample being Néel. Allowing for a surfactant thickness of 2 nm, the result for sample 3 is a good approximation to the actual mean particle radius. For sample 4, as the dominant relaxation mechanism is Néel, Eq. (1.27) and not Eq. (1.16) is applicable and applying Eq. (1.27) with a value of $K = 4 \times 10^4$ J m^{-3} results in an approximate magnetic radius of 5 nm.

2. Case Study 2

The next set of data relate to a water-based ferrofluid with an average particle radius of 5 nm whose particle size distribution is given in Fig. 4. These data, measured over the frequency range 50 Hz–10 MHz, is of particular interest for, as shown in Fig. 12, the susceptibility plot has two loss peaks at frequencies of 250 Hz (f_{m1}) and 2.2 MHz (f_{m2}), respectively [49]. The lower frequency loss peak is due to Brownian rotation, and application of Eq. (1.16) results in an average particle size of 59.4 nm; this size of hydrodynamic radius is indicative of the formation of aggregates. The higher frequency loss peak has contributions from both Brownian and Néel relaxation mechanisms, and application of Eq. (1.27) results in an effective relaxation time of 7.2×10^{-8} s, indicating that the Néel mechanism is dominant at this frequency with an approximate particle radius of 5 nm.

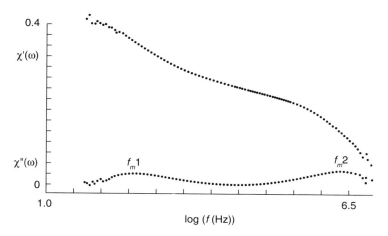

Figure 12. Plot of $\chi'(\omega)$ and $\chi''(\omega)$ vs. $\log f$ for sample 5 with loss peaks at frequencies of $f_{m1} = 250$ Hz and $f_{m2} = 2.2$ MHz.

Treating the data as two finite distributions enables the separate entities to be analyzed by means of the Cole–Cole and approximate ellipse techniques. Fitting the data to two Cole–Cole plots, as shown in Fig. 13, one obtains values of 0.4 and 0.44, which are indicative of a wide range of particle sizes in each separate distribution. Also, if each distribution is fitted to an ellipse, as shown in Fig. 14, the weight function is found to be

$$f(\tau) = \frac{1}{10.2\tau} \qquad 3.9 \times 10^{-6} < \tau < 10^{-1} \text{ s} \quad \text{for ellipse with } \alpha = 0.4$$

and

$$f(\tau) = \frac{1}{11.7\tau} \qquad 2 \times 10^{-10} < \tau < 2.4 \times 10^{-5} \text{ s} \quad \text{for ellipse with } \alpha = 0.44$$

thus giving an overall range of relaxation times of approximately $2 \times 10^{-10} < \tau < 10^{-1}$ s.

The contribution which the Néel component makes to the high-frequency loss peak can be determined by freezing the test sample (thereby preventing the bulk rotation of the Brownian-type particles) and observing the effect which this has on the susceptibility profile. Figure 15 illustrates the result of such an exercise as a result of freezing the test sample to 268 K.

The following points emerge:

The low-frequency loss peak has disappeared.

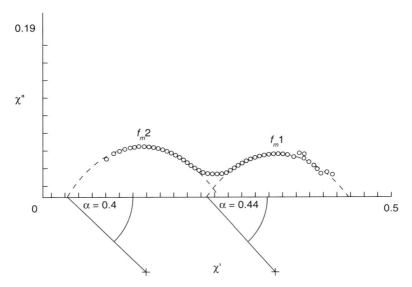

Figure 13. Cole–Cole plot of χ' against χ'' for sample 5 with values of $\alpha = 0.4$ and $\alpha = 0.44$.

The high-frequency loss peak has been attenuated and f_{m2} has been shifted to a lower frequency of 1.3 MHz.

The susceptibility at 50 Hz has been reduced to approximately 0.375 of its room temperature value.

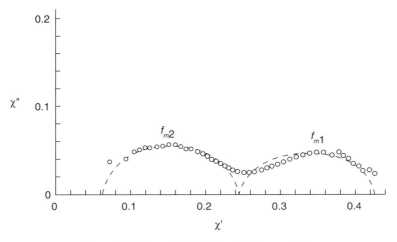

Figure 14. Approximate ellipse plot for sample 5.

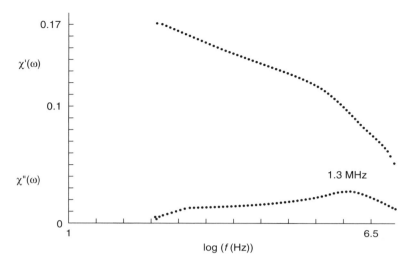

Figure 15. Plot of $\chi'(\omega)$ and $\chi''(\omega)$ against log f for frozen sample 5 with loss peak at frequency of approximately 1.3 MHz.

The elimination of the low-frequency loss peak can be attributed to the prevention of the bulk rotation of the Brownian-type particles of the test sample, and this is reflected in the fall in the susceptibility to 0.15. The reduction in amplitude of the high-frequency loss peak is indicative of the contribution made by the Brownian particles to this loss while the downward shift in the high-frequency loss peak is due to the temperature change and is as predicted by Eq. (1.27).

3. Case Study 3

The next two examples to be presented are of the complex susceptibility data of two magnetic fluid samples, namely fluids 6 and 7, measured over the approximate frequency range 1 kHz–10 MHz [50]. The fluids consisted of suspensions of (1) cobalt ferrite in isopar M and (2) cobalt ferrite in hexadecene, with median diameters of approximately 10 nm and viscosity values of 3.5×10^{-3} and 1×10^{-3} Ns m^{-2}, respectively, being assumed. Here we highlight the application of the Cole–Cole and approximate ellipse techniques and also the determination of the aftereffect functions.

The plot of $\chi'(\omega)$ and $\chi''(\omega)$ versus log f for fluid 6, as shown in Fig. 16(a), is found to have a Debye-type profile with a maximum occurring at a frequency of approximately $f_{max} = 14.6$ kHz. This fluid was analyzed in terms of the Cole–Cole and approximate ellipse plots, as illustrated in Fig. 16(b). The value of α was determined from the Cole–Cole plot while from the

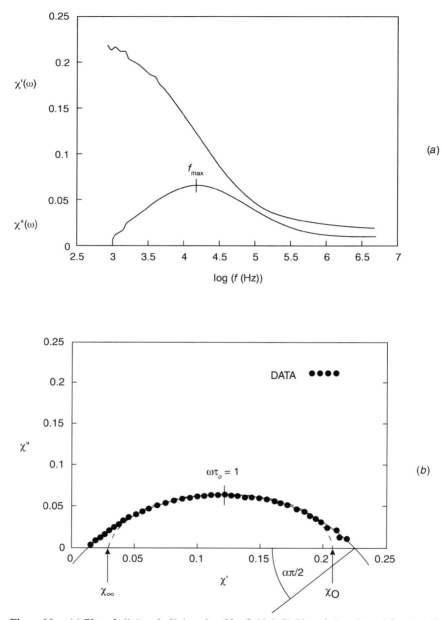

Figure 16. (a) Plot of $\chi'(\omega)$ and $\chi''(\omega)$ vs. $\log f$ for fluid 6. (b) Plot of χ' against χ'' fitted to (1) circular arc and (2) an ellipse from which values of α, $\chi''(\omega_m)$ χ'_0, and χ_∞ of sample 6 are determined.

susceptibility components $\chi''(\omega_{max})$, χ_0, and χ_∞ of the ellipse, the corresponding values of the Fröhlich parameters τ_1 and τ_2 [49] were determined; the results obtained are shown in Table II. Application of Eqs. (1.59) and (1.60) results in a ratio of $r_{max}/r_{min} = 17.9$.

Plots of Cole–Cole and Fröhlich distribution functions as determined by means of Eqs. (1.45) and (1.51), respectively, were then obtained and plotted versus $\ln(\tau/\tau_0)$ in Fig. 17, while the corresponding aftereffect functions as determined by Eqs. (1.67), (1.70), and (1.71), respectively, are shown in Fig. 18, with the noncontinuous nature of the response corresponding to the Cole–Cole case being quite apparent. The Fröhlich distribution function results in a continuous decay function with the fastest initial rate of decay when compared to the corresponding response obtained with the application of the Cole–Cole data, with the decay curves overlapping at position a

TABLE II
Data for Fluids 6 and 7

Fluid	f_m (kHz)	τ_0 (µs)	τ_{od} (µs)	τ_1 (ms)	τ_2 (µs)	α	β
6	14.6	10.8	—	0.8	0.14	0.445	—
7	7.0	22.74	4.55	—	—	0.36	0.47

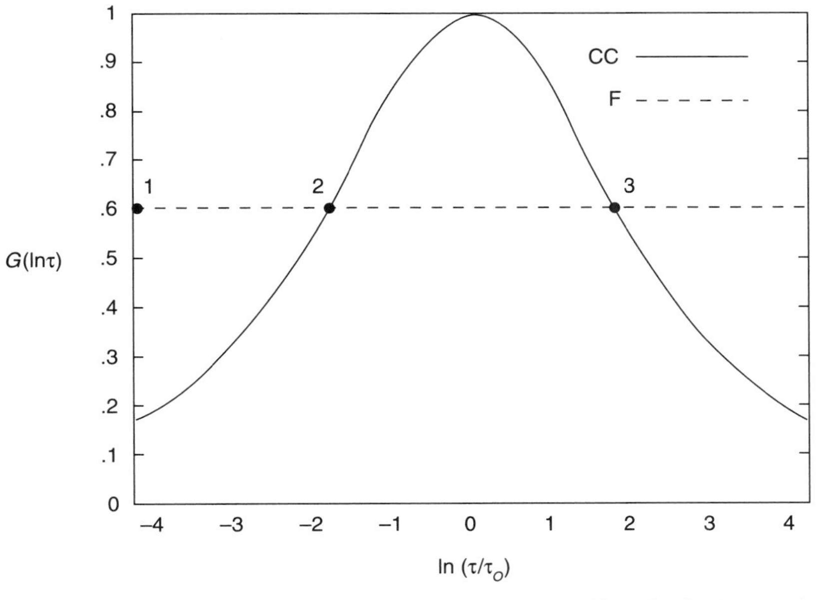

Figure 17.   Plot of $G(\ln \tau)$ against $\ln(\tau/\tau_0)$ for Cole–Cole and Fröhlich distribution functions using values of α, τ_1, and τ_2 determined for sample 6.

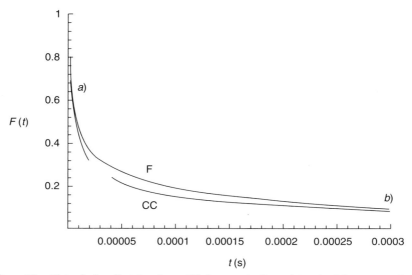

Figure 18. Plot of aftereffect functions $F(t)$ for sample 6, as determined by means of the Cole–Cole and the Fröhlich distribution function.

of the curve. Furthermore from Fig. 18 it is noted that decay curves again overlap at point b.

These effects are due to the different natures of the distribution functions studied and can be explained by means of Fig. 17. Over the region 1–2, which represents the region of fastest decay, the Fröhlich function contributes a greater number of smaller particles (i.e., shorter relaxation times) than the Cole–Cole function, with the result that the initial decay of the after effect function is greater in the former. In the region of point 2 and beyond, to point 3, the Cole–Cole distribution is greater, and as a result the rate of decay associated with this distribution is now greater than that for the Fröhlich distribution, with the result that the curves cross, as illustrated in Fig. 17. Using a similar argument, the curves must again cross when in the region of point 3.

The plot of χ' against χ'' for fluid 7 was found to have a Cole–Davidson profile, as indicated in Fig. 19, with the plot cutting the $\chi'(\omega)$ axis at the high-frequency side at an angle of $\frac{1}{2}\pi\beta$. From this angle the value of the Cole–Davidson parameter β was determined. The fluid was also analyzed in terms of the Cole–Cole plot as illustrated in the same figure and the value of α determined. The values of χ' and χ'' at which $\omega\tau_{od} = 1$ were determined

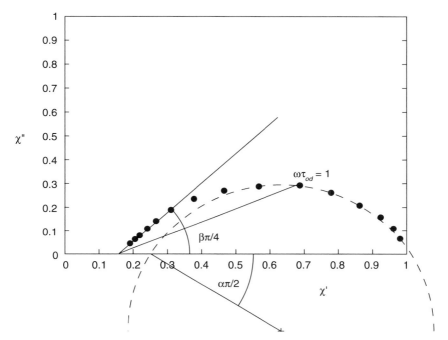

Figure 19. Normalized plot of χ' vs. χ'' fit to Cole–Davidson and Cole–Cole profiles for sample 7.

by projecting forward the bisector of the angle $\frac{1}{2}\beta\pi$ until it cut the plot. These values of χ' and χ'' were then referred to the previously measured complex susceptibility data and the value of ω at which they occurred determined. Since $\omega\tau_{od} = 1$, τ_{od} was easily evaluated. The relevant parameters are shown in Table II.

Plots of the corresponding after effect functions as determined by Eqs. (1.73), (1.76), and (1.77), respectively, are shown in Fig. 20.

These results show the aftereffect plots of the Cole–Davidson and Havriliak–Negami decay functions to be similar, particularly at high and low times; however, the Cole–Davidson had an additional advantage in that a continuous magnetization decay curve could be generated.

4.   Case Study 4

In this case the effects of applying a magnetic polarizing field H to a sample is investigated. Equation (2.8) predicts a reduction in both $\chi'(\omega)$ and $\chi''(\omega)$

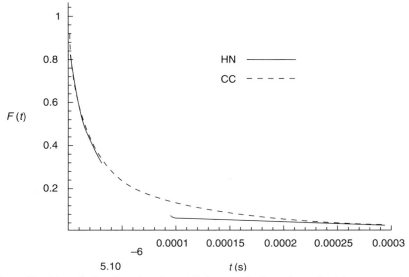

Figure 20. Plot of aftereffect functions $F(t)$ for sample 7, as determined by means of the Cole–Davidson and the Havriliak–Negami distribution function.

with increasing polarizing field, and this is confirmed by the results obtained for a magnetic fluid sample (fluid 8) of cobalt ferrite in hexadecene.

From Fig. 21(a) it is evident that over the polarizing field range of 0–40 kA/m^{-1} χ_0 reduces to approximately 1/20th of its unbiased value, while Fig. 21(b) shows how f_m increases from 300 Hz to 5.6 kHz [51], which, according to Eqs. (1.20) and (1.23), is indicative of a decrease in τ_{eff} from 1.43 to 0.18 ms and an average particle radius decrease of 36–14 nm.

Further information about the distribution of relaxation times and particle radii is found by fitting the data to Cole–Cole and elliptical plots, as shown in Figs. 22(a) and 22(b) respectively.

From Fig. 22(a) it is seen that the effect of the polarizing field on the circular arcs is to vary α from 0.41 to 0.26, which indicates a narrowing of the distribution with polarizing field. Within the limits of experimental error there is little change between $H = 0$ and $H = 15$ kA m^{-1}, but there appears to be a significant change at higher fields between 20 and 40 kA m^{-1}. From Fig. 22(b) it is clear how χ_0 decreases with increasing H.

It is found that the weight function $f(\tau)$ varies as follows:

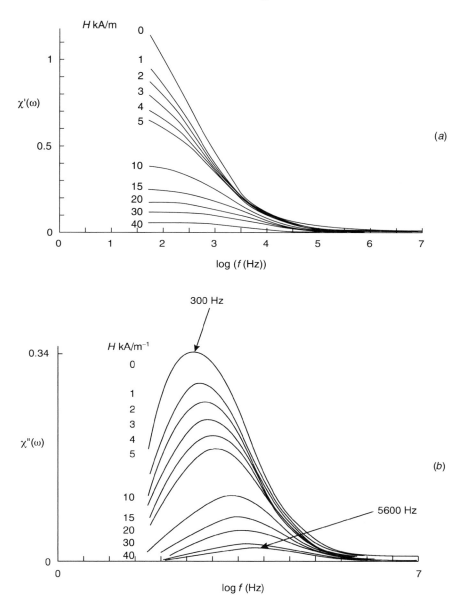

Figure 21. (*a*) Plot of $\chi'(\omega)$ vs. $\log f$ for sample 8 for polarizing fields H of 0, 1, 2, 3, 4, 5, 10, 15, 20, 30, and 40 kA m^{-1}. (*b*) Plot of $\chi''(\omega)$ against $\log f$ for sample 8 for polarizing fields H of 0, 1, 2, 3, 4, 5, 10, 15, 20, 30, and 40 kA m^{-1}, with loss peaks varying over the approximate frequency range 300–5600 Hz.

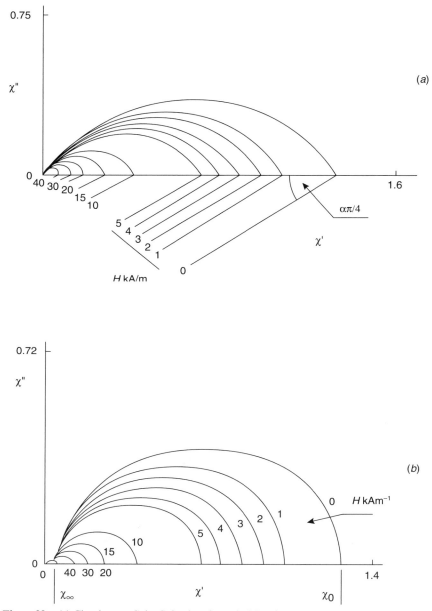

Figure 22. (*a*) Circular arc Cole–Cole plots for polarizing field *H* of 0, 1, 2, 3, 4, 5, 10, 15, 20, 30, and 40 kA m⁻¹. (*b*) Ellipse Cole–Cole plots for polarizing fields *H* of 0, 1, 2, 3, 4, 5, 10, 15, 20, 30, and 40 kA m⁻¹.

1. For $H = 0$

$$f(\tau) = \frac{1}{11.6\tau} \quad \text{for} \quad 2 \times 10^{-6} < \tau < 1.4 \times 10^{-1}\ \text{s} \quad \text{and} \quad \frac{r_{max}}{r_{min}} = 41.3$$

2. For $H = 40\ \text{kA m}^{-1}$

$$f(\tau) = \frac{1}{8.8\tau} \quad \text{for} \quad 3.4 \times 10^{-7} < \tau < 2.3 \times 10^{-3}\ \text{s} \quad \text{and} \quad \frac{r_{max}}{r_{min}} = 19$$

Both methods of analysis portray a situation in which the average particle radius decreases with increasing polarizing field. The apparent decrease in particle size is due to the greater degree of alignment exerted by the polarizing field on large particles compared to smaller ones, because the particle moment is directly proportional to its volume. Thus, as the polarizing field increases, the larger particles are progressively inhibited from contributing fully to the relaxation process.

5. Case Study 5

Data from polarized measurements provide a convenient method for investigating the behavior of the aftereffect function $F(t)$ as a function particle size distribution. Here [52] four polarized measurements of a water-based sample (fluid 9) were taken as being representative of four fluids of different particle size distributions.

Each of the fluids' $\chi'(\omega)$ and $\chi''(\omega)$ susceptibility profiles were fitted to Fröhlich susceptibility equations (1.54) and (1.55) and the corresponding values of τ_1 and τ_2 were determined in each case, as shown in Table III; an example of the data and fit obtained for sample 3 is shown in Fig. 23(b).

This technique provides four samples which effectively have different particle size distributions which should result in four different aftereffect functions.

Figure 24 shows the aftereffect functions of the four samples. It is apparent that each sample has a fast and a slow decay component, with the sample with the narrowest distribution of particles (sample 4) having the fastest decay component with a time constant of approximately 100 μs compared to approximately 610 μs for sample 1).

TABLE III
Values of τ_1 and τ_2

Sample	1	2	3	4
$\tau_1 \times 10^{-3}$ s	5	4	1	0.7
$\tau_2 \times 10^{-5}$ s	5	2	1.5	1

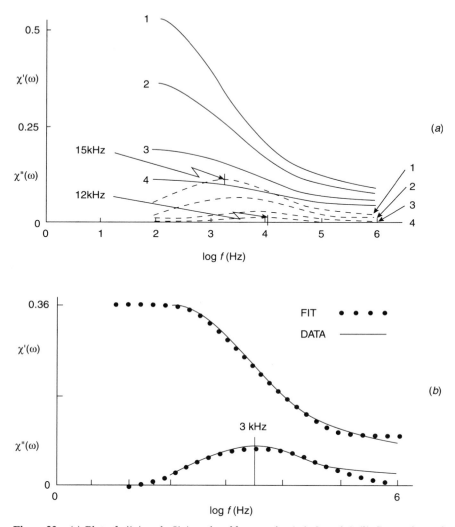

Figure 23. (*a*) Plot of $\chi'(\omega)$ and $\chi''(\omega)$ vs. log f for samples 1, 2, 3, and 4. (*b*) Comparison of experimental and theoretical spectra of $\chi'(\omega)$ and $\chi''(\omega)$ vs. log f for sample 3 and fitting curve obtained by use of the Frohlich susceptibility equations (1.54) and (1.55).

D. Application of Signal Processing Techniques

With the advent of fast computing facilities it has become possible to apply signal transform techniques to assist in the determination of various properties of the complex susceptibility of ferrofluids [53].

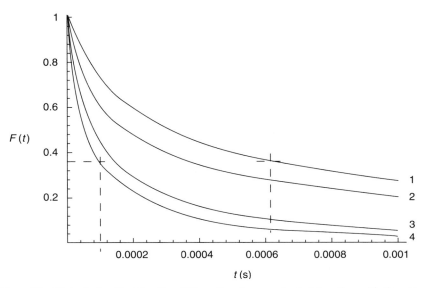

Figure 24. Plot of aftereffect function $F(t)$ vs. time for samples 1–4 together with time constant for samples 1 and 4.

Two such examples are given here based on the use of the fast Fourier transform (FFT) and the Hilbert transform.

1. Application of the Fourier Transform Technique

In this example the Fourier transform technique is used as an alternative method of determining the aftereffect function of ferrofluids.

The Fourier transform $F(\omega)$ of a continuous function $f(t)$ is defined as

$$F(\omega) = \int_{-\infty}^{\infty} f(t) \exp(-i\omega t) \, dt \qquad (2.9)$$

and its inverse function as

$$f(t) = \frac{1}{2\pi} \int_{-\infty}^{\infty} F(\omega) \exp(i\omega t) \, d\omega \qquad (2.10)$$

As an example, consider the case where $f(t)$ is a one-sided exponential pulse $A \exp(-t/\tau)$. From eq. (2.9), the Fourier transform of $f(t)$ is

$$F(\omega) = \frac{A\tau}{1 + i\omega\tau} \qquad (2.11)$$

which has the same form of Eq. (1.16), the Debye equation. So immediately it is known that the inverse FT (IFT) of $\chi(\omega)$ should be an exponential function of the form $A \exp(-t/\tau)$. In reality $\chi(\omega)$ is not a continuous function but a sampled function, and so the discrete Fourier transform (DFT) must be used.

If the number of samples is of the power of 2, then the DFT may be implemented using the FFT algorithm. Where $\chi(\omega)$ consists of N samples, the implementation of the FFT requires approximately $N \log_2 N$ operations. In a practical measurement, $\chi'(f)$ and $\chi''(f)$ are measured at evenly spaced intervals, Δf hertz, from a start frequency of Δf to a frequency f_∞, the latter being the frequency where χ_∞ is deemed to have been reached. Also, the start frequency must be equal to or a multiple of the incremental frequency. The proposed technique also assumes that the zero-frequency data are the same as obtained at the first measurement point. Furthermore, since FFT techniques require the signal to be transformed to be periodic, the values of $\chi'(f)$ and $\chi''(f)$ from 0 to $-f_\infty$ must be generated. This is a simple matter since $\chi'(f)$ and $\chi''(f)$ are even and odd functions of f and are easily determined in software.

A factor which should be considered when choosing the incremental measurement frequency Δf and the final frequency f_∞ is the fact that the limits of integration of the integral in Eq. (2.9) are from $\pm \infty$. In practice this is limited to $\pm f_\infty$, which is equivalent to multiplying $\chi'(f)$ by a rectangular window. Other types of windows, for example, Hamming or triangular, may be used in order to minimize the effects of the rectangular window.

Application of Inverse FFT Transform. To test the validity of the proposed technique, it was first applied to theoretical Debye curves, where one knew [from Eq. (2.11)] what the resultant inverse FFT (IFFT) should be. Initially data of the Debye plots of $\chi'(\omega)$ and $\chi''(\omega)$ for Eq. (1.17), with $\chi_0 = 1$ and $\chi_\infty = 0$, maximum $\omega = 5 \times 10^8$ rad/s, $N = 8192$ points, and $\omega_m = 10^6$ rad/s, were determined and the corresponding transform plot of $F(t)$ versus time obtained as shown in Fig. 25(a); it is found to be of the exponential form predicted. One notes that its magnitude falls to $1/e$ of its peak value in a time of τ seconds. This corresponds to the time constant $\tau = 1/\omega_m$, thus establishing the fact that the "time-constant time" of the IFFT corresponds to the average relaxation time of the particles under examination. Figure 25(b) shows the result for the case where $\chi_\infty = 0.2$ with the time constant being the same as in the previous case. This is as expected since τ is a function of ω_m, which remains unchanged.

The technique is then applied, with $N = 8192$ points, to a profile having two absorption peaks occurring at ω_{m1} and ω_{m2}, corresponding to approx-

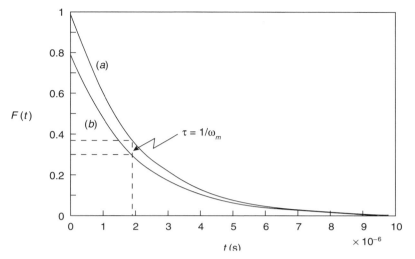

Figure 25.   Plot of $F(t)$ vs. t for Debye case with (a) $\chi_\infty = 0$ and (b) $\chi_\infty = 0.2$.

imately 10^5 and 1.8×10^7 rad/s, respectively, as illustrated in Fig. 26. This profile, which has two distinct slopes in the $\chi'(\omega)$ plot, can be described by the expression

$$\chi(\omega) = \chi_{\infty 2} + \frac{\chi_0 - \chi_{\infty 1}}{1 + i\omega\tau_1} + \frac{\chi_{\infty 1} - \chi_{\infty 2}}{1 + i\omega\tau_2} \tag{2.12}$$

where $\omega_{m1} = 1/\tau_1$ and $\omega_{m2} = 1/\tau_2$. Now the IFFT of Eq. (2.12) $v(t)$, has the approximate form

$$v(t) = \frac{1}{\tau_1}(\chi_0 - \chi_{\infty 1}) \exp\left(-\frac{t}{\tau_1}\right) + \frac{1}{\tau_2}(\chi_{\infty 1} - \chi_{\infty 2}) \exp\left(-\frac{t}{\tau_2}\right) \tag{2.13}$$

so provided ω_{m1} and ω_{m2} are well separated, $v(t)$ should display two distinct exponential curves.

The loglinear plot of Fig. 27(a) shows the result of transforming the data of Fig. 26. It clearly shows the two distinct decaying exponentials B and C which intercept the y axis at approximate values of $\log(5.55)$ and $\log(3.85)$, respectively, as determined by Eq. (2.13) with $(\chi_{\infty 1} - \chi_{\infty 2}) = 0.02$ and $(\chi_0 - \chi_{\infty 1}) = 0.07$.

Having established that the technique is successful with theoretically generated data, it is now applied to susceptibility data obtained for a ferrofluid sample (fluid 10) consisting of a colloidal suspension of single-

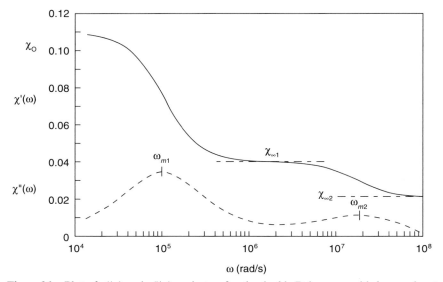

Figure 26. Plot of $\chi'(\omega)$ and $\chi''(\omega)$ against ω for the double-Debye case with loss peaks at $\omega_{m1} = 10^5$ rad/s and $\omega_{m2} = 17.8 \times 10^6$ rad/s.

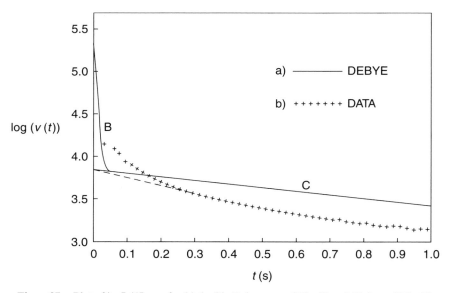

Figure 27. Plot of $\log[v(t)]$ vs. t for (a) double-Debye case of Fig. 26 and (b) data of Fig. 28.

domain magnetite particles with a lognormal volume distribution of medium diameter of 12.1 nm and a standard deviation of 0.51 dispersed in a water carrier.

Measurements were performed over the frequency range 400 Hz–3.28 MHz in steps of 400 Hz and the resulting plots of $\chi'(\omega)$, $\chi''(\omega)$ are shown in Fig. 28, with a first loss peak of $\chi''(\omega)$ occurring at $\omega_{m1} = 10^5$ rad/s and an incomplete second peak occurring at approximately $\omega_{m2} = 1.8 \times 10^7$ rad/s. These parameters correspond to those chosen in the theoretical example of Fig. 26 and enable a comparison to be made between its transformation, Fig. 27(b), and the Debye transformation shown in Fig. 27(a). Again it should be noted that the ferrofluid sample consists of a distribution of particle sizes and hence a distribution of relaxation times, while the theoretical case consists of only two relaxation times represented by two Debye-type profiles, this fact being manifest by (a) the difference in slopes of the corresponding $\chi'(\omega)$ curves and (b) the loglinear plots for the two exponential profiles no longer being linear, reflecting the presence of a distribution of relaxation times. Notwithstanding these differences between the theoretical and dynamic profiles, the similarity between the corresponding transformed

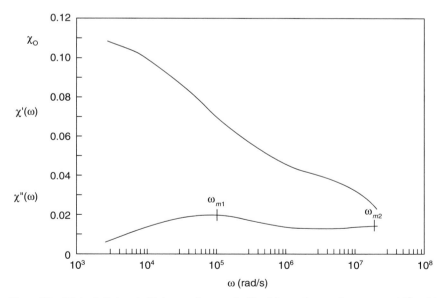

Figure 28. Plot of $\chi'(\omega)$ and $\chi''(\omega)$ vs. ω for sample 10, with two loss peaks at $\omega_{m1} = 10^5$ rad/s and $\omega_{m2} = 17.8 \times 10^6$ rad/s, respectively.

waveforms is quite satisfactory, with the lower frequency exponential cutting the y axis at almost the same point as its theoretical counterpart while the second exponential intersects at approximately log(5.5), compared to the theoretical intercept of log (5.55).

This example clearly demonstrates that the use of FFT techniques enables one to process data, measured in the frequency domain, to generate data in the time domain; here we have effectively performed the equivalent time measurement over the approximate range 0.3×10^{-6}–2.5×10^{-3} s.

2. Generation of Complex Susceptibility Data through the Use of the Hilbert Transform

On occasion, due to the existence of poor susceptibility data, it may only be possible to obtain good-quality data for $\chi'(\omega)$ and not $\chi''(\omega)$. However, $\chi'(\omega)$ and $\chi''(\omega)$ are a Hilbert transform pair, so in knowing $\chi'(\omega)$, one can determine $\chi'(\omega)$ [54,55]. On a point of interest, for the case of spin glass, the literature [56,57] reports on the determination of $\chi''(\omega)$ from a knowledge of $\chi'(\omega)$, with measurements being a function of temperature at spot frequencies.

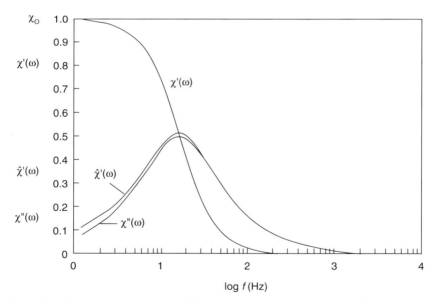

Figure 29. Plot of $\chi''(\omega)$, $\chi'(\omega)$, and its Hilbert transform $\hat{\chi}'(\omega)$ vs. log f for the Debye case with $\chi_\infty = 0$ and $n = 8192$ points.

The Hilbert transform of a function $x(t)$ is defined as

$$\hat{x}(t) = \frac{1}{\pi} \int_{-\infty}^{\infty} \frac{x(\tau)}{(t - \tau)} \, d\tau \tag{2.14}$$

$$= \left(\frac{1}{\pi t}\right) * x(t) \tag{2.15}$$

where * represents the convolution operation.

This operation corresponds to phase shifting all frequency components of $x(t)$ by 90°, and it may be represented by a linear system with an impulse response of

$$h(t) = \frac{1}{\pi t} \tag{2.16}$$

and transfer function

$$H(f) = -i \, \text{sgn}(f) \qquad \left(f = \frac{\omega}{2\pi}\right) \tag{2.17}$$

As convolution in the time domain is equivalent to multiplication in the frequency domain, Eq. (2.15) may be written as

$$\hat{X}(f) = H(f)X(f) \tag{2.18}$$

where $\hat{X}(f)$, $H(f)$ and $X(f)X(f)$ are the Fourier transforms of $\hat{x}(t)$, $h(t)$, and $x(t)$, respectively. The terms $\hat{X}(f)$ and $\hat{x}(t)$ are related by the expression

$$\hat{x}(t) = \int_{-\infty}^{\infty} \hat{X}(f) \, e^{j2\pi f t} \, df \tag{2.19}$$

and from Eq. (2.18), assuming that $X(f)$ is an even function, it follows that

$$\hat{x}(t) = 2 \int_{0}^{\infty} X(f) \sin 2\pi f \, df \tag{2.20}$$

Consider the example of the function [58]

$$x(t) = \frac{1}{1 + t^2} \tag{2.21}$$

This function has a Fourier transform of

$$X(f) = \pi e^{-|2\pi f|} \tag{2.22}$$

and from Eq. (2.20) the corresponding Hilbert transform is

$$\hat{x}(t) = \frac{t}{1 + t^2} \tag{2.23}$$

It is of interest to note that Eqs. (2.21) and (2.23) have the same form as the frequency-dependent term of $\chi'(\omega)$ and $\chi''(\omega)$ in Eq. (1.17). The components χ_0 and χ_∞ are constants, and since the Hilbert transform of a constant is zero, it is clear that $\chi''(\omega)$ is indeed the Hilbert transform of $\chi'(\omega)$.

Realization of the Hilbert Transform. The Hilbert transform may be implemented in the frequency domain by direct computation of Eq. (2.15). For discrete samples, this becomes a matrix multiplication:

$$X[\hat{K}] = h[K] \times [K] \tag{2.24}$$

is an $n \times n$ matrix whose elements are given by $1/[(u + v' - 1)\pi \, \Delta f]$, where u and v' are the row and column indices, respectively, and Δf is the frequency increment, which cannot be greater than the start frequency. The computation of this formula is very time consuming, since it requires of the order of n^2 multiplications and additions. Such a computational burden is greatly reduced by implementing Eq. (2.18). This involves calculating the IFFT of $\chi'(f)$, multiplying by $-i \, \mathrm{sgn}(t)$ and calculating the Fourier transform of the result, as illustrated in the following flow chart:

$$|\chi'(f)| \xrightarrow{\text{IFFT}} |\chi'(t)| \xrightarrow{-i \, \mathrm{sgn}(t)} |\hat{\chi}'(t)| \xrightarrow{\text{FFT}} |\hat{\chi}'(f)|$$

The above method was used to obtain the Hilbert transform data presented here.

Application of Hilbert Transform Technique. To test the accuracy of the technique, it is initially applied to theoretical Debye curves [59] where it is known what the resultant Hilbert transform should be. Initially the technique is applied to Eq. (1.17) with $\chi_\infty = 0$ and $n = 8192$; the corresponding plots of $\chi''(\omega)$, $\chi'(\omega)$, and its Hilbert transform $\hat{\chi}'(\omega)$ are shown in Fig. 29 and the Hilbert transform $\chi'(\omega)$ proves to be a good approximation to $\chi''(\omega)$, with some error existing in the low-frequency region of the plots.

To account for the situation where a χ_∞ component exists, the technique is shown to be equally successful when applied to the case where $\chi_\infty = 0.2$ and $n = 8192$; the resultant plots of $\chi''(\omega)$, $\chi'(\omega)$, and its Hilbert transform $\hat{\chi}'(\omega)$ are shown in Fig. 30.

The technique is now applied to a dynamic situation, the data being obtained for a cobalt ferrite magnetic fluid sample (fluid 11).

Measurements were made on a ferrofluid consisting of a colloidal suspension of single-domain cobalt ferrite particles with a lognormal volume distribution of median diameter of 10.3 nm and a standard deviation of 0.54 dispersed in a perfluoro carrier. Measurements were performed over the approximate frequency range 50 Hz–500 kHz in steps of 50 Hz and the resulting plots of $\chi''(\omega)$, $\chi'(\omega)$, and $\hat{\chi}'(\omega)$ are shown in Fig. 31. The parameter $\hat{\chi}'(\omega)$ proved to be a very good approximation to $\chi''(\omega)$ with ω_{max} for both curves being identical. Some error does exist, particularly in the low-frequency region; however, this error can be reduced by reducing the size of incremental frequency Δf. At high frequencies the onset of the "window" effect can be seen with $\hat{\chi}'(\omega)$ crossing and becoming smaller than $\chi''(\omega)$.

This example illustrates the usefulness of the Hilbert transform in determining complex susceptibility components once one component is known,

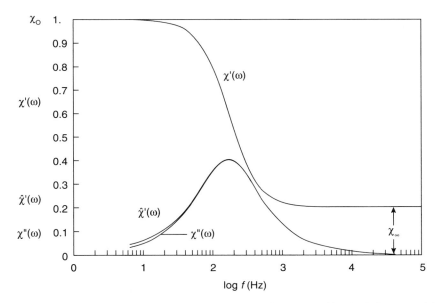

Figure 30. Plot of $\chi''(\omega)$, $\chi'(\omega)$, and its Hilbert transform $\hat{\chi}'(\omega)$ vs. log f for the Debye case with $\chi_\infty = 0.2$ and $n = 8192$ points.

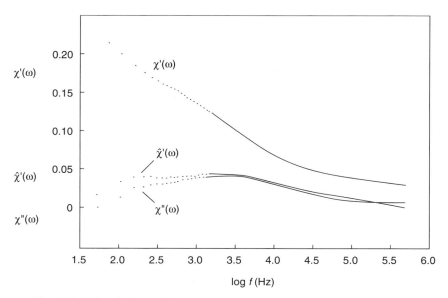

Figure 31. Plot of $\chi''(\omega)$, $\chi'(\omega)$, and its Hilbert transform $\hat{\chi}'(\omega)$ vs. $\log f$ for sample 11.

particularly in a dynamic situation. Of course, as already mentioned, the technique can also be used to generate data on $\chi'(\omega)$ from a knowledge of $\chi''(\omega)$, simply by taking the inverse Hilbert transform of $\chi''(\omega)$. This is due to the fact that the Hilbert transform of a Hilbert transform returns the original signal with a change in sign [60].

III. RESONANT MECHANISMS

In Section I the complex susceptibility $\chi(\omega) = \chi'(\omega) - i\chi''(\omega)$ of a random assembly of single-domain particles was described in terms of its parallel, $\chi_{\parallel}(\omega)$, and perpendicular, $\chi_{\perp}(\omega)$, components, with

$$\chi(\omega) = \tfrac{1}{3}(\chi_{\parallel}(\omega) + 2\chi_{\perp}(\omega)) \tag{1.15}$$

The parallel susceptibility $\chi_{\parallel}(\omega)$ was taken to be purely relaxational in character; here we consider $\chi_{\perp}(\omega)$ the transverse susceptibility.

The transverse susceptibility $\chi_{\perp}(\omega)$ is associated with resonance, and under equilibrium conditions, the magnetic moment m and the internal field H_A of a particle are parallel and any deviation of the magnetic moment from the easy-axis direction results in the precession of the magnetic moment about this axis. In the case where the polar angle θ is small, the angular resonant frequency ω_0 is given by

$$\omega_0 = 2\pi f_0 = \gamma H_A \tag{3.1}$$

where γ is the gyromagnetic ratio and H_A is the internal field. For a particle with uniaxial anisotropy,

$$H_A = \frac{2K}{M_s} \tag{3.2}$$

In the work of Raĭkher and Shliomis [15] the resonance phenomenon is treated in terms of equations proposed by Landau and Lifshitz [61] modified to include stochastic terms. They show that when a radio frequency field is applied perpendicular to H_A, the motion of the magnetic moment has a typical resonant character, with the real and imaginary components of the AC susceptibility having an approximate Lorentz form. They also show that the normalized susceptibility $\chi_\perp(\omega)$, which is associated with resonance, can be described by the equation

$$\chi_\perp(\omega) = \frac{\omega_0(\omega_0 R_3 + i\omega R_4)}{\omega_0^2 R_1 - \omega^2 + 2i\omega\omega_0 R_2} \tag{3.3}$$

where ω_0 is defined by (3.1). The coefficients R_i are given by

$$R_1 = \frac{[F' - F'']}{[F - F']} + \left(\frac{\alpha}{\sigma}\right)^2 \left[1 + \sigma\left(\frac{[F' - F'']}{[F - F']}\right)\right]$$
$$\times \left[3 - \sigma + 2\sigma\left(\frac{[F'' - F''']}{[F' - F'']}\right)\right] \tag{3.4}$$

$$R_2 = \frac{\alpha}{2\sigma}\left[4 - \sigma + \sigma\left(\frac{[F' - F'']}{[F - F']}\right) - 2\sigma\left(\frac{[F'' - F''']}{[F' - F'']}\right)\right] \tag{3.5}$$

$$R_3 = \sigma\left(\frac{[F - F']}{F}\right) + \alpha^2\left(\frac{[F + F']}{2\alpha F}\right)\left[3 - \sigma + 2\sigma\left(\frac{[F'' - F''']}{[F' - F'']}\right)\right] \tag{3.6}$$

$$R_4 = \alpha\left(\frac{[F + F']}{2F}\right) \tag{3.7}$$

where α is a damping parameter and

$$F = F(\sigma) = \int_0^1 dx \, \exp(\sigma x^2) \tag{3.8}$$

and

$$F' = F'(\sigma) = \frac{d(F(\sigma))}{d\sigma} = \int_0^1 dx \, x^2 \, \exp(\sigma x^2) \tag{3.9}$$

and so on. The frequency at which resonance occurs can be found by equating the real part of Eq. (3.2) to zero. This gives

$$\omega_0^2[R_3(\omega_0^2 R_1 - \omega^2) + (\omega^2 R_2 R_4)] = 0 \tag{3.10}$$

and

$$\omega = \left(\frac{\omega_0^2 R_1 R_3}{R_3 - R_2 R_4}\right)^{1/2} \tag{3.11}$$

For values of α between 0.1 and 0.001 and σ between 1 and 10, which are in the range of values expected for the particles in magnetic fluids. $R_3 \gg R_2 R_4$ in all cases, with the result that

$$\omega = 2\pi f_0 = \omega_0 R_1^{1/2} \tag{3.12}$$

The expression of Raĭkher and Shliomis was simplified by Coffey et al. [62] to give

$$\chi_\perp(\omega) = \chi_\perp(0) \frac{1 + i\omega\tau_2 + \Delta}{(1 + i\omega\tau_2)(1 + i\omega\tau_\perp) + \Delta} \tag{3.13}$$

where $\chi_\perp(0)$ is the static transverse susceptibility,

$$\Delta = \frac{\sigma\tau_2(\tau_D - \tau_\perp)}{\alpha^2\tau_D^2} \qquad \tau_D = \tau_o \sigma = \sigma f_0^{-1} \tag{3.14}$$

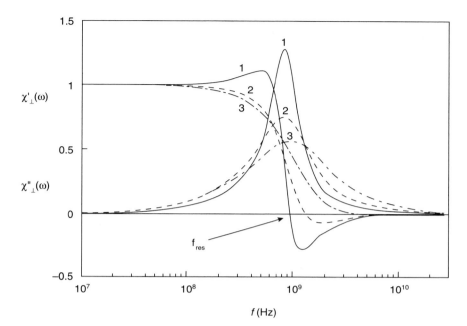

Figure 32. Plots of $\chi'_\perp(\omega)$ and $\chi''_\perp(\omega)$ obtained from Eq. (3.13) for various values of the anisotropy barrier parameter: (1) $\sigma = 1$, (2) $\sigma = 0.5$, and (3) $\sigma = 0.25$, with $\alpha = 0.1$ and $\tau_o = 10^{-9}$ s.

$$\tau_2 = 2\tau_D\left[\sigma - 2 + \frac{2\sigma(\sqrt{\sigma} - D(\sqrt{\sigma}))}{3(\sqrt{\sigma}) - (3 + 2\sigma)D(\sqrt{\sigma})}\right] \qquad (3.15)$$

$$\tau_\perp = 2\tau_D\left[\frac{(2\sigma + 1)D(\sqrt{\sigma}) - \sqrt{\sigma}}{(2\sigma - 1)D(\sqrt{\sigma}) + \sqrt{\sigma}}\right] \qquad (3.16)$$

and $D(x)$ is Dawson's integral, defined by

$$D(x) = e^{-x^2}\int_0^x e^{t^2}\,dt \qquad (3.17)$$

Note that the equation given in Ref. [62] for τ_2 is incorrect [63].

Equation (3.13) enables one to investigate how the transverse susceptibility behaves as a function of the anisotropy energy barrier σ. For the case where $\sigma \to 0$, one sees that $\Delta \to 0$ and $\tau_\perp \to \tau_D$, which reduces Eq. (3.3) to a simple Debye equation, that is,

$$\chi_\perp(\omega) = \frac{\chi_\perp(0)}{1 + i\omega\tau_D} \quad \text{for } \sigma = 0 \tag{3.18}$$

For $\sigma = 0$, $\chi_\perp(0) = \chi_\parallel(0)$, so that both the parallel and transverse susceptibilities are identical for isotropic particles.

Figure 32 shows how $\chi_\perp(\omega)$, calculated from Eq. (3.13), varies with frequency for different values of σ covering the range 0.25–1. The resonant character of $\chi_\perp(\omega)$, for $\sigma \gg \alpha$, becomes deformed with the decrease of σ, and in a continuous fashion the curves become relaxational in character with a Debye response at $\sigma = 0.25$ (curve 3 of Fig. 32).

However, with increase in σ the resonant character of $\chi_\perp(\omega)$ becomes more pronounced, as indicated in Fig. 33, where σ varies from 2 (curve 1) to 20 (curve 3).

The resonant frequency, that is, the frequency at which $\chi'_\perp(\omega)$ goes negative, can be obtained by writing Eq. (3.13) in terms of its real and imaginary

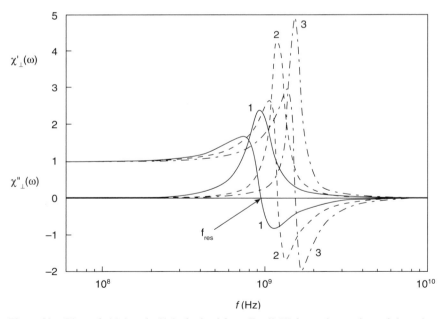

Figure 33. Plots of $\chi'_\perp(\omega)$ and $\chi''_\perp(\omega)$ obtained from Eq. (3.13) for various values of the anisotropy barrier parameter: (1) $\sigma = 2$, (2) $\sigma = 5$, and (3) $\sigma = 20$, with $\alpha = 0.1$ and $\tau_o = 10^{-9}$ s.

components, giving

$$\frac{\chi_\perp(\omega)}{\chi_\perp(0)} = \frac{(1+\Delta)^2 + \omega^2\tau_2^2 - \Delta\omega^2\,\tau_2\,\tau_\perp}{(1 - \omega^2\tau_2\,\tau_\perp + \Delta)^2 + \omega^2(\tau_\perp + \tau_2)^2}$$

$$- i\,\frac{\omega^3\tau_2^2\,\tau_\perp + \omega\tau_\perp + \omega\tau_\perp\,\Delta}{(1 - \omega^2\tau_2\,\tau_\perp + \Delta)^2 + \omega^2(\tau_\perp + \tau_2)^2} \qquad (3.19)$$

Clearly $\chi'_\perp(\omega)$ goes negative when

$$\omega = \omega_{res} = \frac{1 + \Delta}{\sqrt{\Delta\tau_\perp\tau_2 - \tau_2^2}} \qquad (3.20)$$

Substituting for Δ into Eq. (3.20), one finds that, for resonance to occur, the following condition must be satisfied:

$$\frac{\sigma}{\alpha^2} > \frac{\tau_D^2}{\tau_\perp(\tau_D - \tau_\perp)} \qquad (3.21)$$

When α is small, 0.1 say, the condition for resonance is approximately $\sigma > 5\alpha$; that is, if $\alpha = 0.1$, then resonance will occur for $\sigma > 0.5$. Also for $\sigma \gg 1$, $\tau_\perp$, $\tau_2 \approx \tau_D/\sigma = \tau_0$ and the condition for resonance, Eq. (3.21), reduces to $\alpha < 1$.

For very large values of σ, Eq. (3.13) reduces to that of Landau and Lifshitz, namely,

$$\chi_\perp(\omega) = \chi'_\perp(0)\,\frac{(1 + \alpha^2)\omega_0^2 + i\alpha\omega\omega_0}{(1 + \alpha^2)\omega_0^2 - \omega^2 + 2i\alpha\omega\omega_0} \qquad (3.22)$$

where

$$\chi'_\perp(0) = \frac{nM_s^2\,v}{2K\mu_0}$$

and

$$\omega_{res} = \frac{(1 + \alpha^2)\omega_0}{\sqrt{1 - \alpha^2}} \qquad (3.23)$$

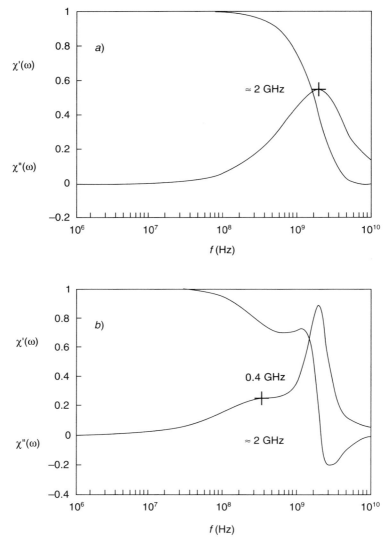

Figure 34. Complex susceptibility components of a monodisperse ferrofluid obtained from Eq. (3.24) for different values of particle radius r: (a) 2 nm, (b) 3 nm, (c) 4 nm, and (d) 5 nm.

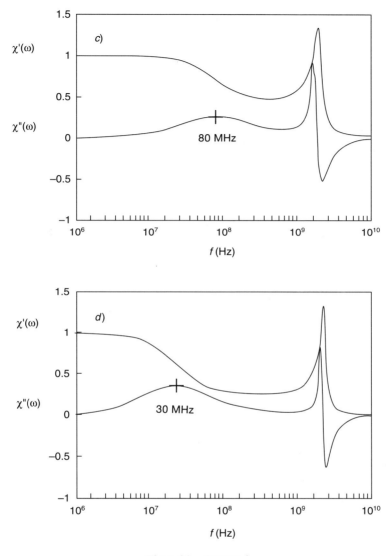

Figure 34. *Continued*

Thus ω_{res} is complex when $\alpha > 1$ and no resonance takes place; hence no matter how large σ is, α has to be less than 1 for resonance to occur. In the case where $\alpha < 1$, σ has to be approximately five times greater than α.

In the case of magnetic tape, particles have very large values of σ, so that the contribution by the relaxation components to the overall susceptibility is negligible, and thus Eq. (3.22) accurately describes the frequency dependence of such materials.

A. Overall Susceptibility

1. Susceptibility of Monodispersed Particles

By combining Eqs. (1.16) and (3.13) for $\chi_{\parallel}(\omega)$ and $\chi_{\perp}(\omega)$, respectively, the overall frequency-dependent susceptibility for monodispersed particles is given by [62]

$$\chi(\omega) = \frac{1}{3}\left[\frac{\chi_{\parallel}(0)}{1 + i\omega\tau_{\parallel}} + 2\chi_{\perp}(0)\frac{1 + i\omega\tau_2 + \Delta}{(1 + i\omega\tau_2)(1 + i\omega\tau_{\perp}) + \Delta}\right] \quad (3.24)$$

An illustration of the frequency dependence of $\chi(\omega)$ predicted by Eq. (3.24) for particle radii of 2, 3, 4, and 5 nm is shown in Figs. 34(a), (b), (c), and (d), respectively. For these plots arbitrary values taken for M_s, α, K, and η were 0.5 T, 0.05, 2 10^4 J m^{-3}, and 10^{-3} Ns m^{-2}, respectively; the number of particles per unit volume, n, was taken to be 5×10^{23} m^{-3}.

For the case of 2-nm particles [Fig. 34(a)], the profile is almost Debye-like, with a loss peak in the $\chi''(\omega)$ component at a frequency of approximately 2 GHz. There is no indication of $\chi'(\omega)$ changing from positive to negative. At a radius of 3 nm [Fig. 34(b)], a distinct resonance is observed at about 2 GHz, with the emergence of a second loss peak at about 0.4 GHz. As the particle radius is increased to 4 nm [Fig. 34(c)], the lower frequency loss peak (corresponding to the relaxation component) shifts down to a frequency of 80 MHz while the resonance effect becomes more pronounced with the frequency of resonance remaining unchanged. Finally, at a particle radius of 5 nm [Fig. 34(d)], two separate and distinct profile components, corresponding to the $\chi_{\parallel}(\omega)$ and $\chi_{\perp}(\omega)$ components of Eq. (3.24), are identifiable, with the lower frequency loss peak occurring at 30 MHz. Thus, overall, the relaxation component is shifted to lower frequencies for larger particles whereas the position of the resonance is largely unaffected by the increase in particle size.

2. Susceptibility of Polydispersed Particles

As has been indicated previously, magnetic fluids have a distribution of particle size and shape giving rise to a distribution in values of relaxation times and resonant frequencies [31]. To take account of the distribution of particle size, Eq. (3.24) has to be modified by an appropriate distribution function suitable for the range of particle radius; one such function is the Nakagami distribution function [64,65] which is used when the variable is positive and nonlimited. For a variable x the Nakagami probability density function is given by

$$p(x) = \frac{2\beta^\beta}{\Gamma(\beta)W^\beta} x^{2\beta-1} \exp\left(-\frac{\beta x^2}{W}\right) \tag{3.25}$$

where β is the width parameter of the distribution and W is a scale parameter equal to the mean square of x:

$$W = \frac{1}{n}\sum_{i=1}^{n} x_i^2 \tag{3.26}$$

The form of the Nakagami distribution for different values of the width parameter β is shown in Fig. 35; the larger is β, the narrower is the distribution. When $\beta = 0.5$ it becomes a normal distribution while for large values of β it becomes a lognormal distribution. Similarly, if x has a Nakagami distribution, then x^2 has a gamma distribution.

Plots 1, 2, and 3 of Fig. 36 display the normalized complex susceptibility components yielded from Eq. (3.24) for three distributions of particle size corresponding to β values of 2, 3, and 4, respectively, with $M_s = 0.5$ T,

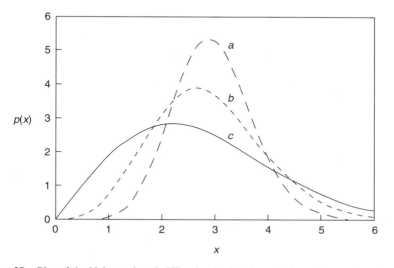

Figure 35. Plot of the Nakagami probability density function $p(x)$ for various values of width parameter β: (a) 3, (b) 2, and (c) 1. The scaling parameter $W = 9$.

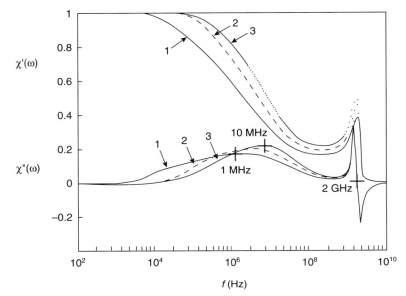

$\chi'(\omega)$

$\chi''(\omega)$

f (Hz)

Figure 36. Plot of normalized complex susceptibility components $\chi'(\omega)$ and $\chi''(\omega)$ vs. f for a polydispersed ferrofluid for various distribution widths of particle size: (1) $\beta = 2$, (2) $\beta = 3$, and (3) $\beta = 4$.

$\alpha = 0.05$, $K = 2 \times 10^4$ J m^{-3}, $\eta = 10^{-3}$ Ns m^{-2}, and $\bar{r} \approx 4.5$ nm. Resonance occurs at an approximate frequency of 2 GHz, while f_{max}, the frequency of the first maximum in $\chi''(\omega)$, lies in the range 1–10 MHz. The figure illustrates that the wider the distribution (small β), the broader the relaxation profile of the complex susceptibility. Also, as expected, it demonstrates that the width of resonance loss peak and indeed the resonant frequency is largely unaffected by the distribution in particle size.

To complete the susceptibility profile of a ferrofluid, one also has to introduce a distribution of resonant frequencies which may be accounted for by a distribution of anisotropy constants.

This is demonstrated in Fig. 37, where a normal distribution of the anisotropy constant K and a Nakagami distribution of r with $\bar{r} = 2.5$ nm and $\beta = 4$ are used in the calculation of the total susceptibility. For the distribution of K, a mean value of 2×10^4 J m^{-3} and a standard deviation of 1×10^4 J m^{-3} were used. The calculations clearly show, in comparison with Fig. 36, that the effect of averaging over K broadens and dampens the resonance peak in the $\chi''(\omega)$ component and smooths the resonance which $\chi'(\omega)$ exhibits.

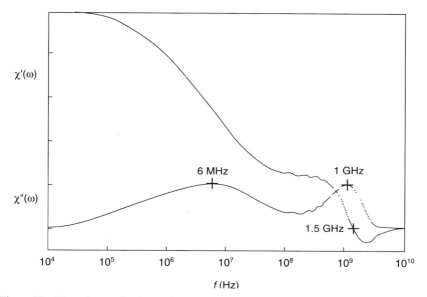

Figure 37. Plot of normalized complex susceptibility components $\chi'(\omega)$ and $\chi''(\omega)$ for a poly-dispersed ferrofluid with a distribution of particle size and a distribution of the effective anisotropy constant.

An example of the application of fitting combinations of distribution functions in a dynamic situation is illustrated in Fig. 38 [65], where plot *a* represents the measured data with a resonant frequency of about 1.7 GHz, as indicated by the transition of $\chi'(\omega)$ from positive to negative and low- and high-frequency loss peaks in the $\chi''(\omega)$ component at 35 MHz and 1 GHZ, respectively.

The data of plot *a* were chosen to determine which combination of distribution functions was the most suitable to use in the fitting procedure. The following combinations were used:

1. No distribution of the anisotropy constant K or particle radius r. The resulting susceptibility profiles are as given by plot *b* in Fig. 38.

2. To take account of possible variation in particle shape, a distribution of anisotropy constant K was introduced. A normal distribution was used but truncated to avoid unrealistic low and negative values of K. The actual form of distribution is not known, and thus the distribution function may be arbitrarily chosen as long as it is not unrealistic. No distribution of r was included in this model. The resulting curves are labelled *c* in Fig. 38.

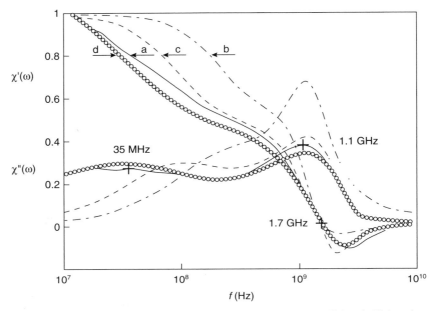

Figure 38. (a) Plot of normalized complex susceptibility components $\chi'(\omega)$ and $\chi''(\omega)$ vs. f over the frequency range 10 MHz–6 GHz for plot (a) measured data. (b) Plot of Eq. (3.24) with no distributions of the effective anisotropy constant K or particle radius r. (c) Plot of Eq. (3.24) fitted with a normal distribution of K but no distribution of r. (d) Plot of Eq. (3.24) fitted with the same normal distribution of K as in (c) and a Nakagami distribution of r.

3. The distribution of K was treated in a manner similar to that described in 2. In addition, a Nakagami distribution of r was included. In this case the resulting curves are labeled d in Fig. 38.

From these curves it is apparent that the best fit is obtained when a distribution in both K and in r are used. The curves for which no distributions were used provide a poor fit at both high and low frequencies. By using a distribution of K, the fit is dramatically improved, particularly in the high frequency region. However, a particle size distribution is necessary to obtain a reasonable fit to the data for the frequencies at which the relaxation mechanisms (Néel and Brownian) operate.

For all three fits a value of $M_s = 0.4$ T and a fluid viscosity of 10^{-3} Ns m^{-2} were used. The damping parameter α and the distribution parameters of the particle radius r and the anisotropy constant K were adjusted to give the best possible fits to the experimental data.

In case 1, where no distributions are used, values of $K = 1.2 \times 10^4$ J m^{-3}, $\alpha = 0.18$, and $r = 4.5$ nm were used.

In case 2, the values $\alpha = 0.07$, $r = 4.5$ nm, and a mean $\bar{K} = 1.2 \times 10^4$ J m^{-3} (with a standard deviation of 7×10^3 J m^{-3}) were used. The fitting procedure involved the summation over 20 values of K in the range 0.5–5×10^4 J m^{-3}.

For case 3 the values $\alpha = 0.07$, $\bar{r} = 4.5$ nm with a width factor of 4, and $\bar{K} = 1.2 \times 10^4$ J m^{-3} (with a standard deviation of 7×10^3 J m^{-3}) were used. Again the fitting procedure involved the summation over 20 values of K in the range 0.5–5×10^4 J m^{-3}.

In case 1, with the absence of distribution functions, a relatively large value of damping parameter $\alpha = 0.18$ is required, a value which is outside the range 0.1–0.01 commonly quoted. In cases 2 and 3, with the use of a distribution of K, the damping parameter required is 0.07 and both the damping and the distribution of K lead to the broadening of the resonance profile shown in Fig. 38, plot c. It can be seen that by using a distribution of K, a reasonably good fit can be obtained. The lower frequency dispersion in the complex susceptibility, which from 10 MHz upward is primarily due to the Néel relaxation, is very sensitive to particle size so that the spread of relaxation times is quite broad. A distribution of particle size is therefore needed to obtain a good fit in this frequency region, as is clear from Fig. 38, plot d.

B. Aftereffect Function in the Nanosecond Time Region

The concept of aftereffect function or magnetic viscosity $F(t)$ has already been referred to in Section I.D; here it will be referred to as $b(t)$ in order to distinguish it from $F(t)$, which was obtained over a very much lower frequency range and where the transformation of susceptibility data measured at 10 MHz, say, gave information in the equivalent time domain at a time of $1/10^7 = 0.1$ µs. Here measurements are performed up to a frequency of 6 GHz, with a corresponding time of 1.6×10^{-10} s. In the case of high-frequency data, it is convenient to use the following theorem [55], which relates $b(t)$ to the frequency-dependent complex susceptibility $\chi(\omega)$.

The aftereffect function $b(t)$ represents the decay of the magnetization after the removal of an external field and, according to Scaife [55], is related to the frequency-dependent complex susceptibility $\chi(\omega)$ via the equation

$$\chi(\omega) = \chi_0 - i\omega \int_0^\infty b(t)[\cos \omega t - i \sin \omega t] \, dt \qquad (3.27)$$

where χ_0 is the static susceptibility.

The complex susceptibility is defined as $\chi(\omega) = \chi'(\omega) - i\chi''(\omega)$, and writing the integral in real and imaginary parts gives

$$\chi'(\omega) - i\chi''(\omega) = \chi_0 - \omega \int_o^\infty b(t) \sin \omega t \, dt - i\omega \int_o^\infty b(t) \cos \omega t \, dt \quad (3.28)$$

Since $b(t)$ is a real function, we can see that

$$\chi''(\omega) = \omega \int_o^\infty b(t) \cos \omega t \, dt \quad (3.29)$$

Now $b(t) \cos \omega t$ is an even function, so that

$$\frac{\chi''(\omega)}{\omega} = \frac{1}{2} \int_{-\infty}^\infty b(t) \cos \omega t \, dt \quad (3.30)$$

which may be written as

$$\frac{\chi''(\omega)}{\omega} = \frac{1}{2} \operatorname{Re}\left\{ \int_{-\infty}^\infty b(t) \exp(-i\omega t) \, dt \right\} \quad (3.31)$$

The component within the brackets in Eq. (3.31) is the Fourier transform of $b(t)$; thus $b(t)$ can be obtained by carrying out an inverse Fourier transform on $\chi''(\omega)/\omega$ with

$$b(t) = 2 \operatorname{Re}\left\{ \frac{1}{2\pi} \int_{-\infty}^\infty \frac{\chi''(\omega)}{\omega} \exp(i\omega t) \, d\omega \right\}$$

$$= 2 \operatorname{Re}\left[F^{-1}\left\{ \frac{\chi''(\omega)}{\omega} \right\} \right] \quad (3.32)$$

Where F^{-1} denotes the inverse Fourier transform.

C. High-Frequency Measurement of the Complex Susceptibility

In order to investigate the resonant properties of magnetic fluids, one requires measuring instrumentation and test cell capable of operating in the gigahertz frequency range. The technique used here is the "coaxial transmission line technique", a well established method originating from the work of Roberts and von Hippel [12], [66–68].

Basically the technique consists of determining the complex impedance $Z = R + iX$ of a magnetic fluid sample enclosed in a short-circuited coaxial cell. The complete analysis is given in Appendix B.

The short circuit produces a maximum magnetic field and a minimum electric field at the sample, thus making the technique particularly suited to the measurement of the magnetic properties of test samples. Fluids have a particular advantage over solids in that they readily fill the coaxial test cell which has an almost radial electric field and concentric magnetic field. In the case of an inductive load, from the measurement of the resistive component R and the reactive component X, the complex permeability $\mu(\omega) = \mu'(\omega) - iu''(\omega)$ and hence the complex susceptibility $\chi(\omega) = \chi'(\omega) - i\chi''(\omega)$ are readily determined by means of the equations

$$\chi'(\omega) = \frac{\lambda}{2\pi d} \left\{ \frac{R_{in}^2 \tan \beta x + (X_{in} - R_0 \tan \beta x)(X_{in} \tan \beta x + R_0)}{(R_{in} \tan \beta x)^2 + (X_{in} \tan \beta x + R_0)^2} \right\} - 1 \quad (3.33)$$

and

$$\chi''(\omega) = \frac{\lambda}{2\pi d} \left\{ \frac{R_{in}(X_{in} \tan \beta x + R_0) - (X_{in} - R_0 \tan \beta x)(R_{in} \tan \beta x)}{((R_{in} \tan \beta x)^2 + (X_{in} \tan \beta x) + R_0)^2} \right\} \quad (3.34)$$

where $R_{in} + iX_{in}$ is the input impedance of the Hewlett–Packard (HP) line at a distance x from the terminating load, $R_0 = 50\ \Omega$, $\beta = 2\pi/\lambda$ is the phase change coefficient, λ is the operating wavelength in the line, and d is the sample depth.

The measurement system also easily lends itself to facilitate polarized measurements. These are realized by placing the coaxial cell between the pole faces of an electromagnet and then varying the polarized field H over the required range.

1. Measurement Technique

The measurements presented here are those made on a HP 50-Ω coaxial line incorporating a coaxial cell with 3 mm inner diameter and 7 mm outer diameter, up to a frequency range of 6 GHz, by means of a HP 8753C network analyzer. Use of standard HP open and short calibrating components enables accurate measurements over the entire frequency range to be obtained in one sweep, without the necessity to disturb the sample during the measurement [67].

This instrument automatically measures the reflection and transmission characteristics of devices by the use of the S, or scattering parameters, which are a measure of the ratio of the power reflected from a device to the power incident on a device. When the instrument is operated in a one-port mode, it measures the S_{11} parameter. Now,

$$S_{11} = \frac{Z_R - Z_0}{Z_R + Z_0} \tag{3.35}$$

where Z_R is the load impedance and Z_0 is the characteristic impedance of the instrument. The instrument has the capability of converting the S_{11} measurements to the real and imaginary components of Z_R by means of the equation

$$Z_R = \frac{Z_0(1 + S_{11})}{1 - S_{11}} \tag{3.36}$$

Thus in the case of an inductive load it automatically measures the resistive component R_L and the reactive component X_L.

In order to avoid dimensional resonance [69], care has to be taken in choosing the sample depth and to satisfy the requirement that the sample depth is much less than the wavelength of electromagnetic radiation in the sample medium; measurements over the frequency range 100–400 MHz were obtained using a sample depth d of 10 mm while a sample depth of 1.5 mm was used over the frequency range 100 MHz–6 GHz.

D.   Wideband High-Frequency Measurements

The following examples of measurements of the frequency-dependent, complex susceptibility $\chi(\omega) = \chi'(\omega) - i\chi''(\omega)$ taken up to an approximate frequency of 6 GHz is again representative of the work undertaken by the author and fellow workers. The examples are chosen to illustrate the usefullness of susceptibility measurements in determining data on a number of magnetic parameters, including, the resonant frequency f_{res}, the internal field H_A, the anisotropy constant K, the gyromagnetic ratio γ, and the aftereffect function $b(t)$.

1.   Case Study 6

This case study illustrates the following:

1. how the resonant frequency f_{res} and the susceptibility $\chi'(\omega)$ vary with particle concentration and
2. how f_{res} is influenced by materials with different values of anisotropy constant K.

The complex susceptibility of four magnetic fluids, namely samples 12, 13, 14 and 15, of magnetite in isopar M (a hydrocarbon) were determined. Identical particles of median diameter 9.6 nm of a lognormal volume distribution were used in each sample but with different concentrations, so that the saturation magnetization of the four samples were of (1) 0.09 T, (2) 0.06 T, (3) 0.04 T, and (4) 0.02 T.

Figure 39 shows the results obtained for the four samples; these plots of $\chi'(\omega)$ and $\chi''(\omega)$ versus frequency in hertz show that all the samples display ferromagnetic resonance with the $\chi'(\omega)$ components going negative at a frequency $\omega_{res/2\pi}$ of approximately 2.0 GHz for the fluid with the highest concentration of particles and 1.7 GHz for the lower concentration.

The plots also show how the $\chi''(\omega)$ component is constant up to a frequency of approximately 400 MHz while at the same time the corresponding $\chi'(\omega)$ component decreases in proportion to particle concentration. These effects are a manifestation of the contribution of the relaxation Néel components $\chi_{\parallel}(\omega)$ to the susceptibility. Beyond 400 MHz a relaxation-to-resonance transition occurs.

Here measurement of the complex magnetic susceptibility of three ferrofluid samples (samples 16, 17 and 18) of (1) cobalt in diester, (2) manganese ferrite in isopar M, and (3) magnetite in a perflouro polyether carrier

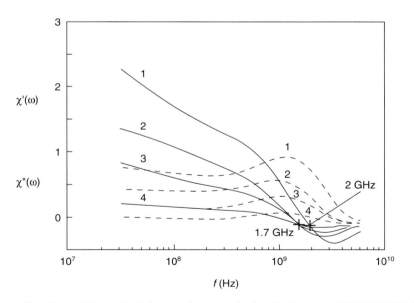

Figure 39. Plots of $\chi'(\omega)$ and $\chi''(\omega)$ versus frequency for ferrofluid samples: (1) 0.09 T, (2) 0.06 T, (3) 0.04 T, and (4) 0.02 T.

(PFPE), with average particle diameters of 7.8, 10.5, and 11.9 nm and approximate magnetizations of 0.04, 0.036, and 0.03 T, respectively, are investigated. Ferromagnetic resonance is observed for all three samples [70].

Figure 40 shows the complex susceptibility results obtained for the three samples; these plots of $\chi'(\omega)$ and $\chi''(\omega)$ versus f show that the $\chi''(\omega)$ component is approximately constant up to a frequency of approximately 400 MHz for samples 17 and 18 while in the case of sample 16 there is a gradual decrease in $\chi''(\omega)$ up to an approximate frequency of 800 MHz; at the same time the corresponding $\chi'(\omega)$ components decrease. These effects are again a manifestation of the contribution of the relaxation components $\chi_{\parallel}(\omega)$ to the susceptibility.

Beyond the 400- and 800-MHz points, a relaxation to resonance transition occurs with the $\chi''(\omega)$ components peaking at approximate frequencies of 0.8 and 1.2 GHz for samples 17 and 18 and at 2.8 GHz for sample 16 and the corresponding $\chi'(\omega)$ components going negative at approximately 1.6, 1.8, and 4.1 GHz, respectively. This difference in values f_{res} is indicative of the difference in the K and hence H_A values of the par-

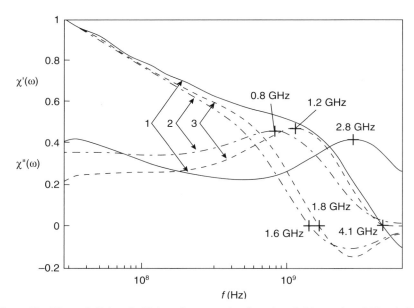

Figure 40. Plots of $\chi'(\omega)$ and $\chi''(\omega)$ vs. frequency for three ferrofluid samples of (1) cobalt in diester, (2) manganese ferrite in isopar M, and (3) magnetite in a PFPE carrier.

ticles, with cobalt being recognized as having the higher values of these parameters.

In the case of the magnetite particles (sample 18), using values of $M_s = 0.4$ T and $K = 2 \times 10^4$ J m^{-3}, Eq. (3.2) gives an H_A value of 10^5 A m^{-1}, and from Eq. (3.1) a corresponding frequency $f_0 = \omega_0/2\pi$ of 3.5 GHz is obtained. Similarly, for the cobalt sample (sample 16) using values of $M_s = 1$ T and $K = 2 \times 10^5$ J m^{-3}, the same exercise yields an H_A value of 4×10^5 A m^{-1} and f_0 of 14 GHz. The value of K for sample 17 is not known, and thus a similar exercise is not possible; however, one notes that in this case the resonant frequency is only slightly higher than that of sample 18.

Thus there is a discrepancy between the measured and calculated frequencies f_{res} and f_0 with the ratio f_{res}/f_0 being 0.49 and 0.3 for samples 16 and 18, respectively. However, Eq. (3.1) assumes a small polar angle θ, which may not be the case in practice, and in fact the equations of Raïkher and Shliomis predict a ratio of $f_{res}/f_0 < 1$ over a wide range of σ values. This is illustrated in Fig. 41, which plots f_{res}/f_0 versus σ for values of damping parameter $\alpha = 0.1$ and $\alpha = 0.05$, respectively. For $\alpha = 0.05$ the ratio f_{res}/f_0 is seen to be approximately equal to 0.5 for $\sigma = 0.5$ and a value of 0.75 for $\sigma = 5$.

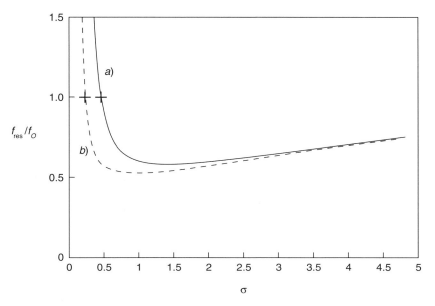

Figure 41.   Plot of f_{res}/f_0 vs. σ for $\alpha = 0.05$ (curve a) and $\alpha = 0.1$ (curve b).

2.   Case Study 7

Here equation 3.32 is used to determine the aftereffect function of samples 12, 13, and 14 of case study 6 where the particles had an average particle diameter of 10 nm and magnetization of 0.09, 0.06, and 0.04 T, corresponding to packing fractions of 0.22–0.10, respectively.

The corresponding normalized aftereffect functions $b(t)/b(0)$ are shown in Fig. 42 and it is clear that there is little difference between the three decay profiles, with the curves, after an initial rapid decay up to a time of approximately 0.4 ns, having a less rapid decay and merging at a time of approximately 2×10^9 s. The rapid decay is attributed to the relaxation of the transverse relaxation component $\tau_\perp$, while the slow component represents the relaxation of the parallel component $\tau_\parallel$. This result indicates that, over the time range concerned, $b(t)/b(0)$ is little affected by the packing fraction and hence particle–particle interactions.

Due to the similarity of the decay curves, it was only necessary to further analyze one sample, as these results are then also representative of all three samples. The 0.04-T magnetic fluid (sample 14) was chosen for this purpose. The resonant frequency f_{res} occurs at approximately 1.7 GHz, while f_{max} is approximately 1.1 GHz.

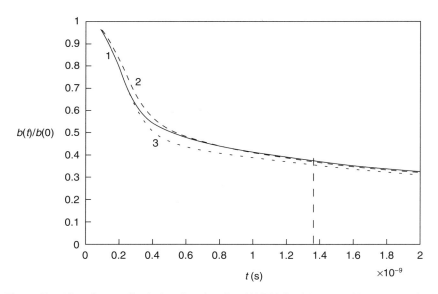

Figure 42. Plot of normalized aftereffect function $b(t)/b(0)$ for (1) 0.09-T, (2) 0.06-T, and (3) 0.04-T magnetic fluids.

To obtain an initial estimate of the decay times involved, the aftereffect function of this sample was fitted to a simple equation, $b(t)_1$ consisting of the sum of two decaying exponentials with different time constants, with

$$b(t)_1 = \left[A \exp\left(-\frac{t}{\tau_1} \right) + B \exp\left(-\frac{t}{\tau_2} \right) \right] \tag{3.37}$$

and where $A = 1$, $B = 0.5$, $\tau_1 = 1.2 \times 10^{-10}$ s, and $\tau_2 = 7 \times 10^{-9}$ s.

The plot of this equation, with an approximate time-constant time of 2.4×10^{-9} s gives a reasonable fit (curve b of Fig. 43) to the experimental data shown in curve a of Fig. 43. The time constants t_1 and t_2 have frequencies $f_1 = 1.3$ GHz and $f_2 = 22$ MHz, respectively. The first frequency, f_1, coincides roughly with the position of the high-frequency loss peak (at 1.1 GHz) in $\chi''(\omega)$, while the lower frequency, f_2, corresponds to the frequency of maximum Néel dispersion. However, to account for a distribution of both anisotropy and particle size, a more accurate representation of $b(t)$ than that obtained by Eq. (3.37) is required. To realize this, the original data of curve a of Fig. 44 were fitted to theoretical profiles generated by Eq. (3.24) and covering the frequency range 10–100 MHz, both modified to

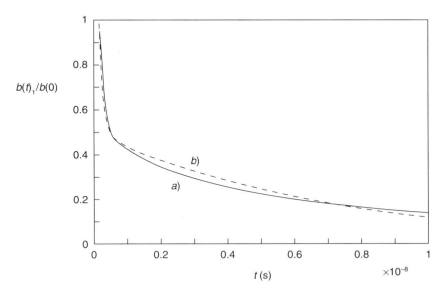

Figure 43. Plot of normalized aftereffect function $b(t)/b(0)$: (a) 0.04-T magnetic fluid and (b) normalized aftereffect function $b(t)_1/b(0)$ for fit equation of the form $b(t)_1 = [A \exp(-t/\tau_1) + B \exp(-t/\tau_2)]$, with $\tau_1 = 1.2 \times 10^{-10}$ s and $\tau_2 = 7 \times 10^{-9}$ s and where $A = 1$, $B = 0.5$.

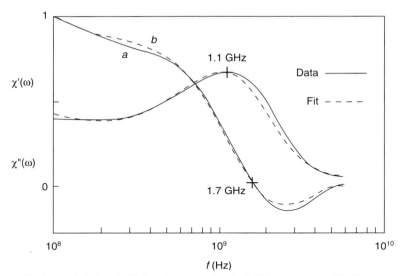

Figure 44. Plot of $\chi'(\omega)$ and $\chi''(\omega)$ vs. frequency for the 0.04-T magnetic fluid: (*a*) corresponding to measured data and (*b*) corresponding fit obtained by use of Eq. (3.24) adapted to include
a normal distribution of anisotropy constant K and a Nakagami distribution of radii r.

include a normal distribution of anisotropy constant K with a mean $\bar{K} =$
1.1×10^4 J m^{-3} with a standard deviation of 8×10^3 and a Nakagami
distribution of radii r with a width factor $\beta = 4$ and a mean particle radius
$\bar{r} \approx 4.5$ nm. A saturation magnetization of 0.4 T and a damping parameter
$\alpha = 0.08$ were also used.

The data fit is shown in curve b of Fig. 44 while the corresponding aftereffect function $b(t)_2$ is shown in curve b of Fig. 45. As anticipated, from
curves a and b of Fig. 45, it is readily seen that $b(t)_2$ is a better representation of $b(t)$ than that obtained by $b(t)_1$ (curve b of Fig. 43) with the decay
curve having an overall time-constant time of approximately 1.8×10^{-9} s
compared to a value of 2.4×10^{-9} s. Here, $b(t)$ contains a contribution not
only from $\tau_\perp$ but also from $\tau_\parallel$, which is taken as being equal to τ_N; thus in
order to obtain a more accurate estimate of $\tau_\perp$, it is necessary to reduce the
influence of the Néel relaxation in the $\chi_\parallel(\omega)$ component of the susceptibility.
This may be realized by applying a polarized field to the sample, an action
which has the effect of increasing the anisotropy field. This results in an
increase in the energy barrier Kv, thus resulting in less spontaneous flipping
of the magnetic moment over the potential barrier during the time of mea-

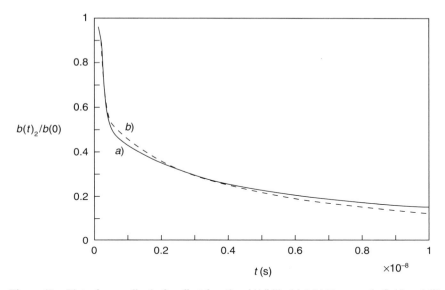

Figure 45. Plot of normalized aftereffect function $b(t)/b(0)$: (a) 0.04-T magnetic fluid and (b) normalized aftereffect function $b(t)_2/b(0)$ for fit data of Fig. 43.

surement. By increasing H, from Eq. (3.1) the form of ω_0 becomes

$$\omega_0 = \gamma(H_A + H) \tag{3.38}$$

with the result that a further increase in H results in an increase in ω_0 and a corresponding change in $\tau_\perp$. The change in $\tau_\perp$ with polarizing field H is investigated by the equation [41]

$$\tau_\perp = 2\tau_N \frac{\dfrac{-\sigma}{T(\xi,\sigma)} (\xi L(\xi) + 1 - \xi^2/2\sigma) - \xi^2/4\sigma + 1/2 + \sigma}{\dfrac{\sigma}{T(\xi,\sigma)} (\xi L(\xi) + 1 - \xi^2/2\sigma) + \xi^2/4\sigma + 1/2 + \sigma} \tag{3.39}$$

where

$$T(\xi, \sigma) = \sqrt{\sigma} \left\{ [\xi L(\xi) + 1 + \xi] D\left(\sqrt{\sigma} + \frac{\xi}{2\sqrt{\sigma}}\right) \right.$$

$$\left. + [\xi L(\xi) + 1 - \xi] D\left(\sqrt{\sigma} - \frac{\xi}{2\sqrt{\sigma}}\right) \right\} \tag{3.40}$$

$$\xi = \frac{mH}{kT} \tag{3.41}$$

and

$$L(\xi) = \coth \xi - \frac{1}{\xi} \quad \text{(Langevin function)} \tag{3.42}$$

The 0.04-T sample was subjected to eight polarizing fields over the range 0–68 kA m^{-1} and the corresponding imaginary $\chi''(\omega)$ susceptibility components obtained. The parameter f_{max} was found to encompass the approximate frequency range 1.1–4.8 GHz. Upon application of equation 3.32, the corresponding normalized aftereffect functions $b(t)_3/b(0)$ were determined and are shown in Fig. 46, labeled 1–8. These plots clearly demonstrate that with increasing polarizing field, $b(t)_3$ gradually becomes oscillatory in nature, as illustrated by curve 8 of the figure. An increase in H results in a

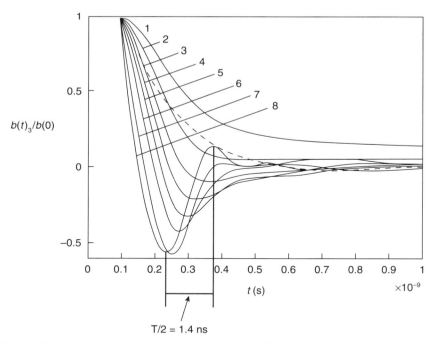

Figure 46. Plot of normalized aftereffect function $b(t)_3/b(0)$ for the 0.04-T magnetic fluid for eight values of polarizing field of approximately (1) 0, (2) 9.6, (3) 20, (4) 25.6 (5) 32, (6) 44, (7) 56, and (8) 68 kA m^{-1}. The envelope of curve 8 is represented by the dashed curve.

higher frequency of oscillation [Eq. (3.38)], and ultimately a point is reached where the $\chi_{\parallel}(\omega)$ component becomes negligible and $\chi(\omega)$ reduces to the transverse susceptibility component $\chi_{\perp}(\omega)$, which is given by Eq. (3.3). Using parameters $M_s = 0.4$ T and $\alpha = 0.07$, the corresponding aftereffect function of Eq. (3.3), say $b(t)_4$, is shown in Fig. 47 and is found to be oscillatory, thus confirming the oscillatory form of the curves of Fig. 46.

Furthermore, the theoretical behavior of $\tau_{\perp}$, with increasing H, may be seen from a plot of Eq. (3.39). Such a plot is shown in Fig. 48 for values of σ of 1, 2, 3, 5, and 10, and it is clearly demonstrated that (1) $\tau_{\perp}$ decreases with increasing σ and (2) for a fixed value of σ, $\tau_{\perp}$ decreases with increasing H (corresponding to a shorter time constant of the envelope of the oscillations). It should be borne in mind that in the case of low values of σ and of ξ (low applied field), which is the situation for the case of the experimental measurements carried out with no applied external field, Néel relaxation will be important so that the aftereffect function will involve both components of $\tau_{\parallel}$ and $\tau_{\perp}$, as illustrated in Fig. 42.

3. Case Study 8

Here it is demonstrated that, by means of the simple expedient of fitting the measured complex susceptibility profiles to suitably adapted classical

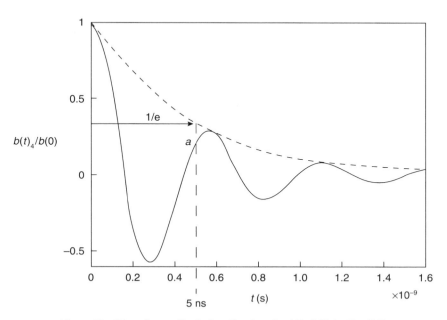

Figure 47.   Plot of normalized aftereffect function $b(t)_4/b(0)$ for Eq. (3.3).

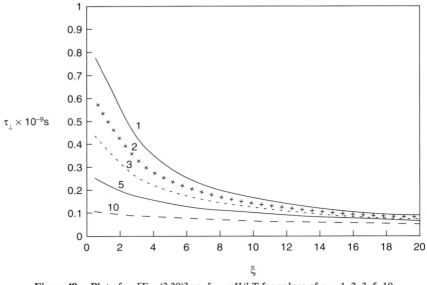

Figure 48.   Plot of $\tau_\perp$ [Eq. (3.39)] vs. $\xi = mH/kT$ for values of $\sigma = 1, 2, 3, 5, 10$.

models, data in the 10–100-GHz frequency band can be determined from measurements made in the frequency range 100 MHz–6 GHz [71]. Transformation of the fitted data thus enables $b(t)$ to be investigated in the 10^{-10}–10^{-11} s time regions, a time region which is generally associated with the precessional decay time τ_0, which is a prefactor of Brown's equations for Néel relaxation.

The results of applying this technique to two magnetic fluid samples— the 0.09-T (sample 12) and 0.04-T (sample 14) samples of case study 6 and a magnetic tape sample (sample 19)—are illustrated; in the case of the ferrofluids the complex susceptibility data $\chi(\omega)$ are again fitted to the equations of Raĭkher and Shliomis and of Debye, suitably modified to include a distribution of particle size r and anisotropy constant K, while for the magnetic tape sample, the fit is realized by means of the Landau–Lifshitz equations suitably modified to cater for a distribution of K. It is demonstrated that a more accurate aftereffect function is obtained in the cases where the fitted profiles are transformed and that, by application of a varying polarizing magnetic field to the magnetic fluid samples, $b(t)$ of the fitted data is shown to be oscillatory in form with the transverse relaxation time $\tau_\perp$ having a periodic time approximately equal to the time obtained from the resonant frequency $f_{res} = \omega_0/2\pi$, f_{res} corresponding to the frequency at which the $\chi'(\omega)$ component goes from a $+ve$ to a $-ve$ value.

Complex magnetic susceptibility measurements over the frequency range 100 MHz–6 GHz were determined for both samples for seven values of polarizing field, with, as an example, typical results of the $\chi'(\omega)$ component of the sample being shown in Fig. 49. This figure shows how f_{res} varies from approximately 1.7 to 4.2 MHz, while the magnitude of $\chi'(\omega)$ below the giga-hertz region decreases with increasing polarizing field. The complex suscep-tibility components for all polarizing fields were then fitted to Eq. (3.24) suitably modified to include a distribution of particle size r and anisotropy constant K. A typical example of this is shown in Fig. 50 for the case of the unpolarized sample where the fit is seen to extend to 10 GHz.

The normalized aftereffect functions for the measured data are shown in Fig. 46. It is clearly demonstrated that with an increasing polarizing field $b(t)_3/b(0)$ gradually becomes oscillatory in nature with curve h, correspond-ing to the strongest value of the polarizing field having an approximate time $\frac{1}{2}T = 1.4$ ns, corresponding to a frequency $1/T = 3.6$ GHz.

However, the oscillatory action is more clearly pronounced in the plots of Fig. 51, which are the aftereffect functions $b(t)_2'/b(0)$ obtained from the corresponding fitted profiles. In this case, for curve g of the figure, the

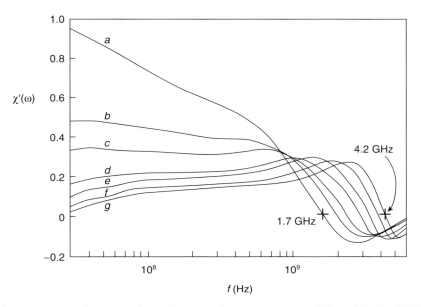

Figure 49. Plot of normalized complex susceptibility component $\chi'(\omega)$ vs. f for the 0.04-T magnetic fluid (sample 14) of magnetite in Isopar M for seven values of polarizing field of approximately (*a*) 0, (*b*) 9.6, (*c*) 20, (*d*) 32, (*e*) 44, (*f*) 56, and (*g*) 68 kA m^{-1}.

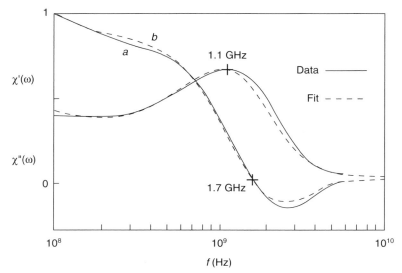

Figure 50. Plot of normalized complex susceptibility components $\chi'(\omega)$ and $\chi''(\omega)$ vs. f for sample 14 of magnetite in Isopar M: (a) measured data over the frequency range 10^8–10^9 Hz; (b) fitted profile over the frequency range 10^8–10^{10} Hz.

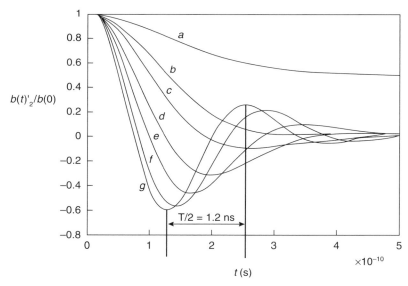

Figure 51. Plot of normalized aftereffect function $b(t)'_2/b(0)$ fitted to data of sample 14 for seven values of the polarizing field of approximately (a) 0, (b) 9.6, (c) 20, (d) 32, (e) 44, (f) 56, and (g) 68 kA m^{-1}.

approximate period of oscillation is 0.25 ns, corresponding to a precession or resonant frequency of 4 GHz. This value compares favorably with measured resonant frequency f_{res} of 4.2 GHz and is more accurate than that obtained from Fig. 46.

In the case of the 0.09-T sample (sample 2), f_{res} was found to encompass the frequency range 2–4.8 GHz when subjected to the polarizing magnetic field. The corresponding aftereffect functions $b(t)'_3/b(0)$ obtained for the measured data are shown in Fig. 52 (left), while those obtained from the fits $b(t)'_4/b(0)$ are shown in Fig. 52 (right): from Fig. 52 (left) the periodic time of curve g is 2.8 ns, corresponding to a frequency of 3.6 GHz. Again the aftereffect function of the fits in Fig. 52 (right) give a more pronounced oscillation with the periodic time for curve g being 2.2 ns, corresponding to a resonant frequency of 4.55 GHz. This value of 4.55 GHz has to be compared to the measured f_{res} frequency of 4.8 GHz and again demonstrates the advantage of transforming the fitted data.

An interesting observation regarding the decay time of the aftereffect function as a function of polarizing field is that the time-constant time, or $1/e$ value, increases with increasing polarizing field: the reason for this effect is as follows.

In a magnetic fluid the existence of particles of different size and shape gives rise to a distribution of precession or resonant frequencies, and thus, when the external field is removed, the magnetic moments of the particles precess back into the directions of the easy axes with different precession frequencies. The moments start off with their position of greatest alignment but soon became increasingly out of phase, canceling each other out and causing the average transverse component of the magnetization, $\tau_\perp(\omega)$, to decay rapidly to zero. This effect is clearly demonstrated in Fig. 53, where a comparison is made between the aftereffect function $b(t)'_5/b(0)$ calculated from Eq. (3.13), the equation of Coffey et al. [15] for the transverse Néel susceptibility for a distribution of particles with different resonant frequencies, as represented by curve b, and that for a single particle of 10 nm diameter (corresponding to an average particle size of a distribution), as represented by curve a. It is demonstrated that the time taken for $b(t)'_5/b(0)$ to decrease to $1/e$ of its initial value is much less for the assembly of particles (curve b) than for the single average particle (curve a). For the assembly of particles $\tau_\perp(\omega)$ is approximately 1.25×10^{-10} s while for the case of the average particle $\tau_\perp(\omega)$, as represented by the decay time of the envelope of curve a, is clearly much longer at approximately 5×10^{-10} s. In this example the oscillatory-type behavior of the single particle is indicative of the fact that its magnetic moment precesses about the anisotropy field H_A several times before it is aligned with the easy axis. In contrast, for the case of the assembly of particles, $b(t)'_5/b(0)$ goes negative only once so that it

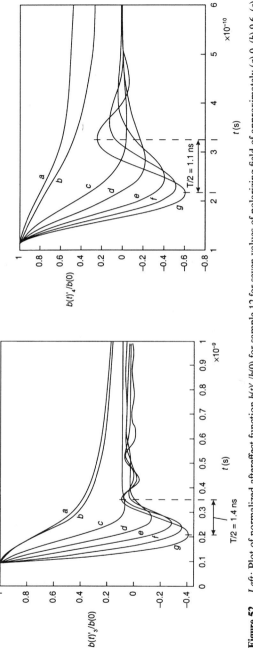

Figure 52. *Left*: Plot of normalized aftereffect function $b(t)'_3/b(0)$ for sample 12 for seven values of polarizing field of approximately (*a*) 0, (*b*) 9.6, (*c*) 20, (*d*) 32, (*e*) 44, (*f*) 56, and (*g*) 68 kA m^{-1}. *Right*: Plot of normalized aftereffect function $b(t)'_4/b$ (0) fit to data of sample 14 for seven values of polarizing field of approximately (*a*) 0, (*b*) 9.6, (*c*) 20, (*d*) 32, (*e*) 44, (*f*) 56, and (*g*) 68 kA m^{-1}.

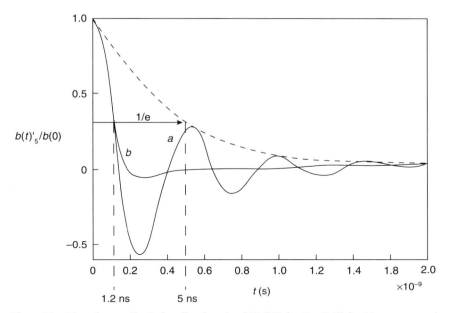

Figure 53. Plot of normalized aftereffect function $b(t)'_5/b(0)$ for Eq. (3.13) for (a) average particle radius of 5 nm and (b) assembly of particles with a distribution of resonant frequencies.

appears that the component of the magnetization transverse to H_A also goes negative only once. This particular example highlights the fact that the results obtained for a single particle, representing an average particle of the assembly, may be completely different to that obtained for an assembly of particles.

Plot a of Fig. 54 displays the complex susceptibility component obtained for the magnetic tape (sample 19). It is immediately apparent that there is insufficient data for a realistic value of aftereffect function to be determined. However, by means of the expedient of using the fitting technique presented here, this obstacle can be overcome as indicated by the aftereffect function $b(t)'_6/b(0)$ (obtained from the fit of plot b of Fig. 54). Here the equations of Landau and Lifshitz were adapted to cater for a Nakagami distribution of K, and in contrast to the ferrofluid cases, due to insufficient measured data, the fitting profile was computed up to a frequency of 100 GHz, a frequency sufficiently high enough to ensure that the susceptibility has neared zero.

Figure 55 shows that $b(t)'_6/b(0)$ has a faster decay rate than the unpolarized ferrofluid samples, going negative at approximately 0.5×10^{-10} s and then tending back to zero without any further oscillation. This fast decay time arises because with this system of particles there is essentially no

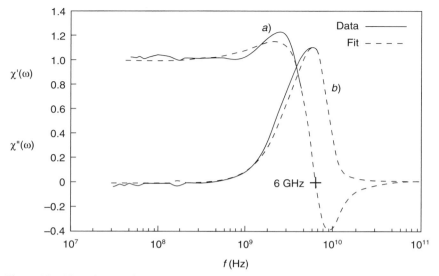

Figure 54. Plot of normalized complex susceptibility components $\chi'(\omega)$ and $\chi''(\omega)$ vs. f for magnetic tape (sample 19): (a) measured data and (b) fitted with a distribution of K.

$\chi_{\parallel}(\omega)$ component and the decay time represents the transverse relaxation time $\tau_{\perp}(\omega)$ of the particles.

4. Case Study 9

In this study the dependence of f_{res} on the presence of an external magnetic field H in the approximate range $0-68$ kA m^{-1} is examined for colloidal suspensions of magnetite particles (sample 19) and cobalt particles in isopar M (sample 20) and enables average values of the internal anisotropy field $\bar{H}_A$ to be measured, from which mean values of the magnetic anisotropy constant of the particles, $\bar{K}$, and the magnetogyric ratio γ are derived [72]. The experimental profiles of $\chi'(\omega)$ and $\chi''(\omega)$ for the magnetite particle system as a function of H are fitted to modified equations of Raĭkher and Shliomis suitably adapted to include a normal distribution of particle energy barriers, $K_{eff} v$, to the rotation of the magnetic moments.

The magnetic fluids used consisted of (i) magnetite particles in isopar M and stabilized with oleic acid and (ii) cobalt metal particles also dispersed in a similar hydrocarbon but stabilized with MOT [di(ethyl-2-hexyl)] sodium sulfosuccinate). The saturation magnetization of both suspensions was 0.04 T.

The magnetite particles were of median diameter 10 nm and standard deviation 0.4 of a lognormal volume distribution and an M_s of 0.04 T while

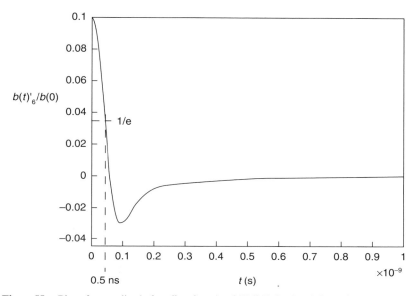

Figure 55. Plot of normalized aftereffect function $b(t)'_6/b(0)$ for fitted data of magnetic tape.

the cobalt metal particles had a median diameter of 7.8 nm, a standard deviation of 0.2, and an M_s of 1.25 T. The real, $\chi'(\omega)$, and imaginary, $\chi''(\omega)$, susceptibility components of the two ferrofluid samples were measured over the stated frequency range for eight values of external fields H between 0 and 68 kA m^{-1}.

The results of the susceptibility measurements obtained are displayed in Fig. 56 for sample 19 and in Fig. 57 for sample 20. In both cases resonance is observed, at a frequency f_{res}, which increases with increase of applied field. It is shown in Fig. 56 that in the case of sample 19 the frequency of resonance spans the approximate frequency range of 1.7–4.2 GHz while the data of Fig. 57 demonstrate that in the case of sample 20 the final relevant frequencies are outside the range of the instrumentation; however, if an extrapolation is made of the $\chi'(\omega)$ component (as indicated by the dashed portion of the plot), f_{res} is estimated as spanning the approximate frequency range 4.2–7 GHz.

Plots of f_{res} versus H for both samples are shown in Fig. 58, which shows that a linear relationship exists between f_{res} and H with slopes of 2.26×10^5 and 2.28×10^5 s^{-1} A^{-1} m, which, within experimental uncertainty, equal the magnetogyric ratio of $\gamma = 2.21 \; 10^5$ s^{-1} A^{-1} m. Thus the frequency

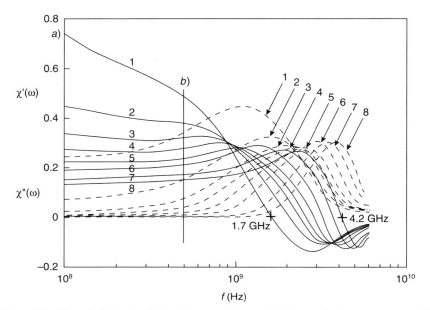

Figure 56. Plot of $\chi'(\omega)$ and $\chi''(\omega)$ vs. frequency for sample 19 over the frequency range 100 MHz–6 GHz for eight values of polarizing field of approximately (1) 0, (2) 9.6, (3) 20, (4) 25.6, (5) 32, (6) 44, (7) 56, and (8) 68 kA m^{-1}.

dependence of ω_{res} can be described by

$$\omega_{res} = 2\pi f_{res} = \gamma(H + \bar{H}_A) = \frac{2\gamma K_{eff}}{M_s + \gamma H} \qquad (3.43)$$

where $\bar{H}_A$ represents some mean value of the anisotropy field and K_{eff} represents some mean value of the anisotropy $\bar{K}$.

The values of $\bar{H}_A$ as determined from Fig. 58 are 44 kA m^{-1} (H_{A1}) for sample 19 and 118 kA m^{-1} (H_{A2}) for sample 20, corresponding to mean values of the intrinsic magnetic anisotropy constant $\bar{K}$ at room temperature of 8.4×10^3 and 7.4×10^4 J m^{-3}, respectively. In the cases of magnetite and cobalt particles, the values of $\bar{K}$, determined from measurements of the low-temperature decay of remanence [16,17], are typically larger by roughly a factor of 2.

From Figs. 56 and 57, it is seen that, for both samples, on increasing the external field, the magnitude of $\chi'(\omega)$ at the lower frequency end of the plots is progressively reduced (regions a–b in Fig. 56 and c–d in Fig. 57), while toward the higher frequencies the profile takes on more of a resonance-type

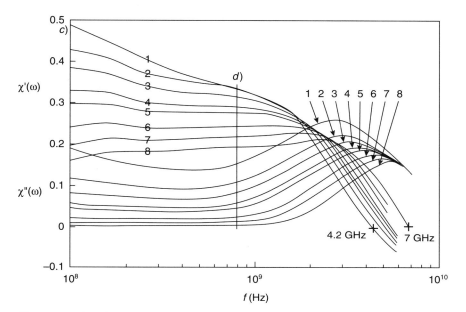

Figure 57. Plot of $\chi'(\omega)$ and $\chi''(\omega)$ vs. frequency for sample 20 over the frequency range 100 MHz–6 GHz for eight values of polarizing field of approximately (1) 0, (2) 9.5, (3) 20, (4) 25.0, (5) 32, (6) 44, (7) 56, and 98) 68 kA m^{-1}.

character. The explanation of this reduction in $\chi'(\omega, H)$ is that the effect of the external field is to effectively increase the value of the anisotropy field $\bar{H}_A$ and hence $\bar{K}$, which is written as K_{eff}. The barrier to rotation of the magnetic moments now becomes $E = K_{eff} v$. This increase in the potential barrier reduces the rate at which Néel relaxation takes place, leading to a reduction in the magnitude of $\chi'(\omega)$. In the corresponding plots of $\chi''(\omega)$, a similar reduction in magnitude is observed at lower frequencies, while at higher frequencies the frequency of the maximum of the loss peak, $f_{max} = \omega_{max}/2\pi$, increases with increase of the external field H. The explanation of this behavior has already been previously explained.

Two processes are responsible for the appearance of the $\chi''(\omega)$ loss peak, namely the Néel relaxation of the smallest particles in the distribution and resonance. With increase in H, the contribution to the loss peak from the Néel component is reduced while the contribution from the resonance component increases.

In zero applied field the easy axes of magnetization have an isotropic distribution, and the observed resonant frequency f_{res} which arises from the

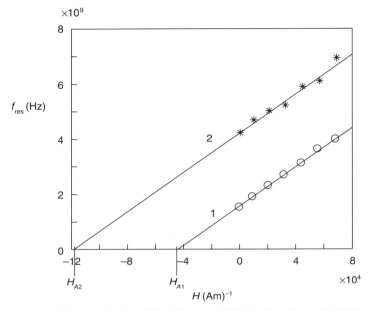

Figure 58. Plot of f_{res} vs. polarizing field H for sample 19 (plot 1) and sample 20 (plot 2) with corresponding values of $\bar{H}_{A1} = 44$ kA m^{-1} and $\bar{H}_{A2} = 118$ kA m^{-1}.

larger particles (large σ) of the distribution is independent of volume. However, f_{res} does depend on the distributions of shape (elongations) of the particles, since shape contributes to K and hence H_A.

Theoretical profiles of $\chi'_{\parallel}(\omega)$ and $\chi''_{\parallel}(\omega)$ are generated using the model of Raĭkher and Shliomis [7], with the model extended to include a normal distribution function to present the distribution of energy barriers, $K_{eff} v$.

A typical example of the fit obtained corresponding to an external field of 20 kA m^{-1} applied to the magnetite-based colloid is given in Fig. 59. To generate this profile, values $M_s = 4.6 \times 10^5$ A m^{-1} and $\alpha = 0.07$ were used. Values of $K_{eff} v$ and the standard deviation (SD) of $K_{eff} V$ resulting from this fit and other fits not illustrated here are shown in Figs. 60 and 61, respectively, for the magnetite-based fluid. A linear relationship between $K_{eff} v$ and the applied field H is observed from Fig. 60, while a monotonic decrease in the standard deviation of $K_{eff} v$ occurs as indicated in Fig. 61.

The value of the barrier to rotation of the magnetic moment E_b with increase in H for a system of uniaxial particles with easy axis aligned can be

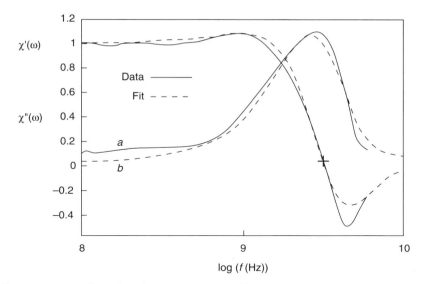

Figure 59. Normalized plot of $\chi'(\omega)$ and $\chi''(\omega)$ vs. f for sample 19 with polarizing field of 20 kA m^{-1}: (a) corresponding to data and (b) corresponding fit obtained by use of the equations of Raĭkher and Shliomis adapted to include a distribution of energy barriers, $K_{\text{eff}}\, v$.

written in the form [18]

$$E_b = K_{\text{eff}}\, V\left(1 + \frac{H}{\bar{H}_A}\right)^2 \tag{3.44}$$

Whereas the increase in f_{res} with increase in external field depends on $\bar{H}_A + H$, the barrier to rotation depends on $(\bar{H}_A + H)^2$ for systems in which the easy axes are aligned. Although the systems studied here do not have their easy axis aligned, nevertheless one might expect a more rapid increase in the value of E_b with increase in external field than that expected for K_{eff}, as was found to be the case in this study. For the case of magnetite particles, over the same external field range, K_{eff} increases by a factor of 2.6, while the corresponding increase in E_b is 3.2.

One possible reason for the reduction in the standard deviation of E_b with increasing H is that as H increases, more and more of the larger particles of the distribution contribute to resonance, thus leaving a narrower distribution of particles to contribute to the relaxation process, with the consequential reduction in the magnitudes of the $\chi_{\parallel}(\omega)$ components, as indicated in Figs. 56 and 57.

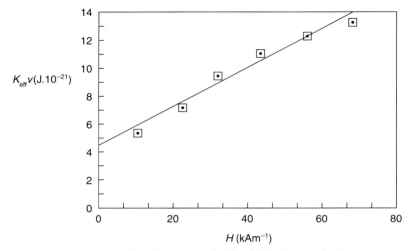

Figure 60. Plot of $K_{eff}v$ vs. polarizing field H for sample 19.

5. Case study 10

In this case measurement of the magnetic field dependence of the complex susceptibility $\chi(\omega)$ over a polarizing range of 0–116 kA m^{-1} of four colloidal suspension of $Mn_xFe_{1-x}Fe_2O_4$ particles in isopar M enables the depen-

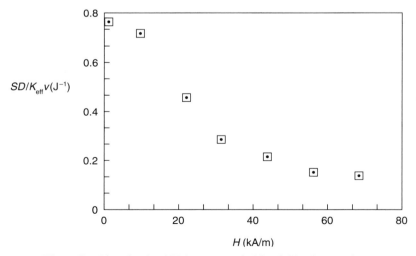

Figure 61. Plot of ratio of $SD/K_{eff}v$ vs. polarizing field H for sample 19.

TABLE IV
Data for Four $Mn_xFe_{1-x}Fe_2O_4$ Samples

	Sample	D_{vm} (Å)	σ_p	M_s (T)
21	$Mn_{0.1}Fe_{0.9}Fe_2O_4$	101	0.40	0.29
22	$Mn_{0.3}Fe_{0.7}Fe_2O_4$	105	0.47	0.27
23	$Mn_{0.5}Fe_{0.5}Fe_2O_4$	99	0.46	0.27
24	$Mn_{0.7}Fe_{0.3}Fe_2O_4$	94	0.46	0.22

dence on x of anisotropy constant K, internal field H_A, magnetogyric ratio γ, and resonant frequency f_{res} to be determined for x having values of 0.1, 0.3, 0.5, and 0.7 [73].

Complex magnetic susceptibility measurements over the frequency range 100 MHz–6 GHz were made for four ferrofluids, namely samples 21, 22, 23, and 24, of $Mn_xFe_{1-x}Fe_2O_4$ particles in isopar M (see Table IV), where D_{vm} is the median diameter of the lognormal volume distribution in angstroms, σ_p is the standard deviation of the distribution and M_s is the bulk saturation magnetization at 300 K.

The real, $\chi'(\omega)$, and imaginary, $\chi''(\omega)$, components of the four magnetic fluid samples were measured over the stated frequency range for 12 values of external fields H between 0 and 116 kA m^{-1}. For the unpolarized case, the results of the measurements are shown in Fig. 62, where it is seen that, for all four samples, resonance is observed. The spread of resonant frequencies $f_{res} = \omega_{res}/2\pi$, being over the approximate frequency range 1.2–1.96 GHz. The effect of an increase in applied polarizing field on the $\chi'(\omega)$, and $\chi''(\omega)$ components is illustrated by the case of sample 24 ($x = 0.7$) in Figs. 62 and 63, respectively, for 12 field values of (1) 0, (2) 6, (3) 20, (4) 32, (5) 44, (6) 56, (7) 68, (8) 74, (9) 80, (10) 92, (11) 104, and (12) 116 kA m^{-1}.

Figure 63 shows that the resonance frequency f_{res} spans the approximate frequency range of 1.2 GHz to >6 GHz while in Fig. 64 it is demonstrated that the corresponding loss peak frequencies f_{max} span the range 0.75 GHz to approximately 5.5 GHz.

From Fig. 63 it is seen that on increasing the external field the magnitude of $\chi'(\omega)$ at the lower frequency end of the plots is progressively

TABLE V
H_A, K, and γ for $Mn_xFe_{1-x}Fe_2O_4$ Samples

Sample	$f_r l$	$f_r h$	$f_m l$	$f_m h$	H_A	K	γ
1	1.9	>6.0	1.1	>6.0	40	5.8	2.36
2	1.7	>6.0	1.0	>6.0	37	5.0	2.48
3	1.4	>6.0	0.8	>6.0	40	5.4	2.43
4	1.2	>6.0	0.74	5.5	29	3.2	2.23

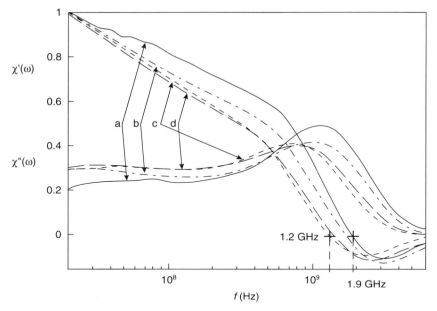

Figure 62. Plot of $\chi'(\omega)$ and $\chi''(\omega)$ vs. f for four $Mn_xFe_{1-x}Fe_2O_4$ samples: (a) $x = 0.1$, (b) $x = 0.3$, (c) $x = 0.5$, and (d) $x = 0.7$.

reduced, while toward the higher frequencies the profile takes on more of a resonance-type character. This effect was observed in all four samples.

In the corresponding plots of $\chi''(\omega)$ in Fig. 64, a similar reduction in magnitude is observed at lower frequencies, while at higher frequencies the frequency of the maximum of the loss peak, $f_{max} = \omega_{max}/2\pi$, increases with an increase of external field H.

A plot of f_{res} versus H for the same sample is shown in Fig. 65, and this shows that a linear relationship exists between f_{res} and H, with a value of 2.23×10^5 s^{-1} A^{-1} m being obtained for the slope multiplied by 2π, which is close to the free electron value of the magnetogyric ratio $\gamma = 2.21$ 10^5 s^{-1} A^{-1} m. This again demonstrates that the frequency dependence of ω_{res} can be described by Eq. (3.43), namely

$$\omega_{res} = 2\pi f_{res} = \gamma(H + \bar{H}_A) \tag{3.43}$$

where $\bar{H}_A$ again represents some mean value of the anisotropy field.

The value of $\bar{H}_A$ in a zero applied field can be calculated by using the value of γ obtained from the slope of ω_{res} versus H and the expression

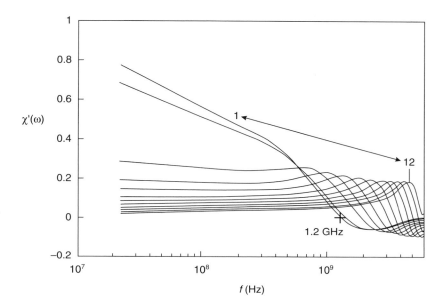

Figure 63. Plot of $\chi'(\omega)$ vs. f for $Mn_{0.7}Fe_{0.3}Fe_2O_4$ for 12 values of polarizing field over the range $0-116$ kA m^{-1}.

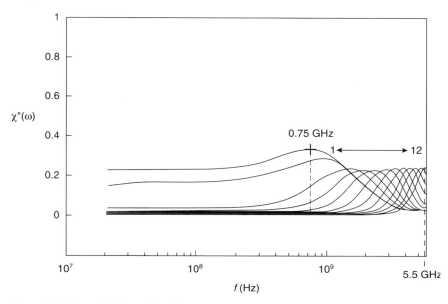

Figure 64. Plot of $\chi''(\omega)$ vs. f for $Mn_{0.7}Fe_{0.3}Fe_2O_4$ for 12 values of polarizing field over the range $0-116$ kA m^{-1}.

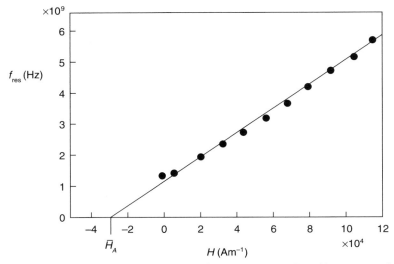

Figure 65. Plot of f_{res} vs. polarizing field H for $Mn_{0.7}Fe_{0.3}Fe_2O_4$ with average value of anisotropy field $\bar{H}_A = 29$ kA m^{-1}.

$\omega_{res} = \gamma H_A$. Alternatively, $\bar{H}_A$ can be calculated from the intercept in Fig. 65. In both methods $\bar{H}_A$ has an approximate value of 29 kA m^{-1}, which corresponds to a mean value of anisotropy constant $\bar{K}$ at room temperature and bulk M_s of 0.22 T of 3.2×10^3 J m^{-3}.

The same procedures were repeated for the remaining three samples, and the complete details for all four samples are given in Table V, where $f_m l$ and $f_m h$ represent the minimum and maximum values of the loss peak frequency while $f_r l$ and $f_r h$ represent the minimum and maximum values of the resonant frequency in gigahertz. The units of H_A, K, and γ are kA m^{-1}, 10^3 J m^{-3}, and 10^5 s^{-1} A^{-1} m.

The results displayed in Table V indicate that, in the unpolarized case, the smaller the value of x, the greater the resonant frequency f_{res} together with the corresponding value of f_{max}. Here, $\bar{H}_A$ is found to be approximately constant at 40 kA m^{-1} for $x = 0.1, 0.3, 0.5$ while decreasing in value by almost 25% for $x = 0.7$. As predicted by Eq. (3.2), the value of $\bar{K}$ is also found to behave in a similar manner. Furthermore, for all the samples the values of the magnetogyric ratio γ are greater than the value of γ for the free electron, and they correspond to γ values in the range 2.02–2.25 $\pm$ 0.03.

E. BROADBAND MEASUREMENT OF THE PERMITTIVITY OF MAGNETIC FLUIDS

Due to its relevance and the simple way it can be realized, it is appropriate to briefly indicate how the complex permittivity $\varepsilon(\omega) = \varepsilon'(\omega) - i\varepsilon''(\omega)$ can be determined by a variation of the technique used in determining the complex susceptibility components $\chi(\omega) = \chi'(\omega) - i\chi''(\omega)$.

The coaxial transmission line technique is ideally suited to the measurement of the complex permittivity of magnetic fluids. In contrast to the short-circuit method used in the determination of $\chi(\omega)$, here the open-circuit technique, whereby impedance measurements are made of the coaxial cell terminated in an open circuit, is used to determine the complex permittivity [74].

The open circuit produces a maximum electric field and a minimum magnetic field at the sample, thus making this technique well suited to the measurement of the dielectric properties of test samples.

From measurements of the real and the reactive components of the test samples, the complex permittivity $\varepsilon(\omega) = \varepsilon'(\omega) - i\varepsilon''(\omega)$, where the components $\varepsilon'(\omega)$ and $\varepsilon''(\omega)$ are given by

$$\varepsilon'(\omega) = \frac{\lambda}{2\pi d} \left\{ \frac{(R_0 + X_{\text{in}} \tan \beta x)(R_0 \tan \beta x - X_{\text{in}}) - R_{\text{in}}^2 \tan \beta x}{(R_0 \tan \beta x - X_{\text{in}})^2 + R_{\text{in}}^2} \right\} \quad (3.45)$$

and

$$\varepsilon''(\omega) = \frac{\lambda}{2\pi d} \left\{ \frac{(R_0 + X_{\text{in}} \tan \beta x)R_{\text{in}} + (R_{\text{in}} \tan \beta x)(R_0 \tan \beta x - X_{\text{in}})}{(R_0 \tan \beta x - X_{\text{in}})^2 + R_{\text{in}}^2} \right\}$$

$$(3.46)$$

where $R_{\text{in}} + iX_{\text{in}}$ is the input impedance of the HP line at a distance x from the terminating load, $R_0 = 50 \ \Omega$, $\beta = 2\pi/\lambda$ is the phase change coefficient, λ is the operating wavelength in the line, and d is the sample depth. These equations were developed similar to Eqs. (3.33) and (3.34) [3] with the analysis being given in Appendix B.

The permittivity results for the four magnetic fluids, namely samples 12, 13, 14, and 15 of case study 6, together with that of isopar M, the carrier fluid, obtained using this technique are shown in Fig. 66; for clarity only the $\varepsilon''(\omega)$ components for the highest and lowest concentrations are shown; however, the $\varepsilon''(\omega)$ component for all four fluids were found to be negligible up to approximately a frequency of 2 GHz, after which a small dispersion peak was exhibited. It can be seen that in general $\varepsilon'(\omega)$ for all the samples is

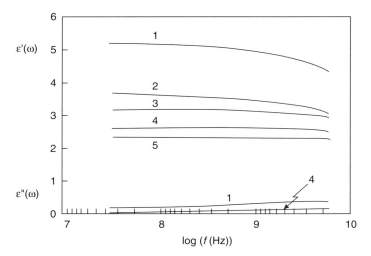

Figure 66. Plots of $\varepsilon'(\omega)$ and $\varepsilon''(\omega)$ vs. frequency in for five fluids consisting of depths of 9 and 1.5 mm over the frequency ranges 30–500 MHz and 100 MHz–6 GHz. For clarity only the $\varepsilon''(\omega)$ component obtained for the 0.09-T and 0.02-T fluids are displayed.

approximately constant up to a frequency of 1 GHz with the magnitude of $\varepsilon'(\omega)$ being a function of particle concentration and with the fluid of highest concentration of particles having the largest permittivity value.

The permittivity of the four ferrofluid samples was then calculated by means of the magnetic analogue of the Günther and Heinrich [75] formula for the complex dielectric constant of a composite dielectric containing a disordered distribution of spheres, which may be written as

$$\varepsilon_{\text{eff}}(\omega) = \varepsilon_2 \left(\frac{C - 2q[(\varepsilon_2 - \varepsilon_1)/(2\varepsilon_2 + \varepsilon_1)]}{C + q[(\varepsilon_2 - \varepsilon_1)/(2\varepsilon_2 + \varepsilon_1)]} \right) \tag{3.47}$$

where $0.75 < C < 1$, ε_1 and ε_2 are the permittivities of magnetite and isopar M, respectively, and q is the packing fraction. Here, $C = 0.75$ corresponds

TABLE VI
Measured and Calculated Permittivities

Fluid	M_s (T)	$\varepsilon'(\omega)$	$\varepsilon_{\text{eff}}(\omega)$
1	0.09	5.07	5.05
2	0.06	3.55	3.8
3	0.04	3.14	3.4
4	0.02	2.6	2.7

to a disordered array of spheres while ε_2 has been determined as having a value of 2.25; these are the values used here. Under the condition that $\varepsilon_1 \gg \varepsilon_2$, the calculated permittivities shown in Table VI were obtained at a frequency of 100 MHz and are found to be in close agreement with the measured components.

It should be noted that this exercise assumes that $\varepsilon(\omega) \approx \varepsilon'(\omega)$, which is a valid assumption over the frequency range where the $\varepsilon''(\omega)$ component is negligible.

Finally, it is a well-established fact that the velocity of propagation in a sample is inversely proportional to $\sqrt{\varepsilon(\omega)\ \mu(\omega)}$ where $\mu(\omega) = \chi(\omega) + 1$; thus, a development of the technique presented which can perform automatic, wideband measurements of both $\varepsilon(\omega)$ and $\mu(\omega)$ parameters has an obvious potential for application in this interesting area of research.

ACKNOWLEDGMENTS

I am indebted to B. K. P. Scaife for introducing me to the topic of ferrofluids and for the time, advice, and encouragement generously given over a number of years. I also express my gratitude to S. W. Charles for his contribution toward this work, not the least for his provision of many of the samples reported on here, and to his wife, Valery, for the hospitality shown me during my visits to Bangor. Thanks are also due R. Dempsey and C. Nolan of the Department of Electronic and Electrical Engineering, Trinity College, for their assistance with the many software problems encountered over the years and also to T. Relihan, formerly of the department. Finally I thank W. T. Coffey for encouraging me to write this work and Yu. Kalmykov for many useful and helpful discussions. The support from Forbairt, the Irish Science and Innovation Agency, is also acknowledged.

APPENDIX A.   DERIVATION OF EQUATIONS FOR THE SLIT-TOROID TECHNIQUE

Consider the toroid shown earlier in Fig. 10:

$$A = \text{cross-sectional area, m}^2$$

$$R_m = \text{mean radius, m}$$

$$L_1 = \text{gap width, m}$$

$$L_2 = \text{mean magnetic path length in toroid}$$

$$\mu_0 = \text{absolute permeability of free space}$$

$$\mu_{\text{eff}} = \text{effective permeability of cut toroid}$$

$$\mu_r = \text{relative permeability of toroid}$$

For N turns a current i (in amperes) produces a magnetizing force

$$H = \frac{Ni}{L_1 + L_2} \quad (\text{AT/m}) \tag{A.1}$$

If the flux is Φ webers, then

$$\Phi = B_{\text{toroid}} A = B_{\text{gap}} A = \mu_0 H_{\text{gap}} A \tag{A.2}$$

Hence

$$B_{\text{toroid}} = \mu_r \mu_0 H_{\text{toroid}} = \mu_0 H_{\text{gap}} \tag{A.3}$$

To produce H_{gap}, we require $H_{\text{gap}} L_1$ ampere-turns and to produce H_{toroid}, we require $H_{\text{toroid}} L_2$ ampere-turns:

Total ampere-turns

$$= Ni = H_{\text{gap}} L_1 + H_{\text{toroid}} L_2 = H_{\text{gap}}\left(L_1 + \frac{L_2}{\mu_r}\right) \tag{A.4}$$

Flux density is given as

$$B = \frac{\Phi}{A} = \mu_{\text{eff}} \mu_0 \times (\text{total ampere-turns}) \tag{A.5}$$

$$= \mu_0 \frac{iN}{L_1 + L_2/\mu_r} \tag{A.6}$$

Therefore,

$$\mu_{\text{eff}} \mu_0 \left[\frac{iN}{L_1 + L_2}\right] = \frac{\mu_0 iN}{L_1 + L_2/\mu_r} \tag{A.7}$$

and

$$\mu_{\text{eff}} = \frac{1 + (L_1/L_2)}{(1/\mu_r) + (L_1/L_2)} \tag{A.8}$$

If $L_1 \ll L_2$

$$\mu_{\text{eff}} = \frac{\mu_r}{1 + \mu_r(L_1/L_2)} = \frac{1}{1/\mu_r + L_1/L_2} \tag{A.9}$$

Let L_0 be the inductance of the excitation winding without any material in the toroid, that is, the inductance of N turns uniformly around the periphery of the toroid; then

$$L_0 = \frac{\mu_0 N^2 A}{L_1 + L_2} \approx \frac{\mu_0 N^2 A}{L_2} \tag{A.10}$$

Let L_s be the inductance of the toroid without a cut; then

$$L_s = \frac{\mu_0 \mu_r N^2 A}{L_1 + L_2} \approx \frac{\mu_0 \mu_r N^2 A}{L_2} \tag{A.11}$$

If L_{cut} denotes the inductance of the toroid with an empty cut,

$$L_{\text{cut}} = \frac{\mu_0 \mu_{\text{eff}} N^2 A}{L_1 + L_2} \approx \frac{\mu_r \mu_0 N^2 A}{(1 + \mu_r L_1/L_2)(L_1 + L_2)} \tag{A.12}$$

$$= \frac{L_s}{1 + \mu_r L_1/L_2} \tag{A.13}$$

Therefore

$$\text{Inductance with a cut} = \frac{\text{inductance without a cut}}{1 + \mu_r L_1/L_2} \tag{A.14}$$

So, from (A.10)

$$L_{\text{cut}} = \frac{L_0}{1/\mu_r + L_1/L_2} \tag{A.15}$$

Thus, the impedance of the coil on the core with a cut is

$$Z_{ce} = R_w + i\omega L_{\text{cut}} \tag{A.16}$$

$$= R_w + \frac{i\omega L_0}{\left[1/(\mu_r' - i\mu_r'') + \dfrac{L_1}{L_2}\right]} \tag{A.17}$$

Therefore

$$Z_{ce} = R_w + \frac{i\omega L_0(\mu_r' - i\mu_r'')}{1 + \gamma_g(\mu_r' - i\mu_r'')} \qquad (A.18)$$

where

$$\gamma_g = L_1/L_2 \qquad (A.19)$$

Insertion of a fluid of relative permeability μ_1, $(\mu_1 = \mu_1' - i\mu_1'')$ into the gap gives

$$L_{\text{cutfull}} = \frac{\mu_0 N^2 A}{[L_1/\mu_1 + L_2/\mu_r]} \qquad (A.20)$$

$$= \frac{\mu_0 N^2 A}{(L_2/\mu_r)(1 + L_1\mu_r/L_2\mu_1)} \qquad (A.21)$$

$$= \frac{L_0(\mu_r' - i\mu_r'')}{1 + L_1(\mu_r' - i\mu_r'')/L_2(\mu_1' - i\mu_1'')} \qquad (A.22)$$

Now

$$Z_{\text{cutfull}} = R_{\text{wire}} + i\omega L_{\text{cutfull}} \qquad (A.23)$$

$$= R_w + \frac{i\omega L_0(\mu_r' - i\mu_r'')}{1 + L_1(\mu_r' - i\mu_r'')/L_2(\mu_1' - i\mu_r'')} \qquad (A.24)$$

From (A.18)

$$(Z_{ce} - R_w)[1 + \gamma_g(\mu_r' - i\mu_r'')] = i\omega L_0(\mu_r' - i\mu_r'') \qquad (A.25)$$

From (A.24)

$$Z_{cf} - R_w = \frac{i\omega L_0(\mu_r' - i\mu_r'')}{1 + \gamma_g(\mu_r' - i\mu_r'')/(\mu_1' - i\mu_r'')} \qquad (A.26)$$

Therefore

$$\frac{(\mu_1' - i\mu_1'') + \gamma_g(\mu_r' - i\mu_r'')}{\mu_1' - i\mu_1''} = Z[1 + \gamma_g(\mu_r' - i\mu_r'')] \qquad (A.27)$$

where

$$Z = \left(\frac{Z_{cf} - R_w}{Z_{ce} - R_w}\right)^{-1} \tag{A.28}$$

Hence

$$(\mu_1' - i\mu_1'') + \gamma_g(\mu_r' - i\mu_r'') = Z(\mu_1'' - i\mu_1'')[1 + \gamma_g(\mu_r' - i\mu_r'')] \tag{A.29}$$

and

$$\mu_1' - i\mu_1'' = \frac{-\gamma_g(\mu_r' - i\mu_r'')}{1 - Z[1 + \gamma_g(\mu_r' - \mu_r'')]} \tag{A.30}$$

From (A.18)

$$\mu_r' - i\mu_r'' = \frac{Z_{ce} - R_w}{(i\omega L_0 + \gamma_g R_w - \gamma_g Z_{ce})} \tag{A.31}$$

Consequently

$$\mu_1' - i\mu_1'' = \frac{-\gamma_g\left[\dfrac{Z_{ce} - R_w}{i\omega L_0 + \gamma_g R_w - \gamma_g Z_{ce}}\right]}{1 - Z[1 + \gamma_g(\mu_r' - i\mu_r'')]} \tag{A.32}$$

$$= \frac{-\gamma_g[Z_{ce} - R_w]}{i\omega L_0 + \gamma_g R_w - \gamma_g Z_{ce} - Zi\omega L_0} \tag{A.33}$$

Substituting for Z from (A.28) yields

$$\mu_1' - i\mu_1'' = \{-\gamma_g[Z_{ce} - R_w][Z_{cf} - R_w]\}$$
$$\times \{(Z_{cf} - R_w)(i\omega L_0 + \gamma_g R_w - \gamma_g Z_{ce}) - (Z_{ce} - R_w)i\omega L_0\}^{-1} \tag{A.34}$$

Consider the denominator of Eq. (A.34):

$$(Z_{cf} - R_w)(i\omega L_0 + \gamma_g R_w - \gamma_g Z_{ce}) - (Z_{ce} - R_w)i\omega L_0$$
$$= [(R_{cf} - R_w - iX_{cf})(i\omega L_0 + \gamma_g R_w - \gamma_g R_{ce} - i\gamma_g X_{ce})$$
$$- (R_{ce} - R_w + iX_{ce})i\omega L_0] \tag{A.35}$$

The real part of this expression is

$$[-\omega L_0 X_{cf} + \gamma_g R_w R_{cf} - \gamma_g R_w^2 - \gamma_g R_{ce} R_{cf}$$
$$+ \gamma_g R_w R_{ce} + \gamma_g X_{cf} X_{ce} + \omega L_0 X_{ce}] \quad \text{(A.36)}$$

and the imaginary part

$$[i\gamma_g R_w X_{cf} - i\gamma_g R_{ce} X_{cf} - i\gamma_g R_{cf} X_{ce}$$
$$+ i\gamma_g R_w X_{ce} + i\omega L_0 R_{cf} - i\omega L_0 R_{ce}] \quad \text{(A.37)}$$

Let

$$A = \gamma_g[R_{ce} R_{cf} + R_w^2 - R_w R_{cf} - R_w R_{ce} - X_{ce} X_{cf}] \quad \text{(A.38)}$$

$$B = \gamma_g[-R_w X_{ce} - R_w X_{cf} + R_{cf} X_{ce} + R_{ce} X_{cf}] \quad \text{(A.39)}$$

$$U = [\gamma_g R_w R_{cf} + \gamma_g R_w R_{ce} + \gamma_g X_{cf} X_{ce} - \gamma_g R_w^2]$$
$$- \gamma_g R_{ce} R_{cf} + \omega L_0 X_{ce} - \omega L_0 X_{cf}] \quad \text{(A.40)}$$

and

$$V = [\gamma_g R_w X_{cf} - \gamma_g R_{ce} X_{cf} - \gamma_g R_{cf} X_{ce}$$
$$+ \gamma_g R_w X_{ce} + \omega L_0 R_{cf} - \omega L_0 R_{ce}] \quad \text{(A.41)}$$

$$\mu_1' - i\mu_1'' = \frac{[-A - iB][U - iV]}{U^2 + V^2} \quad \text{(A.42)}$$

$$= -\frac{[AU + BV] - i[BU - AV]}{U^2 + V^2} \quad \text{(A.43)}$$

Therefore,

$$\mu_1' = \frac{-[AU + BV]}{U^2 + V^2} \qquad \mu_1'' = \frac{[BU - AV]}{U^2 + V^2} \quad \text{(A.44)}$$

Since

$$\mu_1' = 1 + \chi_1' \quad \text{(A.45)}$$

$$\chi_1' = \frac{-[AU + BV]}{U^2 + V^2} - 1 \quad \text{(A.46)}$$

and

$$\mu_1'' = \chi_1'' = \frac{[BU - AV]}{U^2 + V^2} \quad \text{(A.47)}$$

APPENDIX B. ANALYSIS FOR THE DETERMINATION OF THE COMPLEX SUSCEPTIBILITY OF MAGNETIC FLUIDS

The input impedance of a transmission line at a distance x from a terminating load Z_R, as shown in Fig. B.1, is given by [76]

$$Z_{in} = \frac{V_{in}}{I_{in}} = \frac{V_R \cosh \gamma_0 x + Z_0 I_R \sinh \gamma_0 x}{I_R \cosh \gamma_0 x + (V_R/Z_0) \sinh \gamma_0 x} \tag{B.1}$$

where Z_0 is the characteristic impedance and γ_0 is the propagation constant, which may be written as

$$\gamma_0 = \alpha + i\beta \tag{B.2}$$

where α is the attenuation coefficient, $\beta = 2\pi/\lambda$ is the phase change coefficient, and λ is the operating wavelength in the line. In the case where the line has a very low loss, it is assumed that $\alpha \approx 0$ and $\gamma_0 = i\beta$ and Eq. (B.1) becomes

$$Z_{in} = Z_0 \left\{ \frac{Z_R + iZ_0 \tan \beta x}{Z_0 + iZ_R \tan \beta x} \right\} \tag{B.3}$$

When the load is a short circuit (i.e., $Z_R = 0$ and $V_R = 0$), the input impedance is

$$Z_{in} = iZ_0 \tan \beta x \tag{B.4}$$

Consider the case of a short-circuited, air-filled coaxial transmission line terminated with a toroidal sample of material of thickness or depth d, as shown in Fig. B.2, with Z_0 and γ_0 being the characteristic impedance and propagation constant of the air-filled line, respectively.

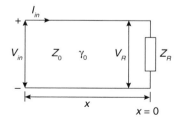

Figure B.1. Equivalent circuit of transmission line terminated in a load impedance Z_R.

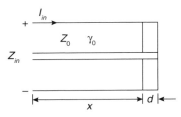

Figure B.2. Model of coaxial transmission line terminated in toroidal sample with short circuit.

The short-load section of Figs. B.2 and B.3 can be modeled, similar to that of Fig. B.1 with Z_R being the input impedance, as illustrated in Fig. B.3, and where Z_1 and γ_1 are the characteristic impedance and propagation constant of the sample-filled line. In this case it cannot be assumed that the line is lossless so the general equation (B.1) has to be used. Since the line is shorted, the input impedance Z_R may be written as

$$Z_R = Z_1 \tanh \gamma_1 d \qquad (B.5)$$

The intrinsic impedance Z of a medium with absolute complex permeability μ is given by

$$Z = \frac{i\omega\mu}{\gamma} \qquad (B.6)$$

while the characteristic impedance Z_1 of a coaxial line containing such a medium as the dielectric is [77]

$$Z_1 = \frac{1}{2\pi} Z \ln \frac{b}{a} \qquad (B.7)$$

where b is the radius of the outer conductor and a is the radius of the inner conductor of the coaxial line. From Eqs. (B.6) and (B.7), Eq. (B.5) then

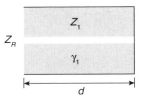

Figure B.3. Model of short-circuited toroidal sample.

becomes

$$Z_R = A \frac{i\omega\mu_1}{\gamma_1} \tanh \gamma_1 d \qquad (B.8)$$

where $A = (1/2\pi) \ln(b/a)$. A significant simplification can now be made if it is assumed that $\gamma_1 d \ll 1$, that is, that the sample depth is much less than the wavelength of electromagnetic radiation in the sample medium.

Under this assumption, $\tanh \gamma_1 d \approx \gamma_1 d$ and Eq. (B.8) becomes

$$Z_R = iA\omega\mu_1 d \qquad (B.9)$$

Substituting in Eq. (B.4) for Z_R, one obtains

$$Z_{\text{in}} = Z_0 \left\{ \frac{iA\omega\mu_1 d + iZ_0 \tan \beta x}{Z_0 - A\omega\mu_1 d \tan \beta x} \right\} \qquad (B.10)$$

which in terms of the permeability of the sample, μ_1, reduces to

$$\mu_1 = Z_0 \left\{ \frac{Z_{\text{in}} - iZ_0 \tan \beta x}{iA\omega d Z_0 + Z_{\text{in}} A\omega d \tan \beta x} \right\} \qquad (B.11)$$

On elimination of the constant factor A, which incorporates the dimensions of the coaxial line, one obtains

$$\frac{\mu_1}{\mu_0} = \frac{\lambda}{2\pi d} \left\{ \frac{(Z_{\text{in}}/Z_0) - i \tan \beta x}{(Z_{\text{in}}/Z_0) \tan \beta x + i} \right\} \qquad (B.12)$$

Noting that $\mu_1 = \mu_0 \mu_r$ and separating into real and imaginary components, that is, $\mu_r = \mu_r' - i\mu_r''$, with $Z_{\text{in}} = R_{\text{in}} + iX_{\text{in}}$ and $Z_0 = R_0$, one obtains

$$\mu_r' = \frac{\lambda}{2\pi d} \left\{ \frac{R_{\text{in}}^2 \tan \beta x + (X_{\text{in}} - R_0 \tan \beta x)(X_{\text{in}} \tan \beta x + R_0)}{(R_{\text{in}} \tan \beta x)^2 + (X_{\text{in}} \tan \beta x + R_0)^2} \right\} \quad (B.13)$$

and

$$\mu_r' = \frac{\lambda}{2\pi d} \left\{ \frac{R_{\text{in}}(X_{\text{in}} \tan \beta x + R_0) - (X_{\text{in}} - R_0 \tan \beta x)(R_{\text{in}} \tan \beta x)}{((R_{\text{in}} \tan \beta x)^2 + (X_{\text{in}} \tan \beta x) + R_0)^2} \right\} \quad (B.14)$$

Now, since

$$\mu'(\omega) = 1 + \chi'(\omega) \tag{B.15}$$

and

$$\mu''(\omega) = \chi''(\omega) \tag{B.16}$$

the relative, complex susceptibility components are readily determined from impedance measurements of the short-circuited line.

APPENDIX C. ANALYSIS OF THE DETERMINATION OF THE COMPLEX PERMITTIVITY OF MAGNETIC FLUIDS

To measure the complex permittivity of a medium, one places the sample at the end of the coaxial line which is terminated by an open circuit. This ensures that the electric field is large and the magnetic field small within the sample, provided that the wavelength within the sample is much greater than the sample thickness.

Following a similar procedure to that used in the measurement of the complex magnetic susceptibility, equations for the evaluation of the complex permittivity components can be determined from measurements obtained while the sample is open circuited.

When the load is an open circuit, that is, $Z_R = \infty$ and $I_R = 0$, then the input impedance is

$$Z_{\text{in}} = Z_0 \coth \beta x \tag{C.1}$$

Since the line is open circuited, the input impedance Z_R may be written as

$$Z_R = Z_1 \coth \gamma_1 d \tag{C.2}$$

The intrinsic impedance Z of a medium with absolute values of complex permittivity ε_1 is given by

$$Z = \frac{\gamma_1}{i\omega\varepsilon_1} \tag{C.3}$$

Again using Eq. (B.7), Z_R is found to be given by

$$Z_R = A \frac{\gamma_1}{i\omega\varepsilon_1} \coth \gamma_1 d \tag{C.4}$$

where $A = (1/2\pi) \ln(b/a)$. We make the assumption that the sample depth is much less than the wavelength in the sample. In this case $\coth \gamma_1 d \approx 1/\gamma_1 d$ and Eq. (C.4) becomes

$$Z_R = \frac{A}{i\omega\varepsilon_1 d} \tag{C.5}$$

which represents the load impedance in terms of the permittivity of the sample under test.

Substituting this into Eq. (B.1) for Z_R one obtains

$$Z_{in} = Z_0 \left\{ \frac{(A/i\omega\varepsilon_1 d) + iZ_0 \tan \beta x}{Z_0 + (A/\omega\varepsilon_1 d) \tan \beta x} \right\} \tag{C.6}$$

which in terms of the permittivity of the sample reduces to

$$\varepsilon_1 = \frac{A}{\omega d Z_0} \left\{ \frac{(Z_0/Z_{in}) - i \tan \beta x}{(Z_0/Z_{in}) \tan \beta x + i} \right\} \tag{C.7}$$

For an air-filled line the capacitance per unit length of line is given by

$$C = \frac{\varepsilon_0}{A} \tag{C.8}$$

and on elimination of the constant factor A, which incorporates the dimensions of the coaxial line, one obtains

$$\frac{\varepsilon_1}{\varepsilon_0} = \frac{\lambda}{2\pi d} \left\{ \frac{(Z_0/Z_{in}) - i \tan \beta x}{(Z_0/Z_{in}) \tan \beta x + i} \right\} \tag{C.9}$$

Resolving for the real and imaginary parts of the relative permittivity, we obtain

$$\varepsilon'(\omega) = \frac{\lambda}{2\pi d} \left\{ \frac{(R_0 + X_{in} \tan \beta x)(R_0 \tan \beta x - X_{in}) - R_{in}^2 \tan \beta x}{(R_0 \tan \beta x - X_{in})^2 + R_{in}^2} \right\} \tag{C.10}$$

and

$$\varepsilon''(\omega) = \frac{\lambda}{2\pi d} \left\{ \frac{(R_0 + X_{in} \tan \beta x)R_{in} + (R_{in} \tan \beta x)(R_0 \tan \beta x - X_{in})}{(R_0 \tan \beta x - X_{in})^2 + R_{in}^2} \right\} \tag{C.11}$$

REFERENCES

1. D. C. F. Chan, D. B. Kirpotin, and A. Bunn, Jr., *J. Magn. Magn. Mater.*, **122**, 374 (1993).

2. K. Raj and R. Moskowitz, *J. Magn. Magn. Mater.*, **85**, 233 (1990).

3. B. Wilson, *Phil. Trans. R. Soc.*, *London*, **69**, 51 (1779).

4. F. Bitter, *Phys. Rev.*, **41**, 507 (1932).

5. W. C. Elmore, *Phys. Rev.*, **54**, 1092 (1938).

6. S. S. Papell, U.S. Patent 3,215,572 (1965).

7. R. E. Rosensweig, *Int. Sci. Tech.*, **55**, 48 (1966).

8. R. E. Rosensweig, U.S. Patent 3,917,538 (1975).

9. M. M. Maiorov, *Magnetohydrodynamics* (Trans. of *Magnitnaia Hidrodinamika*), **2**, 21, 135 (1979).

10. F. Soffge and E. Schmidbauer, *J. Magn. Magn. Mater.*, **24**, 54 (1981).

11. P. C. Fannin, B. K. P. Scaife, and S. W. Charles, *J. Phys. E, Sci. Instrum.*, **19**, 238 (1986).

12. S. Roberts and A. R. von Hippel, *J. Appl. Phys.*, **17**, 610 (1946).

13. J. Frenkel and J. Dorfman, *Nature*, **126**, 274 (1930).

14. C. Kittel, *Phys. Rev.*, **73**, 155 (1948).

15. Y. L. Raĭkher and M. I. Shliomis, *Sov. Phys. JETP*, **40**, 526 (1975).

16. A. Aharoni and S. Shtrikman, *Phys. Rev.*, **109**, 1522 (1958).

17. C. P. Bean, *J. Appl. Phys.*, **26**, 1381 (1955).

18. C. P. Bean and J. D. Livingston, *J. Appl. Phys.*, **30**, 120S (1959).

19. E. Kneller, in *Magnetism*, Academic Press, New York 1963, Vol III. 382.

20. S. Chikazumi, in *Physics of Magnetism*, S. H. Charap, ed., Wiley, New York 1964, p. 128.

21. P. Debye, *Polar Molecules*, Chemical Catalog Company, New York, 1929.

22. W. F. Brown, *Phys. Rev.*, **130**, 1677 (1963).

23. L. Néel, *Ann. Geophys.*, **5**, 99 (1949).

24. V. Schunemann, H. Winkler, H. M. Ziethen, A. Schiller, and A. X. Trautwein, in *Magnetic Properties of Fine Particles*, J. L. Dormann and D. Fiorani, eds., Elsevier Science, Amsterdam, 1992, p. 371.

25. L. Bessais, L. B. Jaffe, and J. L. Dormann, *Phys. Rev. B*, **45**, 14, 7805 (1992).

26. D. P. E. Dickson, N. M. K. Reid, C. A. Hunt, H. D. Williams, M. El-Hilo, and K. O'Grady, *J. Magn. Magn. Mater.*, **125**, 345 (1993).

27. M. I. Shliomis and Yu. L. Raĭkher, *IEEE Trans. Magn.*, **MAG-16**, 237 (1980).

28. A. Aharoni, *Phys. Rev. B*, **46**, 5434 (1992).

29. W. T. Coffey, P. J. Cregg, D. S. F. Crothers, J. T. Waldron, and A. W. Wickstead, *J. Magn. Magn. Mater.*, **131**, L301 (1994).

30. W. T. Coffey, D. S. F. Crothers, Yu. P. Kalmykov, E. S. Massawe, and J. T. Waldron, *Phys. Rev.*, **49**, 1869 (1994).

31. M. I. Shliomis and V. I. Stepanov, in *Relaxation Phenomena in Condensed Matter*, W. T. Coffey, ed., Advances in Chemical Physics, Vol. 87, Wiley, New York, 1994, p. 1.

32. P. C. Fannin, *J. Magn. Magn. Mater.*, **136**, 49 (1994).

33. M. I. Shliomis, *Sov. Phys.-Usp.*, **17**, 53 (1974).

34. K. S. Cole and R. H. Cole, *J. Chem. Phys.*, **9**, 341 (1941).

35. D. W. Davidson and R. H. Cole, *J. Chem. Phys.*, **19**, 341 (1951).

36. S. Havriliak and S. Negami, *Polymer*, **8**, 161 (1967).

37. C. J. F. Böttcher and P. Bordewijk, *Theory of Electric Polarisation*, Elsevier Scientific, Amsterdam, 1978, Vol. 2.

38. H. Fröhlich, *Theory of Dielectrics*, 2nd ed., Clarendon, Oxford, 1958.

39. P. C. Fannin, B, K. P. Scaife, and S. W. Charles, *J. Phys. D: Appl. Phys.*, **21**, 1035 (1988).

40. K. Higasi and A. Matsumoto, *Dielectric Relaxation and Molecular Structure*, Monograph Series No 9, Research Institute of Applied Electricity, Hokkaido University, Sapporo, 1961.

41. W. T. Coffey, Yu. P. Kalmykov, and P. J. Cregg, in *Advances in Chemical Physics*, Vol. 83, Wiley Interscience, New York, 1993, p. 264.

42. B. K. P. Scaife, *Principles of Dielectrics*, Clarendon, Oxford, 1989.

43. M. Abramowitz and I. Stegun, *Handbook of Mathematical Functions*, Dover, New York, 1964.

44. P. C. Fannin and S. W. Charles, *J. Magn. Magn. Mater.*, **136**, 287 (1994).

45. A. Aharoni, *Phys. Rev. B*, **46**(9), 5434 (1992).

46. D. C. Slater, IEE Conference Publication No 177, *Third International Conference on Dielectric Materials, Measurements and Applications*, 1979, p. 132.

47. P. C. Fannin, B. K. P. Scaife, and S. W. Charles, *J. Phys. D: Appl. Phys.*, **21**, 533 (1988).

48. P. C. Fannin and S. W. Charles, *J. Phys. E, Sci. Instrum.*, **22**, 412 (1989).

49. P. C. Fannin and S. W. Charles, *J. Phys. D. Appl. Phys.*, **22**, 187 (1989).

50. P. C. Fannin, *J. Mol. Liq.*, **69**, 39 (1996).

51. P. C. Fannin, B. K. P. Scaife, and S. W. Charles, *J. Magn. Magn. Mater.*, **85**, 54 (1990).

52. P. C. Fannin, Yu. P. Kalmykov, and S. W. Charles, *J. Phys. D: Appl. Phys.*, **27**, 194 (1994).

53. P. C. Fannin, S. W. Charles, and T. Relihan, *J. Phys. D: Appl. Phys.*, **27**, 189 (1994).

54. V. V. Daniel, *Dielectric Relaxation*, Academic, London, 1967.

55. B. K. P. Scaife, *Principles of Dielectrics*, Clarendon, Oxford, 1989.

56. L. Lundgren, P. Svedlindh, and O. Beckman, *J. Magn. Magn. Mater.*, **25**, 33 (1981).

57. C. C. Paulsen, S. J. Williamson, and H. Maletta, *J. Magn. Magn. Mater.*, **54–57**, 209 (1986).

58. M. Schwartz, *Information Transmission, Modulation and Noise*, McGraw-Hill International, New York, 1990.

59. P. C. Fannin, A. Molina, and S. W. Charles, *J. Phys. D: Appl. Phys.*, **26**, 194 (1993).

60. M. S. Roden, *Analog and Digital Communication Systems*, Prentice-Hall International, London, 1991.

61. L. D. Landau and E. M. Lifshitz, *Phys. Zs. Sov. Union*, **8**, 153 (1935).

62. W. T. Coffey, Yu. P. Kalmykov, and E. S. Massawe, in *Advances in Chemical Physics*, Vol. 85, Part 2, S. Kielich and M. W. Evans, eds., Wiley Interscience, New York, 1993, p. 667.

63. Yu. P. Kalmykov, private communication, September 1997.

64. M. Nakagami, in *Statistical Methods of Radio Wave Propagation*, W. C. Hoffman, Ed., Pergamon Press, New York, 1960.

65. P. C. Fannin, T. Relihan, and S. W. Charles, *Phys. Rev. B*, **55**(21), 14423 (1997).

66. M. N. Afsar, J. R. Birch, and R. N. Clark, *Proc. IEEE*, **74**(1), 183 (1986).

67. P. C. Fannin, T. Relihan, and S. W. Charles, *J. Phys. D. Appl. Phys.*, **28**, 2003 (1995).

68. A. R. Von Hippel, *Dielectric Materials and Applications*, M.I.T. Press, Cambridge, Mass. 1954.

69. F. G. Brockman, P. H. Dowling, and W. G. Steneck, *Phys. Rev.*, **77**(1), 85 (1950).

70. P. C. Fannin, in *Proceedings of the 2nd International Workshop on Fine Particle Magnetism*, Bangor, Wales, U. K., D. P. Dickson and S. A. Watson, Eds., Department of Physics, University of Liverpool, 1996.

71. P. C. Fannin, T. Relihan, and S. W. Charles, *J. Phys. D. Appl. Phys.*, **30**, 533 (1997).

72. P. C. Fannin, S. W. Charles, and T. Relihan, *J. Magn. Magn. Mater.*, **162**, 319 (1996).

73. P. C. Fannin and S. W. Charles, *IEEE Trans. Magn.*, **33**(5), 4251 (1997).

74. P. C. Fannin, S. W. Charles, and T. Relihan, *J. Magn. Magn. Mater.*, **167**, 274 (1997).

75. F. Günther and D. Heinrich, *Z. Phys.*, **185**, 345 (1965).

76. M. N. Sadiku, *Elements of Electromagnetics*, Holt, Rinehart and Winston, New York, 1989.

77. C. S. Henson, *Solutions of Problems in Electronics and Telecommunications*, Pitman Press, London, 1969.

AUTHOR INDEX

Numbers in parentheses are reference numbers and indicate that the author's work is referred to although his name is not mentioned in the text. Numbers in *italic* show the pages on which the complete references are listed.

SUBJECT INDEX